Springer Series in
MATERIALS SCIENCE 76

Springer

Berlin
Heidelberg
New York
Hong Kong
London
Milan
Paris
Tokyo

Springer Series in
MATERIALS SCIENCE

Editors: R. Hull   R. M. Osgood, Jr.   J. Parisi   H. Warlimont

The Springer Series in Materials Science covers the complete spectrum of materials physics, including fundamental principles, physical properties, materials theory and design. Recognizing the increasing importance of materials science in future device technologies, the book titles in this series reflect the state-of-the-art in understanding and controlling the structure and properties of all important classes of materials.

Volumes 10–60 are listed at the end of the book.

G.E. Freger
V.N. Kestelman
D.G. Freger

Spirally Anisotropic Composites

With 201 Figures

 Springer

Prof. G.E. Freger
Dr. D.G. Freger
East Ukrainian University
20-A Molodeznaya Street
Lugansk 91034
Ukraine
E-mail: g_freger@yahoo.com
          freger@ukr.net

Prof. V.N. Kestelman
KVN International
632 Jamie Circle
King of Prussia
PA 19406, USA
E-mail: kvnint@earthlink.net

Series Editors:

Professor Robert Hull
University of Virginia
Dept. of Materials Science and Engineering
Thornton Hall
Charlottesville, VA 22903-2442, USA

Professor Jürgen Parisi
Universität Oldenburg, Fachbereich Physik
Abt. Energie- und Halbleiterforschung
Carl-von-Ossietzky-Strasse 9–11
26129 Oldenburg, Germany

Professor R. M. Osgood, Jr.
Microelectronics Science Laboratory
Department of Electrical Engineering
Columbia University
Seeley W. Mudd Building
New York, NY 10027, USA

Professor Hans Warlimont
Institut für Festkörper-
und Werkstofforschung,
Helmholtzstrasse 20
01069 Dresden, Germany

ISSN 0933-033X

ISBN 3-540-21188-8 Springer-Verlag Berlin Heidelberg New York

Library of Congress Cataloging-in-Publication Data.
Freger, G.E. (Garry Efimovich), 1945– .
Spirally anisotropic composites/G.E. Freger, V.N. Kestelman, D.G. Freger.
p.cm.– (Springer series in materials science, ISSN 0933-033X; v. 76)
Includes bibliographical references.
ISBN 3-540-21188-8 (alk. paper)
1. Composite materials. 2. Anisotropy. I. Kestel'man, V.N. (Vladimir Nikolaevich)
II. Freger, D.G. (Dmitry Garrievich), 1974– . III. Title. IV. Series.
TA418.9.C6F745 2004   620.1'18–dc22   2004045624

Typesetting by the authors
Text conversion by Frank Herweg, Leutershausen
Cover concept: eStudio Calamar Steinen
Cover production: *design & production* GmbH, Heidelberg

Printed on acid-free paper   SPIN: 10970218   57/3141/tr   5 4 3 2 1 0

Preface

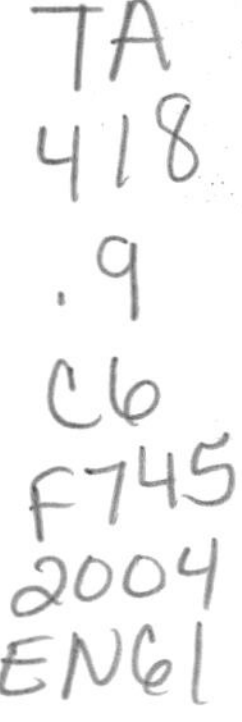

This book is about essentially new composite materials. It contains the results of the authors' studies of problems of increasing of transversal elasticity, strength under compression and other characteristics of polymeric composites.

Most of studies in the world are focused on hybridization and space sheathing for regulation of mechanical performance of composite materials. However hybridization requires random distribution of bundles high-modulus of materials in structure of a part that presents technological difficulty. Use of space sheathing of materials requires complicated equipment and some cases can not be used. This book describes how to combine processes hybridization and space sheathing on a micro-level, with the purpose of reaching random distribution high-modulus fillers in the structure of a material and his spatial arrangement.

The design of the spirally reinforced filling compound allows joining more rigid filaments in bundles, to ensure their regular arrangement in a material at the expense of another reinforcement of a material stacked on a spiral on a surface of a bundle. Thus the overlapping hybridization and space sheathing happens on a micro-level. On the basis of the analysis of microstructure of a material the dependences for definition of a degree of filling of aggregate and structural – geometrical performances of structure of a material are obtained. The representative unit and possible ranges of change of main parameters of structure are defined, that allowed to investigate stress and deformation and to optimize the composition under different aspects of a loading.

The tasks about an elastic strain of composites under the stress perpendicular to a direction of stacking of a main armature are considered. From solution of a plane problem of the elastic theory the dependences for an estimate of stress-deformed states of a material are obtained under transversal loading. Thus, changing properties of components, type of structure, shape of a unit and thickness of a stratum, is possible to develop materials with heightened transversal performances.

The stress-deformation and elastic constants of composites with the spirally reinforced filling compound are defined under longitudinal expansion and shift. The task of optimization of structure of an aggregate is decided (solved) with the purpose of definition of best values of its parameters and properties of making builders for deriving materials with the most favorable combination elastic and strength of performances.

The features of the technological process of deriving of materials with the spirally reinforced filling compound surveyed. The optimum technological parameters of deriving of an intermediate product are defined on the basis of stability of layout of a supplementary armature. On the basis of theoretical and experimental researches the special process equipment for a winding, impregnation, tension and press of a material designed. The complex of experimental researches of mechanical performances of the materials obtained on a basis of spirally reinforced filling compounds is carried out. The influence of main technological parameters and structures of materials on their mechanical properties is researched. Is shown, that at an inappreciable loss in rigidity in a direction of a main sheathing, these materials have excellent performance. The shape of the spirally reinforced units, geometrical sizes of a stratum and the conditions of his formation play significant role in strength and rupture of these composites.

A new modification of the rigid contact conditions for massive bodies with a thin local-orthotropic layer has been studied. The elasticity and thermoelasticity problems for an inhomogeneous spirally-anisotropic cylinder have been solved. The solution for the elasticity and thermo-elasticity problems for the main material layers enables to estimate the stress-strain state, the efficient longitudinal elastic modulus and effective longitudinal linear thermal expansion coefficient by the solution of the algebraic system of equations.

It is shown that introduction of spiral layers into the structure of hybrid composite rods varies their deformation behavior and efficient characteristics. It is recommended to place layers possessing high elasticity modulus outside of the rod elements incorporating several types of the main reinforcement. When designing the manufacturing process of spirally reinforced rods it is important to consider the impact of process factors on the product quality.

The problem of optimal design of hybrid spiral-reinforced rods is solved. It is minimized to the quality function given in a space of control parameters under some complex of limitations. The type of goal function and limitations imposed on the control vector are governed by loading conditions. The considered numerical examples show that the structures of optimal rod elements, the material and relative content of the main components are specified by operational conditions.

Use of aggregates on a basis of the spirally reinforced filling compounds and method of a spiral sheathing in units of constructions of different assignment is illustrated by examples: rod parts for assembly of frames of multidirectional textures of a heat-shielding coverage of flying means, measuring units of a transmitter of an ablation of a material of a heat-shielding, strength of units of power constructions, thin-walled conic envelopes with outside frames. It is shown that the high specific performances of a material with the spirally reinforced filling compound allow to decrease significantly structural weight, increase their mechanical performances, to simplify technology of production.

This book is addressed not only engineers and scientists using these materials, but for students and researchers, who will carry out their studies on new advanced composites, composite structures and manufacturing technologies.

Ukraine, USA

April 2004

G.E. Freger
V.N. Kestelman
D.G. Freger

Contents

1 Hybridization and Spatial Reinforcement of Composite Materials

Composite materials are used in various designs to improve the characteristics of constructions and reduce their weight. The properties of these materials and the problems of obtaining structural elements based upon them have received a great deal of attention in a number of countries. The fields of composite applications are diversified. They include structural elements of flying vehicles, their casings, wings, fuselages, tails and nose cones, jet engine stators, panels for various purposes, main rotors of helicopters, heat – proofing components, construction elements such as panels, racks, shields, backing elements, etc. In shipbuilding, composites are used in the hulls of light boats, shells, cowlings and so on. Composite materials have long been employed in roof panels in railway transport, window frames and flooring. Automakers are using composites in almost all car parts – body, shaft, brake disks, etc. At present, composite implants, canes and crutches for the handicapped, stretchers and many others are in extensive medical use. Problems of design, experimental check-up and production processes of composites and their products have been expounded in a great many investigations. As a result, the physico–mechanical characteristics of the materials and variations in their properties in response to static and dynamic loads have received detailed description, and the fundamentals of designing different purpose composite structures have been elaborated. Rapid expansion of application domains, toughening loading and operational regimes have called for the elimination of the intrinsic drawbacks of composite materials, to which in the first place belong weak transverse strength and stiffness, low longitudinal shear modulus and poor compressive strength in the reinforcement direction. Various means to avoid these defects are currently used, including the development of manufacturing processes with the application of different physical and chemical treatments, hybridization, use of novel structural schemes of materials, etc. Advanced manufacturing processes enable the realization of the original characteristics of components in full, which improves composite transverse and shear parameters to a certain extent but cannot change the situation radically. This is the reason why procedures connected with the development of new design schemes of the material reinforcement have come to the forefront. Analysis of the recent investigations shows that the most promising trends in developing new composites are hybridization and reinforcement.

Analyses of the experimental and theoretical investigations of hybrid composite materials indicate that the most expedient in this respect is the creation of intralayered hybrid composites with regularly disposed bunches of high-modular small-size fibers. There are difficulties in designing these composites, such as provision of a regular location of bunches, their effective separation, improved contact conditions at the interfaces, choice of reinforcing fillers, etc. It should be emphasized that hybridization does not fully overcome the drawbacks. For example, resistance to interlayer shearing and transverse tearing remains at the same level. To improve these characteristics the 3D reinforcement of materials should be exercised. The resultant composites show essentially enhanced stiffness and resistance to interlayer shearing, although this greatly complicates their manufacture.

As stated above, spatial reinforcement and hybridization improve the properties of materials and expand application spheres. In a number of cases mentioned, methods are combined to improve the performance of products. Hybridization and spatial reinforcement can be combined at the filler level to achieve hybrid spatially reinforced filler. Nicely and Davis [1] produced one of the first modifications of a filler, on the base of boron fibers helically wound with glass fibers. Later, Jearinimidis [2] used the method in materials based on organic fibers. Similar actions were undertaken for materials on boron fibers. Along with raising the processability of high-modular fillers, they have solved a number of the problems of hybridization. In this case, regular arrangement of rigid fibers joined in bunches is assisted by another reinforcing material laid over the surface of the main reinforcing bunch. Thus, hybridization and 3D reinforcement at the micro level produce promising hybrid spatially reinforced fillers for composite materials. Fillers of this type have been studied at length only in respect of boron-reinforced plastics, which display improved processability and other elevated properties. However, recommendations developed for the boron plastic structure and its production Process are inapplicable to other plastics, due to essential differences between boron fibers and other fibrous fillers. Thus, further adoption of materials based on hybrid spatially reinforced fillers is constrained in practice, because of the absence of scientifically grounded methods for their design, technology and experimental evidence on their physico-mechanical properties. To attain the indicated objectives, a profound consideration of a series of tasks is required, including design of the material, analysis of structural parameters, investigation of peculiarities of the manufacturing process, elaboration of recommendations on the choice of the main technological parameters, solution of a complex of problems on the Stress-strain state, elastic properties, optimization of material structure, and its physico-mechanical properties. A comprehensive knowledge of these problems has made it possible to develop composite materials based on hybrid spatially reinforced fillers that display refined characteristics under transverse loading, shear and longitudinal compression. Composite materials are employed extensively in various fields of engineering and household application, thanks to a host of positive properties, such as high specific strength and stiffness, monitored heat and electric conductivity, etc. They can be made radio transparent and nonmagnetic to fit construction, heat-proofing and erosion-resistant materials.

Along with glass-reinforced plastics, composites based on boron, carbon, organic and other types of fibrous fillers that considerably reduce the weight of structures and improve their technical characteristics are widely applicable today. The materials are especially efficient in aviation and space applications, as they are capable of significantly reducing the volume of the product. The peculiarities of the materials are their remarkable anisotropy of properties, and high characteristics in the reinforcement direction, which are in contrast to low transverse parameters [3], under shear and longitudinal compression [4]. The degree of anisotropy of unidirectional composites can be judged from Table 1.1, compiled from modulus E_Z/G_{ZX} and stresses σ_Z^+/τ_{XZ} in different directions presented in [5–9].

Table 1.1. Anisotropic Parameters of Unidirectional Composites

Parameter	Glass plastic	Carbon plastic	Boron plastic	Organo-plastic
E_Z/E_X	3–8	20–30	8–13	12–18
E_Z/G_{ZX}	10–35	20–80	20–100	25–40
σ_Z^+/σ_X^+	15–40	15–50	15–40	50–100
σ_Z^+/τ_{XZ}	12–40	13–40	15–50	10–20
σ_Z^+/σ_Z^-	0.5–2.8	0.5–3.5	0.5–2.0	1.5–7.5
σ_Z^+/σ_X^-	6-10	3-10	8-10	8-20

It is evident that the highest anisotropy is displayed by the composites based on high-modular and high-strength fillers, i.e. boron, carbon and organic plastics. Their elastic modulus, tensile strength in the transverse direction, and interlayer shear modulus are much lower compared to the properties in the reinforcing direction. This is because characteristics of unidirectional materials under such types of loading are governed mainly by the polymer matrix properties, and they are one order of magnitude smaller than the filler. Moreover, the provision of a fine adhesive joint at the interface [8, 10–16] plays a significant role in the material's strength [17–22, 287, 288, 301].

The problem of raising transverse and shear parameters of composite materials has become very topical recently [11, 23–25, 283, 296, 302], especially in respect of load-carrying structures. It has been found that the reliability of structures operating under high tangential Stresses improves by an order of magnitude when the shear strength is increased by just 10%. This indicated in [26–30] that low values of these properties are critical during compression and bending of composite- based articles. Poor resistance to transverse separation leads to delamination of thick-walled articles under slight transverse stresses [31, 32] and to buckling of individual layers under compression [33–36] and outer pressure.

The low compressive strength of filament-wound unidirectional composite materials exerts a constraining effect on the effective thickness of the structure. Several standard products, such as rods, plates, shells, etc. assisting in structural stability, often break as a result of interlayer shearing. On the other hand, in contrast to the strict requirements of matrices and of better adhesion between components, there has arisen a necessity of inserting soft, damping layers in materials experiencing impact and vibration loads [37–39].

It is apparent that the bearing capacity of structures of composite materials is interrelated with the structure and properties of the ingredient materials, on which contradictory demands are frequently imposed.

Novel kinds of reinforcing fibers and binders as well as modern technological processes have recently been elaborated to improve the physical and mechanical characteristics of composite materials, to employ the intrinsic properties of the original components fully. Along with this, advanced structural schemes of composite materials and members composed from them have been perfected or developed anew. Very strong and high-modular modern reinforced materials based on boron, silicon carbide, organic or carbon fibers and specific binders have been the bases for producing articles with improved resistance to corrosion, heat strength, and highly specific characteristics [7, 40, 294]. Nevertheless, the materials used in common reinforcing schemes do not yield any proportional growth of all characteristics but only in the reinforcing direction. In this connection, it has been suggested to employ profile fibers and tape-type filler [41–43] to refine properties in a transverse direction and shear. Such fillers are able to attain transverse strength and elastic modulus values constituting about 80–85% of the corresponding characteristics in the reinforcement direction, and reach interlayer shear strengths of 100 to 140 MPa for the unidirectional packing scheme. However, both production and adoption of such fillers are strongly complicated due to the necessity for precision packing, which raises their cost.

Properties and structure of composite materials are accounted for at article production and, consequently, are conditioned by a series of technological parameters, including impregnation and tension regimes, thermal treatment and so on [44, 45, 278, 282–286]. Of importance is the problem of ensuring a high-quality adhesion at component interfaces [28, 46–51, 287–290]. According to [52], increased adhesion at the interface and contact area improves the interlayer shear strength by 10%. As is known [48], the so-called interfacial layer is formed in the course of article formation from a composite material. The extent of the layer size and properties effect on characteristics of the material has been studied in a number of studies [23–25, 30, 53–55]. To change the contact boundaries between components, improve their adhesion strength, reduce flaws in the material structure and improve binder properties, a series of specific technological procedures has been devised. With this aim, certain methods of chemical and physical treatments are used [56, 57]. These include fiber-surface modification, treatment with dressing agents, oiling, processing in electric and magnetic fields, pressure or vacuum impregnation, ultra-sound treatment, high-frequency heating at setting and whiskering. These methods make it possible to realize more fully the initial characteristics of components, but do not resolve the problem of improving the properties of the material as a whole. The most promising trends in the creation of new composite materials are considered to be hybridization and spatial reinforcement, which significantly improve their properties over a wide range.

In historical terms, the first steps in hybridization were the introduction of some discrete fillers into the material composition. This elevated slightly their physico-mechanical characteristics, and improved their technological properties. Later it was established that the combination of continuous fibers of various compositions in one material can be the basis for varying all the properties of the composite. Four possibilities arise [58, 59]: to replace some costly fibers by cheaper ones; to acquire a wide spectrum of properties depending on definal conditions [60, 61]; to incorporate a weak chain into systems that enable to visualize Structural damage at an early stage of operation in extreme conditions; to improve some specific characteristics of the material, including fracture toughness and fatigue life.

1.1 Classification of Hybrid Materials

Classification of modern hybrid composite materials keeps to the technological principle of their production. Multicomponent hybrid materials are subdivided as follows: (i) *discretereinforced*, when one phase is inserted inside another [62], (ii) *intralayered*, with even distribution of both phases [59], (iii) *layerwise*, when layers of different fibers alternate across the packet thickness, (iv) *intralayer-layerwise*, in which alternating layers are composed of intralayered hybrids [63, 64] and (v), *superhybrids*, which are multilayer materials incorporating layers of homogeneous materials, simple composites and hybrid composites [65]. In *discrete-reinforced* materials we speak of the insertion of stiffer reinforcing fibers into the material structure. For example, shell structures can also behave as stiffening Elements. *Intralayered* hybrid composites, based on the principle of homogeneous blends, are produced from primary filaments, inside which fibers of different compositions are arranged. It should be borne in mind that it is not always possible to produce such filaments and evenly distribute component fibers inside them, since the manufacturing processes of the fibers may differ significantly. *Layerwise* materials represent a packet of sequentially laid sheet fillers, having various fibers. So far, it is possible to vary their properties by changing the geometrical parameters of the layers, the sequence of their packing and direction of orientation. *Intralayer-layerwise* hybrid composite materials can be created by either simple combination of the latter two types of materials, or by producing a bicomponent fabric. In the latter case, the fibers are orientated beforehand, which is set by the fabric structure. This is the reason for frequent local distortions of fibers and the resultant impairment of their physic-mechanical properties [66, 67]. Finally, *superhybrids* can be produced by joining, e.g. layers of a composite material with homogeneous metal layers [68]. The main problem arising in this case is the quality of the adhesive joints between layers.

1.2 Hybrid Effect in Multicomponent Materials

The so-called hybrid effect in unidirectional composites [58, 69–74, 279, 300] is exhibited in a deviation from the additivity rule of strength and stiffness, and increased breaking strain of materials in contrast to that in a more rigid component. The additivity rule [75–77] assists in the estimation of strength and elastic characteristics of hybrid composites. The rule turns out to be true only under specific conditions [78]. According to [79], only the upper limit of the properties of unidirectional hybrid materials can be estimated by the additivity rule. As for the lower limit of tensile strength, on compression and bending it is proposed to use a piecewise-linear dependence, consisting of two linear sections that correspond to different fracture modes. Experimental data presented in [79] prove that the characteristics of the class of materials under study differ from theoretical calculations by 12–17%.

It has been shown in [80] that whether or not the properties of a hybrid composite conform to the additivity rule depends on the type of fibers used. For example, glass-reinforced plastic impregnated with organic fibers Kevlar-49 acquires characteristics complying with the additivity rule.

As is known, the additivity rule allows calculating of properties, i.e. modulus of elasticity, of multi component materials based on degree of filling by each of the components. For example, according to this rule modulus of elasticity of carbon-glass plastic composite can be determined as:

$$E_{com} = E_c\varphi_c + E_g\varphi_g + E_b(1 - \varphi_c - \varphi_g),$$

where E_{com}, E_c, E_g, E_b are modulus of elasticity of composite, carbon filler, glass filler, and resin, respectively; φ_c, φ_g are degree of filling of composite by carbon fiber and glass fiber, respectively.

In contrast, impregnation with high-modular carbon fibers brings about a minimum compressive strength of the glass plastic that deviates from the theoretical one when the carbon fiber content is about 20%. Similarly, for the elastic modulus, the critical carbon fiber content in the named material is about 40%. We find that the compressive strength of a laminate hybrid composite, based on carbon-glass-reinforced plastic or carbon-organoplastic, can be calculated by the additivity rule under the following constraints: the components bear the load proportionally to their elasticity moduli, all layers become deformed jointly, and the strength of the monolayer of carbon fibers varies. It is natural to choose a corresponding binder to ensure required adhesion at the interfaces between layers. Moreover, as indicated in [81], differences between the structures of the materials and of binder content in the layers should be accounted for, if the additivity rule is to be employed correctly. For two-layered hybrid composites, the additivity rule applies when the limiting strains of the reinforcing fibers (e.g. organo-glass-reinforced plastics, carbon-boron-plastics) are approximately similar. When limiting strains of, e.g. carbon-glass-reinforced plastics, carbon-organoplastics and so on differ markedly, fibers break simultaneously under tension, insofar as their limiting strain is dependent on that of the fibers, whose volume content is higher and thus, the hybrid effect is exhibited. The apparent breaking strain of carbon fibers in a carbon-glass-reinforced plastic is above 0.7–1.5%, which is attributed to multiple breakages of carbon fibers restricted by the glass plastic matrix. Similar conclusions have been derived in [80]. When breaking strains approach very closely (e.g. high-modular structure HMS fibers show 0.6% breaking strain, while that of the high-tensile structure HTS is 1.2%) there are no observed discontinuities on curves of compressive strength and elastic modulus dependence upon volume share of fibers. As HTS fibers are replaced by glass fibers displaying a 1.9% breaking strain value, discontinuities appear. Similar results yield the dependences of hybrid composite properties on the elasticity moduli ratio of components. A feature of all the investigations is that only the initial elastic modulus value of the hybrid material can be estimated by the additivity rule, and this is true only until the composites train is equal to the stiffest fiber breaking strain combined with the thermal Stress-induced negative strain generated at material setting [77]. Notice that the residual thermal compressive strain cannot be fully considered to be the only reason for strain growth at the onset of breakage. It has been reported in [30, 82, 277] that only 10% strain increment may be explained by this stress.

To explain the hybrid effect for two-layered hybrids, a specific model has been constructed in [82]. By this model, stress redistributions have been evaluated as an individual high-modular fiber tearing off, and overloading factors for neighboring fibers and their inefficient length as well. The composite strength in total was estimated by the Monte-Carlo method. It was established that the breaking strain of the hybrid composite was higher than that of material filled only with fibers with low limiting strain. This suggests that the proposed method gives a theoretical substantiation of the hybrid effect. Experimental results using the acoustic emission procedure are consistent with this cited data.

The behavior of hybrid composites has been studied in [83] by the Stress–strain diagrams proceeding from their resistance to impact loading. If there are only brittle fibers in the composite structure, they break when the limiting strain sufficient to exercise breaking and the generation of new surfaces is achieved. If the composite does not break as the strain sufficient for fiber rupture is attained, it means that there is not enough energy for the breaking

Process and the fibers may be loaded further. Moreover, the fiber stays intact until the moment the strain comes to a limit adequate for the fracture energy. The limiting strain has been determined for two types of unidirectional composites: (i) there is strong bonding between fibers and the matrix, and (ii) fibers adhere to the matrix through sliding friction. For the former an equation for the minimum breaking strain of a fiber in the hybrid composite is

$$\varepsilon_{\min}^2 = \frac{2\gamma_F}{E_f(1 + 1/\alpha)r_f} \left(\frac{G_m}{E_f}\right)^{1/2} \frac{V_f^{1/4}}{(1 - V_f^{1/2})^{1/2}}, \tag{1.1}$$

where γ_F – necessary energy for fiber surface fracture; r_f – fiber radius; $\alpha = E_m V_m / E_f V_f$; E_m, G_m – moduli of normal elasticity and matrix shear; V_m, V_f – volume shares of the matrix and fiber; E_f – elasticity modulus of fibers.

For the latter case:

$$\varepsilon_{\min}^2 = \frac{12\gamma_F \tau E_m V_m^2}{E_c V_f E_f^2 r_f}, \tag{1.2}$$

where τ – breaking strength at shear; E_c – elastic modulus of the composite.

It has been assumed that independently of the fiber rupture, point breakage propagates immediately to other brittle fibers, i.e. multiple rupture takes place, which is not always observed in practice. Reported in [83], data are supported by theoretical and experimental results [84].

When studying the hybrid effect it is suggested in [30] that one should consider material fracture at tension as a statistic process that accounts for the fiber-strength distribution and contact – surface properties. The theoretical analysis proceeds from consideration of the statistical distribution of fiber strength values and their variation lengthwise. Thus, an equation for R_z has been derived expressing the relationship between the lower deformation limit for the hybrid composite failure and the strain of the material containing less rigid fibers only:

$$R_\varepsilon = 2^{1/q} \left(\frac{\bar{\varepsilon}_2}{\bar{\varepsilon}_1}\right)^{1/2} \left(\frac{k}{k_h}\right)^{1/2} \left(\frac{\delta}{\delta_h}\right)^{1/2q}, \tag{1.3}$$

where $\bar{\varepsilon}_1, \bar{\varepsilon}_2$ – mean strain values of fibers with low and high limiting strains; δ, δ_h – inefficient fiber lengths for single-component and hybrid composites; k, k_h – stress concentration factors of single-component and hybrid composites; $q = 20$ – Weibull's distribution parameter.

For a hybrid composite with a uniform distribution of Kevlar-49+Tornel-300 fibers, the theoretical value of $R_\varepsilon = 1.22$, while the experimental value is $R_\varepsilon = 1.04$, i.e. the correlation is satisfactory. As for laminated materials, the critical factor is the sequence of individual packing layers. For example, for the laminate material HMS+E – glass fiber, the theoretical value of $R_\varepsilon = 2.26$, and experimental values for two layers of carbon fibers laying between glass fiber layers are $R_\varepsilon = 1.31$, while with only one layer $R_\varepsilon = 1.83$. It has been also established that multilayered materials with alternating layers show better agreement between experimental data and calculations. In [30] another explanation of the failure mechanism is given for the hybrid composite as compared to [83], according to which the fracture ceases as contact occurs between the crack and less brittle fibers. A similar supposition on the role of bunches was put forward earlier in [62], where the hybrid effect is related to the fixing of ruptured carbon fibers in the glass-plastic matrix, which makes them capable of further load bearing.

Moreover, the hybrid effect is explained in terms of dynamics [85]. The dynamic coefficient of stress concentration has been calculated in materials with alternating layers of fibers with small and large elongation at the rupture. It is reported that two stress waves are generated at fiber rupture whose difference is due to the difference in mass per unit length of fibers. The amplitude of waves and the time of the beginning of destruction depend on differences in elastic modulus for every fiber. Nonsimultaneous rupture brings a phase shift of the waves and a positive hybrid effect.

It has been set forth in [86] that deviation of the elastic modulus at bending, calculated by the additivity law values, is highest when the composite consists of a few layers. Composites behave differently under bending than under strain as all fibers are in different stress conditions. If glass fiber layers are laid outside, the value on the lower and, in the case of carbon fibers, on the upper side. For the former case the value is $E_u = 63\,\mathrm{GPa}$ and in the latter $E_u = 93\,\mathrm{GPa}$, in contrast to the rated one $E_u = 81\,\mathrm{GPa}$. As the number of layers grows, the difference diminishes and with seven or more layers experimental values start to coincide with those found by the additivity rule. Experiments [86] showed that hybrid effect depended on fibers location.

The role of the fiber contact surface with the matrix is crucial for attaining the hybrid effect [87–89, 291]. For instance [90], two different composites were tested. The first composite had fibers HMS and HTS with different mechanical and interfacial properties but with the same surface contact of fibers with the matrix. The second hybrid composite was made from carbon and glass fibers. It was stated that only the second composite was in agreement with the additivity rule although the tensile strength turned out to be lower than that by the additivity rule. The same materials were tested in [91] by the three-point bending procedure, which, in contrast to transverse fracture [90], visualized the interlayer shear-induced failure. In addition, the elasticity moduli at bending appeared higher (61 GPa instead of 55.6 GPa), while fracture toughness and energy decreased. The general properties dependency on the interfacial state turns out to be the same.

Similarly, in [92] fracture toughness K_{1C} and critical rate of strain energy liberation G_{1C} have been studied in hybrid composites based on glass fibers E and carbon fibers TORAYCA T-300. Samples with an incision were tested following the three-point bend procedure. The positive hybrid effect was found to be most prominent for G_{1C}. The most optimum ratio of carbon and glass fibers appeared to be equal to 0.33/0.17, which produces $G_{IC} = 105.7\,\mathrm{KJ/m^2}$ and $K_{IC} = 29.8\,\mathrm{MN/m^{3/2}}$ and is higher than the unidirectional plastic (48.6 and 22.72) and carbon plastic (65.67 and 24.12).

The hybrid effect has been studied in [86, 93–95]. The transverse strength has been observed to reduce at compression in materials with a combination of glass fiber and jute fiber [96], with glass fiber content from 0.3 to 0.55. Small disagreement with the additivity rule has also been found in the elastic modulus in materials consisting of propylene and cellulose fibers [94].

The distribution of fibers of low limiting strain as well as fiber bunches is important for the hybrid effect [97–101]. A microstructural model proposed in [101] takes into consideration the load that the weakest fragments continue to bear after they have failed. It is assumed that each bunch of fibers ruptures upon reaching the breaking strength of the weakest unit, during which the homogeneous mixing of fibers promotes the elevation of the damage threshold and residual strength at bending through regulated fracture energy. In the experiments, 3 mm long carbon fibers constituting 48–52% of the composite were tested. It has been proved that differences in fracture behavior depend upon composite formulation and related fiber distribution. Strength starts to vary if the brittle fiber content is about 20%, which corresponds to the onset of fiber – bunch formation.

Analysis of a series of investigations devoted to the evaluation of the hybrid effect in carbon-glass-reinforced plastic and organo-carbon plastic [102–104, 278, 297] shows that the hybrid effect value can be 20 to 85% of the effect calculated by additivity rule. In order to understand the degree of influence of fiber distribution and amount of low breaking strain, the following types of hybrid composites have been studied: i) laminated, ii) with separate bunches of carbon fibers, and iii) with even fiber distribution. In the first case the composite becomes broken through rupture of the central layers of carbon fibers, followed by crack propagation over layer interfaces. If the hybrid is reinforced with fiber bunches, the delamination length lowers as the size of carbon fiber bunches decreases. Thus, for a bunch consisting of 1620 fibers the delamination length is $L_d = 1$–$1.2\,\mathrm{mm}$, if the bunch contains 10,000 fibers it is $L_d = 8$–$18\,\mathrm{mm}$. Hybrid composites with even distribution of fine bunches fracture chaotically as the bunches rupture one-by-one, not simultaneously, which raises the breaking strain. This suggests that the hybrid effect takes place in cases where more rigid fibers occupy a lesser volume, are organized in small size bunches and are evenly distributed across the bulk.

The fracture mechanism of a hybrid composite based on carbon fiber bunches evenly spread in the glass-reinforced plastic have been treated in [100]. As a result, a statistical failure model for the composite fibers has been constructed to describe the part of the hybrid effect not related to thermal effects. Hybrid material efficiency can be characterized by the ratio of strength at rupture to bunch size. According to the statistical model, the smaller the binder volume in a bunch of carbon fibers the higher its failure strain and until glass fibers hamper damage propagation from bunch to bunch, the material displays a higher mean failure strain than the plastic reinforced with only carbon fibers. Recently developed specific microscopic methods have made it possible to evaluate the effect of bunch size on the material characteristics using model samples. Analysis of the results allows us to conclude that the strength improvement of fibers of low breaking strain with diminishing bunch size can be related to isolation of the bunches by the fibers of higher breaking strain values. So far, it can be stated that the characteristic bunch size and their regular distribution change the material characteristics and govern the stress–strain curve. Test evidence presented in [105–109, 285, 289, 299] prove the significance of the rigid fiber bunch size. Thus, [105] shows that the fracture energy is higher in the layered hybrid composite than in a bunch-type one (laminate $-65\,\mathrm{kJ/m^2}$, bunch-type $-80\,\mathrm{kJ/m^2}$).

The reviewed studies give grounds for a conclusion that the hybrid effect is peculiar to a number of composites, depends on the type of fibers, their volume content, arrangement in the material structure and conditions at the interfaces between components. It has been theoretically derived and experimentally substantiated that better hybrid composites were reinforced with small size bunches of rigid fibers spread evenly within the material bulk.

1.3 Theoretical and Experimental Research on Hybrid Composites

Methods of stress analysis used for hybrid composites proceed from the postulates of the mechanics of composite materials, Stress concentration and fracture mechanics. The linear theory of laminated composites is used most often in studying hybrid materials during which Stresses in the layers are commonly determined and material properties are forecasted [67, 107, 110–116, 275–286, 290, 291, 293, 294]. It should be underlined that the theory of laminated composites is applied to hybrids in cases where they display linear or close to linear stress to strain to failure dependence. Calculated using the theory, stresses can be used along with the fracture criterion of layers to forecast the original and final fracture of hybrids [117]. Moreover, Hill's criterion is commonly employed as the strength criterion [118].

The strength equation can be written as

$$\frac{\sigma_1^2}{F_1^2} - \frac{\sigma_1\sigma_2}{F_1^2} + \frac{\sigma_2^2}{F_2^2} + \frac{\tau_{12}^2}{F_{12}^2} \leq 1,$$

where σ_1, σ_2, τ_{12} – stress in material; F_1 – material strength in direction of reinforcement; F_2 – material strength in transverse direction; F_{12} – material shift strength.

The common strength criterion for anisotropic media is the tensor-polynomial one [75] mostly used in analysis of bicomponent composites. However, it is rarely employed for hybrid materials because of the absence of the required experimental data. The strength theories of conventional composites turn out to be inadequate for predicting breaking stress in hybrid materials. It is possible to use the tensor-polynomial criterion to calculate the strength of hybrid composites, but it necessitates profound experimental investigation.

The stress field in the vicinity of a brittle fiber rupture under tension in the direction of reinforcement has been estimated for a unidirectional poly-fibrous material [119] belonging to an intralayered hybrid composite. It is accepted that a brittle fiber encircled by a quasi-homogeneous anisotropic matrix (polymer binder reinforced with unbreakable fibers) experiences only tension, whereas the matrix transfers shear stresses and experiences part of the tensile load. The derived evidence on stress variations assists in defining the inefficient boron fiber length in the organo-boron plastic. It makes it possible later to determine optimal parameters of the structure of hybrid composites.

Great attention has been paid to the problems of inner stresses and their effect on the properties of hybrid materials [30, 120–125]. For example, in [121] intralayered materials have been compared to interlayer hybrids from the standpoint of stress concentration. It was concluded that the properties of composites will start to decrease at some critical reinforcement degree. It can be explained by changes of interaction at the border of the fiber-matrix.

Strength, toughness and fracture kinetics of hybrids are treated mainly from the viewpoint of linear elastic mechanics of composites [20, 126–131, 275]. For a hexagonal elastic constants of the material have been derived in [132] showing their dependence on properties of its components and their amounts. Boron plastic is shown to increase such characteristics as longitudinal shear modulus and transverse elastic modulus with the boron fiber content growth. They show that at the initial deformation stage the characteristics can be defined using the additivity rule. They state that there exists some critical value of low-modular fiber content at which the material strength is minimal, e.g. for boron plastics it is 52–58% of the total fiber content in the material. They show that the strength of carbon-glass-reinforced plastics at compression, tension and bending has a minimum value at a glass fiber content of 65–80%; for carbon-organoplastics it is 32%.

In [111, 133, 134] elastic properties of intralayered and layerwise hybrid composites have been determined using the methods of boundary estimate [135] and self-consistent field [136–138]. Materials based on glass and Kevlar-49 fibers and a phenolic binder formed by either alternation of layers or mixing strands of constituent fibers, has been studied and perfect agreement between experimental and theoretical data has been reached in [134]. The intralayered hybrid composite on the base of the same components has been studied experimentally [111] during which its elastic characteristics under tension along and across the fibers and during longitudinal shear have been evaluated. It has been established that experimental longitudinal elastic modulus values are slightly greater than the theoretical ones. It should be noted that the transverse elastic modulus for the intralayered hybrids is higher than that of the laminates, which supports the earlier derived conclusion on the demand for more even distribution of brittle fibers within the material.

In order to define the elastic properties of organo-boron plastic, the averaged elastic characteristics of two structural elements of the material have been estimated in sequence. The first element consists of anisotropic fibers in an isotropic binder, the second incorporates isotropic fibers in an anisotropic matrix. The method presumes introduction of brittle fibers into a quasi-homogeneous matrix. For this purpose organo-boron plastics have been studied for variations in the amounts of the constituents. As a result, theoretical data have been found to diverge from experimental evidence as the boron fiber content grows. Note that the greatest divergence has been observed for the transverse elastic modulus and longitudinal shear modulus. Similar data on intralayered hybrids with different combinations of components, including organo-glass-reinforced plastics, organo-carbon plastics and carbon-glass-reinforced plastics, have been obtained. The results show excellent agreement between theory and experiment. Note that all studied materials display monotonic variation of elastic characteristics with changing content of ingredients. The method proposed in the studies is simple, has been computerized and is suitable for determining elastic properties of intralayered hybrids. As shown earlier, in the course of the hybrid composite loading, more brittle fibers rupture as the strain grows and the strain diagram changes its form. Problems of hybrid composite fracture and their behavior in loading conditions have been studied elsewhere [119, 122, 139–145]. Investigations in fracture mechanics of hybrid composites have yielded the dependence of fracture energy on the type of bonds between components [146], their arrangement in the composite [147–149], volume content of fibers with low breaking strain values [54], test conditions [91, 150–152] and operational regimes [65, 290]. It has also been established that the fracture energy of a hybrid composite depends on the fracture energies of each component. According to [153], variation of the glass fiber content from 0 to 100% (with total fiber content 0.55–0.6) in a unidirectional hybrid composite with intralayered arrangement of components augments fracture energy from 70 for the additivity rule to $180\,\text{kJ/m}^2$. In [82, 134] the Monte-Carlo method [154] is used to define tensile strength of materials consisting of E glass fibers and Kevlar-49. In [155], stress concentration factors are calculated around the rupture point of a fiber or a group of fibers based on shear analysis on the assumption that the matrix responds only to shear, and the fibers only to tension.

Experiments with organo-glass plastics [134] and carbon-glass-reinforced plastics [155] were in good agreement with theoretical data. For transverse tension of a unidirectional flexible composite, a stochastic model of composite failure has been constructed in [156], based on the principle of damage accumulation. The calculation makes allowance for a series of assumptions for the critical zone of fracture. The algorithm takes into account many factors, such as scale, irregularities of fiber packing, interfacial strength, etc. When designing hybrid composite materials, it is of importance to solve the problem of Structural and formulation optimization [157, 158, 293]. As the optimum criteria of material characteristics, compressive strength, mass lowering or cost reduction may be used. It is stated in [159] that even if the physico-mechanical characteristics of the material depend almost linearly on the volume ratio of the employed fibrous fillers, then specific parameters and the relation of mechanical characteristics to the unit product mass cost display nonlinear behavior. However, the same materials might have different dependences on their specific cost and mass values. Hence, the problem of optimization of a hybrid composite formulation should be decided individually, for each individual case.

To estimate strength, stiffness and failure kinetics of hybrid composites, linear elastic mechanics of composites is mostly used. By making assumptions, mainly for the intralayered and layerwise composites, stress concentrations in structural elements have been determined and strain diagrams and elastic characteristics have been found. Moreover, it has also been established that the fracture process in hybrid composites is dependent upon variations of the properties of the ingredients, as well as on some other structural and technological parameters of the material. Failure models based on these assumptions enable the forecasting of the strength and deformability of certain types of hybrid composites and their optimization.

1.4 Properties of Hybrid Composite Materials

A vast experimental body reviewed in [84] deals with the physico-mechanical characteristics of hybrid composites, basically laminated hybrids with alternating layers, composites of the sandwich type having glass fiber layers inside and a carbon fiber layer outside, as well as materials with different kinds of fibers in each layer. The original components were commonly glass fibers of E and S grades, carbon fibers HMS, HTS, Thornel and organic fiber Kevlar, with the total fiber content in the plastic being 55 to 65%. The investigators estimated impact toughness, stiffness and tensile, bending and compression strength. There is, however, a lack of fatigue characteristics under tension and bending as well as of strength and stiffness under compression. Analysis of the presented results suggests that elastic, strength and impact characteristics of hybrid composites vary within a wide range and, with the exceptions of compressive and bending strength, occupy an intermediate position between these characteristics of materials reinforced by just one type of fiber. It is reported in [160] that, depending on the orientation of its layers, the strength of carbon plastic and carbon-organoplastic under tension along fibers can vary at tension along fibers from 524 to 689 MPa, bending strength from 951 to 1150 MPa, tensile strength across fibers varies somewhat less from

53.8 to 46.9 and transverse elastic modulus from 13.1 to 11.7 MPa. Similarly, [161] showed that laminated carbon-glass-reinforced plastics with different numbers and thicknesses of layers display 1.23 to 2.57 times higher fatigue strength than carbon plastics.

Further investigations of intralayered and laminated hybrids concern the dependence of their properties on the character of their interactions at interfaces [90], filler type [162, 163], testing mode (e.g. longitudinal compression [164], fatigue [165], thermal effect [98]) and operational conditions [166]. The authors of [167] discussed the possibility of using standard test methods for materials based on carbon, glass and organic fibers with various proportions of components. The authors of [92] found that when estimating the fracture toughness of unidirectional carbon plastics, it varies within 22.7–29.8 MN/m$^{3/2}$, depending on the ratio of components. Moreover, the material elastic constants have been proved to have intermediate values lying between those of material constituents, which corroborate the results obtained for laminated hybrid composites. Note that most of investigations aimed at estimating physico-mechanical characteristics and their dependence on structural and technological parameters are devoid of data on the properties of hybrid composites under transverse compression, tension and longitudinal shear. An exception is the work in [160], in which experimental and theoretical evidence is presented on elastic properties for intralayered hybrid composites. It is shown that hybridization only slightly affects the strength and stiffness characteristics under transverse loading and longitudinal shear of unidirectional composites. For example, the transverse modulus of carbon plastics varies from 86 to 64 MPa when the carbon fiber content increases from 0 to 50% and the transverse shear modulus varies from 37 to 27.8 MPa.

The problem of the hybrid composite warping is important. It is well known that independently of the manufacturing process, the proportions of reinforcing fibers, or differences in their thermophysical characteristics, bring about internal thermal stresses in the material structure. For instance, thermal stresses generated in laminated hybrid composites as a result of cooling to room temperature can be determined by the relation

$$\tau_{xz} = \Delta t(\alpha_{x1} - \alpha_{x2})\sqrt{\frac{G_m E_{x1} E_{x2} F_1 F_2}{b_m \delta_m (F_1 E_{x1} + F_2 E_{x2})}}, \tag{1.4}$$

where Δt – temperature variation; a_{x1}, a_{x2}, E_{x1}, E_{x2}, F_1, F_2 – coefficients of the linear thermal expansion, elasticity moduli and cross-sectional areas of adjoining layers; G_m, b_m, δ_m – shear modulus, width and thickness of the binder layer.

Equation (1.4) shows that dependence stresses in a material consisting of carbon tape and glass fabric layers make up 2 MPa. Similar data on the effect of thermal expansion coefficients and inner Stresses in a layered material containing two glass fiber layers located between carbon fiber layers have been obtained in [168]. It has also been shown that in carbon fibers, cooling produces compressive stresses of about 50 MPa, while cooling of glass fibers produces tensile stresses of about 25 MPa. The stresses bring about 0.029% compressive strain in the carbon fibers and 0.05% tensile strain in the glass fibers. Carbon fibers turn out to be already compressed in the initial state to up to 10% of their breaking strain value. Evidently, by varying critical parameters, the sequence of packing and mutual orientation, it is possible to control to some extent the residual thermal strains in the material [169]. One such method is set forth in [117], which essentially consists of combining hybridization with programmed winding. In this case, the tension force of the carbon fibers during the winding of a laminated carbon-glass-reinforced plastic is equal to zero, that of the glass fibers changes following a definal law. It has been established that radial thermal stresses can be varied within a wide range, depending on thickness ratios of the material layers (12 to 10 MPa). A simpler method of eliminating article warping resulting from inner thermal stresses is variation of the mutual orientation of individual layers. For example, symmetrical anti-phase disorientation of carbon-glass-reinforced plastic, consisting of a glass fabric, Kevlar fiber and a carbon braid, practically eliminates warping at packing angles of $\pm 7°$ and $\pm 15°$.

1.5 Application of Hybrid Composite Materials

Raised specific strength, fracture toughness, impact viscosity and specific stiffness along with reduction of mass, cost and sensitivity to incision make hybrid composites advantageous over other composite materials in such fields of application as aviation, space, ship-building, medicine and transport. Often the structures must meet specific and often contradictory requirements [2, 57, 170–174, 295].

The authors of [64, 149, 292, 295] report on the use of hybrid composites in blades of turbine engines and helicopters with elevated resistance to large-caliber rifle fire, in wing skin and horizontal stabilizers of warplanes and truss supports of spaceships. As a review [175] on the study and use of hybrid composites shows, the level of their design and production technique allows the manufacture of high-quality products competitively with conventional composites. Along with the usual production methods it is expedient to use pultrusion and hydrodynamic mixing, so as to reach an even distribution of components in the hybrid composite structure.

Hybrid composites also appear to be promising in the automobile industry [176]. Some car parts incorporate organic, glass and carbon fibers whose combination assists in raising mechanical characteristics at compression, impacting and fatigue strength. For example, a carbon-glass-reinforced plastic with carbon fiber content of up to 7% was used in a power-shaft design of a truck, almost halving the shaft mass. Hybrid composites are highly promising for use in the flywheels of battery-driven vehicles. Information on the fabrication of a 584 mm diameter and 26.2 kg weight flywheel rotating at a $25{,}000\,\text{min}^{-1}$ frequency is cited in [177]. The flywheel rim was made by winding nine layers of glass fibers and a layer of organic fibers.

Hybridization of materials gives opportunities for varying their properties within a wide range, thus significantly expanding their spheres of application. Hybrid materials used in various products elevate their physic-mechanical and service characteristics, reduce mass and cost, and improve process productivity.

An important parameter of governing hybrid composites is the ratio of the elasticity moduli of the high-modular and low-modular components – β. Its values are presented in Table 1.2.

Figure 1.1 and others "+" means tensile, "−" compression, β-ratio of modulus of elasticity of high-moduli and low-moduli components, E_z^+, σ_z^+-moduli of elasticity and strength of hybrid composite under tensile in direction of reinforcement, E_z^-, σ_z^--moduli of elasticity and strength of hybrid composite under compression in direction of reinforcement, E_x^+-moduli of elasticity of hybrid composite under tensile in direction perpendicular to reinforcement, E_{HC}, σ_{HC}-moduli of elasticity and strength of hybrid composite, E_{IM}, σ_{IM}-moduli of elasticity and strength of hybrid composite based on low-moduli fibers.

In Fig. 1.1, the results of investigations on compressive strength and tensile strength variations in response to different volume share of components for various materials with different β [65, 80, 159]. It is evident that with increasing β, the parameters of the materials also increase. Similar conclusions can be made for elastic characteristics of the composites (Fig. 1.2) where $\beta = 5.8$ corresponds to GY-70/S-glass material and $\beta = 3.6$ to boron/organic

Table 1.2. Parameter β for various hybrid composites

	Hybrid	β
1.	S-glass/Kevlar-49	1.03
2.	E-glass/Kevlar-49	1.38
3.	HMS/HTS	1.46
4.	HTS/Kevlar-49	1.58
5.	HTS/S-glass	1.62
6.	HMS/E-glass	2.14
7.	HTS/E-glass	2.17
8.	HMS/Kevlar-49	2.3
9.	HMS/S-glass	2.36
10.	Boron/Kevlar-49	2.88
11.	Boron/S-glass	2.95
12.	Boron/E-glass	3.96
13.	Graphite GY-70/S-glass	5.8

Reference data make it possible to determine the dependence of material properties on this factor.

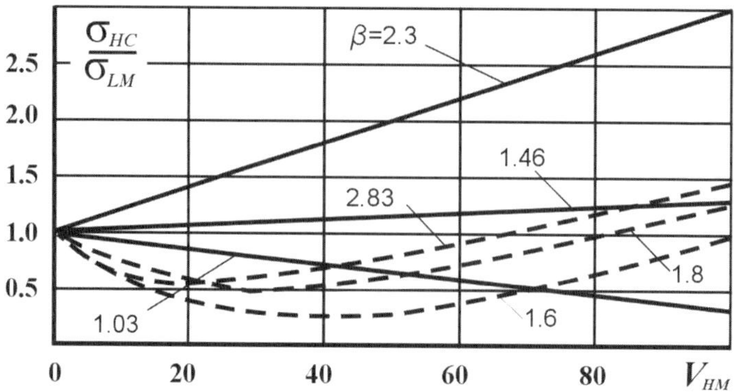

Fig. 1.1. Variations of compressive strength (*solid lines*) and tensile strength (*dotted*) of a hybrid composite as a function of high-modular fiber content $\beta = 1.03$ – Kevlar 49/S-glass, $\beta = 1.46$ – HMS/HTS, $\beta = 2.3$ – HMS/Kevlar 49, $\beta = 1.6$ – Kevlar 49/Thornel, $\beta = 1.8$ – Kevlar 49/S-glass, $\beta = 2.83$ – Thornel/S-glass

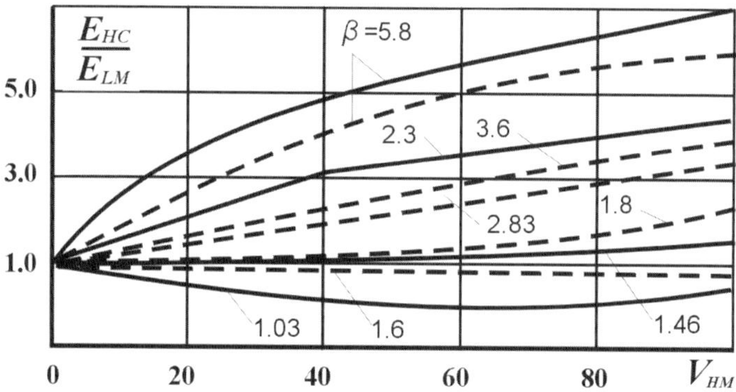

Fig. 1.2. Variations of elastic modulus at compression (*solid lines*) and tension (*dotted*) for hybrid composites depending on high-modular fiber content [65, 80, 159] E_{HC}-moduli of elasticity of hybrid composite, E_{LM}-moduli of elasticity of composite based on low-moduli fibers, V_{HM}-content (%) of high-moduli fibers in reinforcing fibers

fibers. As seen from Fig. 1.3, elastic properties also respond to β variations under transverse loading [65]. Absence of data on transverse strength and shear characteristics of hybrid materials make it difficult to analyze dependence of the properties on the β parameter. Changes in strength and elastic characteristics of hybrid composites with varying volume fraction of high-modular and low-modular components 40/60 are shown in Fig. 1.4 according to [65, 80, 159]. The results show that the elastic properties of the material grow drastically with increasing β.

On the degree of anisotropy of hybrid composites one can judge from Table 1.3, where E_Z – moduli of elasticity in the direction of reinforcement,

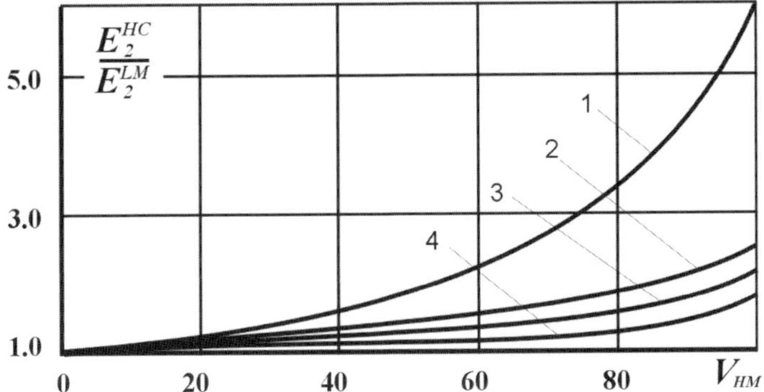

Fig. 1.3. Dependence of transverse elastic modulus on high-modular fiber content under transverse tension in the fiber direction [65] *1* organo-boron plastic ($\beta = 5.97$), *2* organo-glass plastic ($\beta = 2.24$), *3* organo-carbon plastic ($\beta = 1.62$), *4* glass-carbon plastic ($\beta = 1.34$)

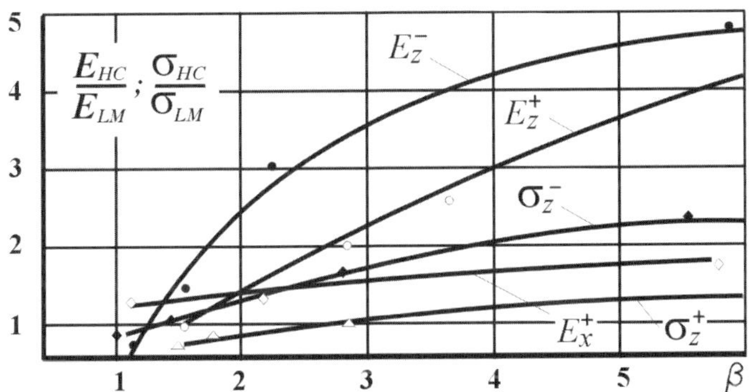

Fig. 1.4. Variations of strength and stiffness of hybrid composites depending on β parameter [65, 80, 159]

E_X – moduli of elasticity in the direction of reinforcement, G_{ZX} – shear moduli in the direction of reinforcement, σ_z^+ – strength of composite under tensile in the direction of reinforcement, σ_z^- – strength of composite under compression in the direction of reinforcement, σ_x^+ – strength of composite under tensile in the direction perpendicular to reinforcement, σ_x^- – strength of composite under compression in the direction perpendicular to reinforcement, τ_{xz} – strength of composite under shear.

When the results are compared with similar results derived for unidirectional materials (Table 1.1), we see that anisotropic parameters of hybrid composites lie within the interval of those for materials reinforced by each component individually.

Table 1.3. Anisotropic Parameters of Unidirectional Hybrid Composites [164, 178]

Composite parameters	E_Z/E_X	E_Z/G_{ZX}	σ_Z^+/σ_X^+	σ_Z^+/τ_{XZ}	σ_Z^+/σ_Z^-	σ_Z^+/σ_X^-
Organo-carbon plastic	14–18	32	55–0	20–30	1.5–4.5	6–15
Carbon-glass plastic	10.5–15	29.5–36.5	35–45	30–35	1.4–2.5	8.5–18
Organo-glass plastic	8–13	15–36.5	95–110	45–50	3–5	16–18
Organo-boron plastic	13–15.5	32–51	25–125	25–50	0.8–5	5–20
Boron-glass plastic	–	–	–	14.5	0.8	–

As shown above, one of the features of the hybrid effect is the departure of the material properties from the additivity rule. The hybrid effect is $[(E_{HC} - E_A)/E_A] \times 100\%$, where E_{HC} – experimental moduli of elasticity of hybrid composite, E_A – moduli of elasticity of hybrid composite calculated by additivity rule. Most investigations have confirmed experimentally and theoretically, that the presence of the hybrid effect and its value, as follows from Fig. 1.5, depends on parameter β.

The structure of hybrid composite materials largely governs their physic-mechanical properties. For example, changes in the sequence of layers [86] might essentially vary the properties of the material as a whole (Fig. 1.6). However, as the number of layers increases, the difference in strength reduces and the hybrid effect grows (Fig. 1.7). This also alters the fracture behavior of the material and the delamination length, as seen from Fig. 1.7, shortens perceptibly in the carbon-glass plastics as the number of layers diminishes.

If one compares layerwise and intralayered hybrid composites, the advantages of the latter become apparent. It turns out that high-modular fibers function better in bunches that are evenly spread throughout the material. Bunch size is influenced significantly by the hybrid effect [99] and this influence is higher than in laminated composites. Data of acoustic emission and fractography also support the fracture behavior in composites containing bunches with different fiber amount.

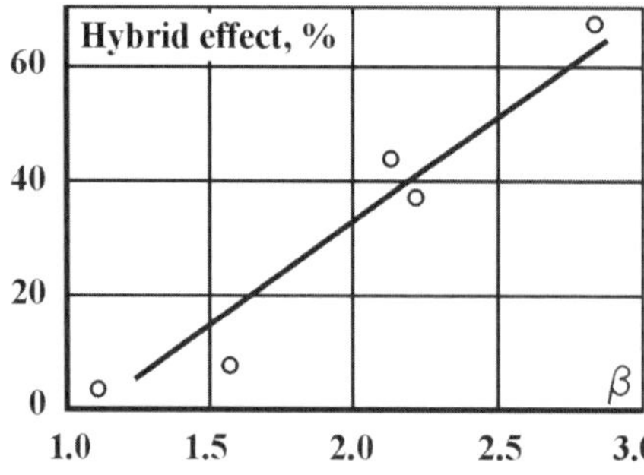

Fig. 1.5. Dependence of the hybrid effect value on β parameter [84, 90, 168]

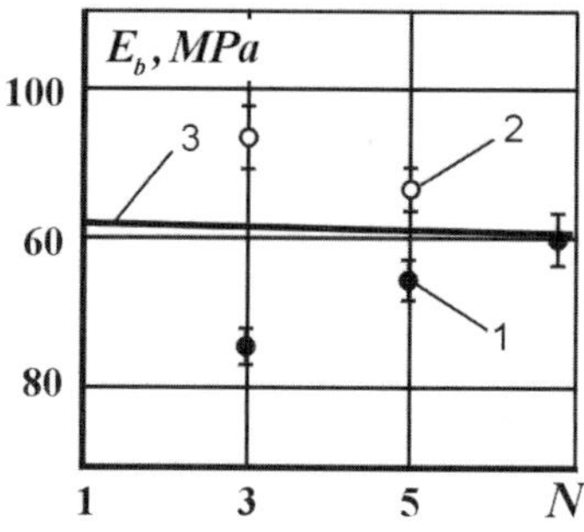

Fig. 1.6. Changes in elastic modulus under bending for a layerwise hybrid composite depending on the number of layers and their sequence [86]. *1* – the external layer – glass fibers; *2* – the external layer – carbon fibers; *3* – the additivity rule; E_i – modulus of elasticity of hybrid composite under bending; N – number of layers in hybrid composite

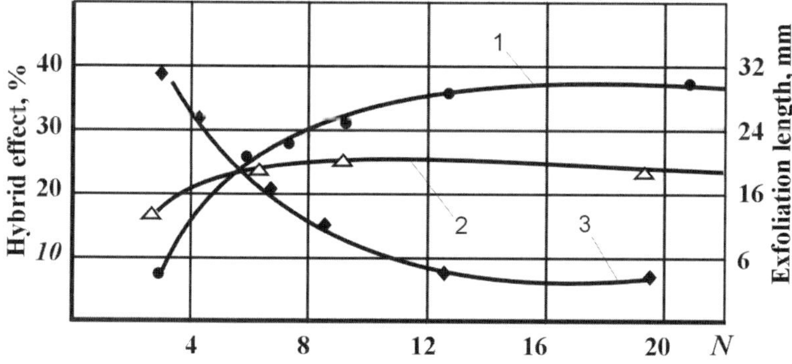

Fig. 1.7. Dependence of hybrid effect and exfoliation length on the number of layers [100]. *1* – hybrid effect in HTS/E-glass ($\beta = 18$), *2* – the same in HMS/E-glass ($\beta = 2.14$), *3* – exfoliation length at composite HTS/E-glass fracture

Hence, it is expedient to form intralayered hybrid composites with regular disposition of high-modular fiber bunches of small size; the (parameter of the material should be as large as possible.

It should be borne in mind that in the course of manufacturing and application of these materials, it is necessary to solve a number of problems, including provision of regular arrangement of high-modular fiber bunches against the low-modular filler, reduction of strength and elastic anisotropy, study of interfacial interaction in the hybrid composite as well as possible surface treatment types of adjoining components, aimed at improving the interfacial bonding quality. Analysis of the technical literature shows that some of these problems can be solved by using spatial reinforcement of articles and their combination with hybridization.

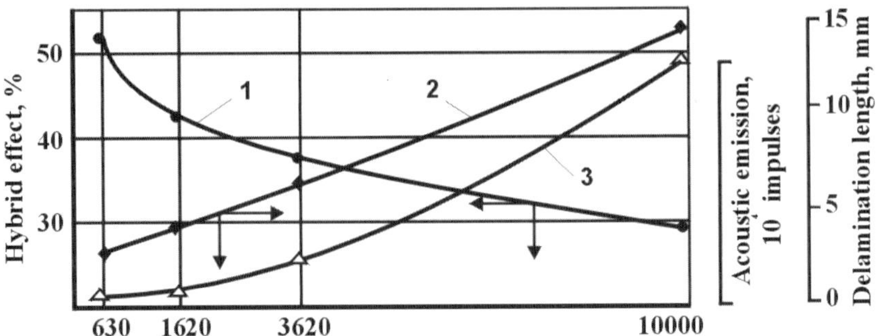

Fig. 1.8. Alteration of hybrid effect depending on bunch size of high-modular fibers [99]. *1* – alteration of hybrid effect, *2* – acoustic emission on rupture of fibers, *3* – delamination length at fracture

1.6 Spatial Reinforcement of Composites

The formation of additional interlayer physico-chemical links in a composite material can contribute to resistance to interlayer shear and transverse separation. This can be attained differently – either by weaving methods, 3D reinforcement or whiskering. Because the delamination of composites under nonstationary temperature loads is most harmful, spatial reinforcement was first used for the heat proofing of aircraft [179, 180].

The presence of reinforcing filler lowers the probability of article breakage as a result of crack propagation caused by micro defects. Later, materials of the carbon-carbon type [179–181], in which the low heat-resistant matrix is substituted by one after pyrolysis, gained popularity. The combination of spatial reinforcement with the pyrolyzed matrix has expanded the temperature range of such composites.

There are many processes for manufacturing spatially reinforced composite materials [7, 182–185, 296–298, 300]. Classification of such materials has been proposed in [9]. The first group includes composites in which spatial links are formed as a result of the distortion of part of the fibers in one direction (a system with two fibers). To the second group belong materials having fibers arranged in the third direction (the system of three fibers), which can be spaced just as mutually orthogonally at an angle in one of the reinforcement planes. Composites in which spatial links are formed through the introduction of thread-like crystals into the interfiber spaces are related to the third group glass fibers and high-modular carbon fibers are used as reinforcing materials for spatially reinforced composites. One more problem that presents great difficulties is the choice of the binder for such materials because impregnation and removal of air and gas inclusions from their structures is complicated.

It is stated in [186] that shaping of materials from a system of two fibers raises its strength on interlayer shear by 2 to 2.5 times, strength under transverse tension 4–5 times and interlayer shear modulus 1.2 times. In such circumstances the degree of flexure of the basic fibers, characterized by the disorientation angle θ, exerts a profound effect on the material properties [9]; where certain properties improve at the expense of others. For example, when θ increases from 0 to 30° the interlayer shear modulus doubles, while the elastic modulus in the packing direction of distorted fibers is reduced by as much as 1.5 times. Materials formed by a two-fiber system used in winding articles operating under pressure considerably reduce the finished article thickness. According to [187], the relative thickness of articles characterized by the outer to inner radius ratio is 1.4 when the described material is used, and on winding with a unidirectional tape it is 2.15, with a similar inner radius and pressure.

In contrast to a two-fiber system, the reinforcing filler in systems with three fibers is introduced in all three directions. With this aim a reinforcing framework can be formed in not only Cartesian and polar co-ordinates but also at alternating packing angles of reinforcing fibers across the packet thickness. Simplest of all is the method of insertion of woven and laminated materials [97, 179, 182]; this can, however, lead to damage of the inserted fiber layers. More refined methods employ complex technological equipment and a special program [184, 188]. The chief problem in this case consists of the introduction of fibers in a third direction. One of the means of lowering fiber damage is the netting of the reinforcing frame in a plane around steel tubes or dowels, which are then dragged through using fibers in the third direction. Dowels can damage the layers being inserted. As for tubes, although they lower fiber damage, they considerably reduce the degree to which the material can be filled.

Mechanical properties of spatially reinforced materials based on carbon or quartz fibers and phenolic binder are cited in [179, 189, 191, 192, 275]. The comparison of materials with similar materials formed from a two-fiber system proves that the spatially reinforced composites acquire higher characteristics. composites on a base of quartz fibers with a system of three fibers improve strength at transverse tension from 19 to 67 MN/m^2 and stiffness from 7.7×10^3 up to 21.7×10^3 MN/m^2. According to [182], the composite based on 3D fabric displays interlayer shear strength 85 MN/m^2 and 3700 MN/m^2 stiffness. For material based on a molded mat (two-fiber system) these values are 40 and 2500 MN/m^2, respectively.

Similar data have been derived for the improvement of fabric- based composite properties in a transverse direction. In this case, the packet is inserted with organic fibers constituting 4.4% of the binder volume, as a result of which strength and modulus of interlayer shear increase by 45% and 13%, respectively.

The method of radial reinforcement of composite shells should be noted where short steel fibers are used and embedded in the material in the course of the shell winding. As a result, the strength of the interlayer shear increases by 17–45% and strength under transverse tension by 12–88%. This, however, increases the weight of the structure, and its durability is conditioned by the corrosion resistance of the steel fibers.

Links between layers in composites reinforced by fibers are created by thread-like crystals either grown on the fiber surface or introduced into the binder. Mechanical properties of such materials depend upon the type and number of crystals, binder composition, type and volume share of fibers, and production procedures. Introduction of thread-like crystals of silicon carbide in amounts of 5.2% weight into, e.g. the carbon plastic based on Modmor-11 fibers, raises its strength on interlayer shear from 16 to 82 MPa at invariable elastic modulus in the main direction. It is noted in [193] that introduction of 4% by volume of thread-like crystals during winding of glass-reinforced plastic rings increases their interlayer strength by 38% and compressive strength by 17%.

For an arbitrary spatial reinforcement there are rather accurate methods for defining elastic characteristics proceeding from known properties of constituent components. Evaluation of such materials is usually made by averaging their stiffness [9, 194–196]. For example, in [9], elastic characteristics estimated by the principle of summing up the reiterating elementary layers and elastic properties of each layer are found in two steps. First, the properties of the reduced matrix are defined through averaging the elastic properties of fibers laid perpendicularly to the layer and binder plane. Secondly, the properties of the layer are estimated proceeding from the elastic properties of fibers laid parallel to the layer plane and those of the modified matrix. The conducted experiments showed good agreement with theoretical results.

As can be seen, spatial reinforcement is an efficient means of improving the characteristics of composites under transverse loading and longitudinal shear. Yet, it is to be borne in mind that the considerable intricacy of the manufacturing Process may often result in a cost increase of the product. In addition, as soon as there is a gain in one property, there is loss in another property in the main reinforcement direction.

1.7 Combination of Hybridization and Spatial Reinforcement

As shown above, the physico-mechanical properties of an article can be changed by altering the sequence and direction of individual packing layers. For the materials of the layers, different reinforcing fillers can be used. For the production of a number of standard structures of aircrafts, composites formed of glass and carbon fiber layers are often employed. Note that to eliminate warping and improve transverse characteristics, glass fiber layers are laid at $\pm 7°$ or $\pm 15°$ angles to the packing direction of carbon fibers.

This suggests that during the manufacture of such products the processes of hybridization and spatial reinforcement should be united in one cycle.

Model materials based on quartz and silicone fibers and reinforced by a system of three fibers have been described in [9]. Such hybrid spatially reinforced composites display higher characteristics in the transverse direction and at shear than layered materials, even with insignificant silicone fiber content (up to 13%) in the third direction. The interlayer shear strength of a composite with the filling degree 45% constitutes $84\,MN/m^2$, which is higher than that of the laminate. A similar combination of spatial reinforcement and hybridization is encountered in laminated materials on a base of carbon fibers inserted with organic fibers [97] or in radially reinforced materials.

The authors of [9, 197] show that the introduction of slight amounts of fibers forming spatial orientation reinforcing much improves the material properties. An example is the metal-composite on the basis of a copper matrix reinforced with spirally tungsten fibers [198]. It has been shown that variation of the spirally winding angle within 15–$76°$ essentially alters the limiting strain and fracture energy of the matrix. Similar data on fracture mode variation are cited in [38]. Here, the reinforcing fibrous elements include two components, the core consisting of straight fibers, and the shell formed of other fibers, which envelop the core over a spirally. By varying the core-shell contact conditions the straight fibers can be made unbreakable, which also contributes to raising characteristics at cyclic loads through additional energy dissipation.

Thus, along with the hybridization and spatial reinforcement of a structure as a whole, these methods can be used to manufacture reinforcing fillers as well.

One of the first spatially reinforced hybrid fibrous fillers was developed and produced on the basis of boron and organic fibers as specified in [199]. To raise the productivity of the method it has been proposed to wind, at a certain pitch, bunches of boron fibers with glass fibers. A similar filler structure was devised in [234], where organic fibers were the main reinforcement on whose surface glass fibers were spirally wound. Material thus produced resembles vegetable structures.

The method of combined hybridization and spatial reinforcement has been extended to spirally reinforced fillers based on boron fibers. As the reinforcing filler, complex boron-glass fibers have been used, in which seven parallel boron fibers are braided with a glass fiber at 4 mm pitch. Materials with this filler and uniaxial or orthogonal reinforcement schemes show high compressive, shear and bending strength and fatigue strength surpassing by 25–30% the similar characteristics of the usual plastics.

Complex boron-glass fibers are easily impregnated with a binder and, when compacted at forming, ensure the needed filler content in the material. Application of named fillers has made it possible to reduce manufacturing time by 10 times. The substitution of the monofiber in the filler for the spatially reinforced one impairs (by 10–15%) the mechanical characteristics in the reinforcement direction (elasticity modulus in the main) but improves the shear characteristics by 10–15% and substantially rises – by 35–65% transverse strength and impact toughness. Hence, by joining seven fibers together one can obtain complex boron-glass braids for which high transverse strength values are characteristic.

Boron-carbon braids have been designed in which boron fibers, complex glass fibers and carbon ones in various combinations are wound with glass fibers at 3.5–4.1 mm pitch. Notice that with the increasing number of carbon fibers the material properties at bending are not impaired, i.e. the weight effectiveness of the material improves.

Application of spatially reinforced fillers does not exclude other means of raising the transverse and shear characteristics of the composites. For instance, chemical activation of complex boron-glass fibers increases the shear strength of the boron-plastics from 67.5 to 104 MPa.

Application of spatially reinforced fillers also assists in regulating mechanical properties of the composite. Reinforcing fibers laid in a spiral unite the main reinforcing fibers into bunches, which makes them operate uniformly and increases fracture energy at the bunch/matrix interface.

1.8 Conclusions

The combination of hybridization and spatial reinforcement methods yields promising spatially reinforced fillers for composite materials. By using fibers of different strength, deformation, geometrical and other parameters and by varying their correlation and mutual position one can obtain fillers able to regulate properties of composites within a wide range and to refine their manufacturing process. Yet there is a lack of investigations into materials of the type, whilst studies available in this field chiefly deal with boron plastics and their modification. Nevertheless, interest in the development and investigation of materials based on carbon and organic fibers with high specific mechanical characteristics and essential anisotropy of physico-mechanical properties is gradually growing.

2 Major Principles
of Developing Material Structures
with Hybrid Spirally Reinforced Filler

2.1 Analysis of Structural Schemes of Materials

The design of materials with hybrid spirally reinforced filler is a system of specifically arranged filler elements consisting of bundles of straight-line fibers, threads, and plaits wound by cockling filaments of some other fibrous material (Fig. 2.1). Later, such elements would be called spirally reinforced elements. Since the major force is perceived at loading by the inner fiber bundles, called the element core, they will be considered as the main reinforcement, whereas the winding filaments are an auxiliary one. The layers of auxiliary reinforcement that separate individual bunches of the main reinforcing material we will call intermediate layers, or simply layers.

Goods made from materials with the spirally reinforced filler are produced by the winding or spreading out of a prepreg consisting of spirally reinforced elements, followed by molding. During the process either a unidirectional or layerwise with crossing, or spatial schemes of reinforcement and others can be realized. It is emphasized that the presence of the auxiliary reinforcement allows for spatial reinforcement of both the spirally reinforced element and the material as a whole even in unidirectional composites. Its structure and parameters are strongly conditioned by the type of fibrous materials used for the main and auxiliary reinforcement.

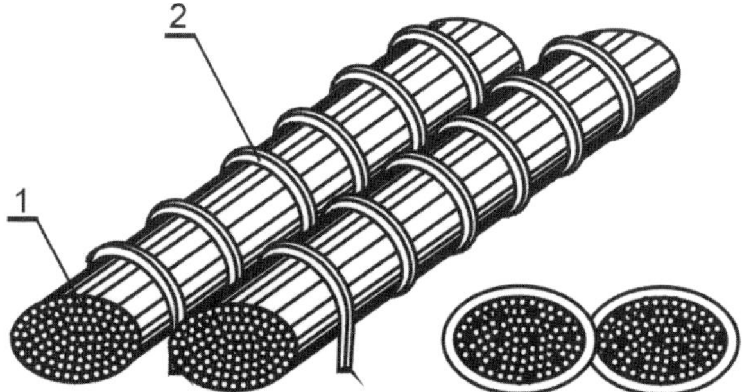

Fig. 2.1. Structure of the hybrid spirally reinforced filler

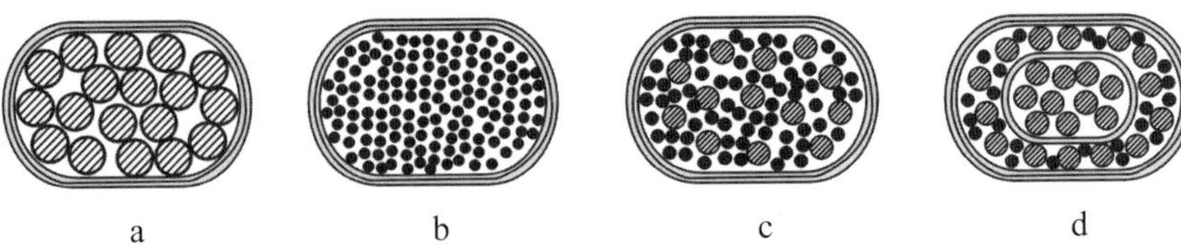

Fig. 2.2. Sectional view of spirally reinforced elements used for composite materials

When the fibers are of comparatively large diameter (above $100\,\mu$m), e.g. boron, they can be laid in a hexagonal manner inside the spirally reinforced elements. This ensures structural regularity and maximum rated filling degree. The target element acquires close to circular form and displays good dimensional stability, preserved at all stages of machining (Fig. 2.2a). When there are small amounts of fibers inside the element, the reinforcement is better saturated with various types of polymer binders, without attracting any special impregnation procedures.

In contrast, production of spirally reinforced elements from fine fibers ($d_a = 5$–$20\,\mu$m), including glass, carbon, organic filaments and braids has the following peculiarities. The small diameter of the elementary fibers and their large number within the element composition hampers the perfect arrangement of the fibers and ensures a regular structure, similarly to a unidirectional composite (Fig. 2.2b). The attained element is inclined to dimensional variations during processing and might acquire an elliptical or polygonal shape while forming. Impregnation of the structural element core is impeded due to a high enough filling degree, so it is recommended to preimpregnate the main reinforcement followed by winding with an auxiliary reinforcement.

To reach a spirally reinforced structure of the elements, bunches of a polyfibrous filler can be used as the main reinforcement (see Fig. 2.2c). Elements of such a structure are strongly filled and display properties characteristic of both the above-considered types of spirally reinforced elements.

It is possible to form structures of spirally reinforced fillers of higher orders, based on coaxial or group winding of simple elements. These elements are, in their turn, structural units of composite materials and, very often, are finished parts of the rod type or supporting members of power structures (Fig. 2.3).

In Fig. 2.4, the main types of spirally reinforced elements are presented. According to the classification proposed, there are three chief types of elements: (i) fillers of composite materials with different types of fibers used and different kinds of spirally reinforced element, (ii) semi-finished products, and (iii) members of articles.

In the latter case one can isolate them as finished products, such as rods and hollow elements produced by melting of the mandrel, as insets employed at assembly. These types of elements are subdivided by the principle of manufacturing articles on the basis of spiral reinforcement and hybridization.

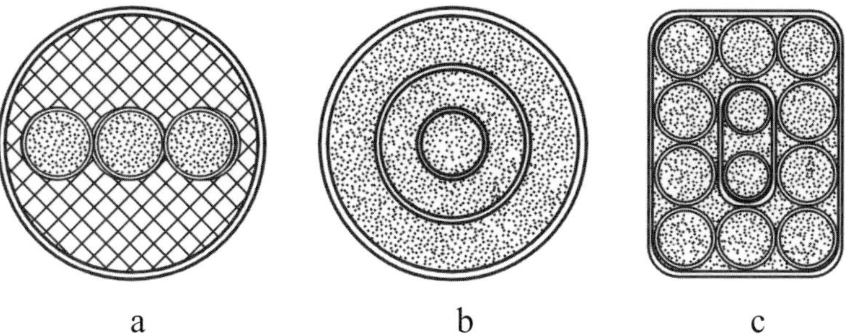

Fig. 2.3. Types of spirally reinforced elements of higher orders: (**a**) complex spirally reinforced element, (**b**) coaxially reinforced element, (**c**) – combined

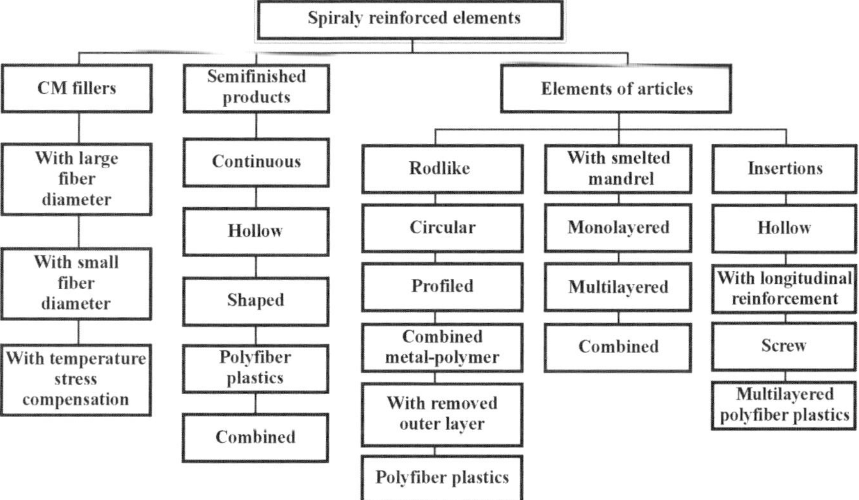

Fig. 2.4. Classification of spirally reinforced elements

Composites with spirally reinforced filler are referred to as materials of a complex structure of two levels, each of which is formed at different stages of the technological process. The major structural unit, i.e. the spirally reinforced element, conditions the first level characterizing the material macrostructure. Thus, the material structure is dependent on both mutual arrangement and type of elements. The second level characterizes the material microstructure and is specified by the parameters of the spirally reinforced element itself. A diagram of the major structural parameters of the material with spirally reinforced filler is illustrated in Fig. 2.5.

Microstructural parameters of the elements are realized during processing of the fibrous reinforcement into spirally reinforced braids. They are dependent upon impregnation of the main reinforcing material, processing regimes,

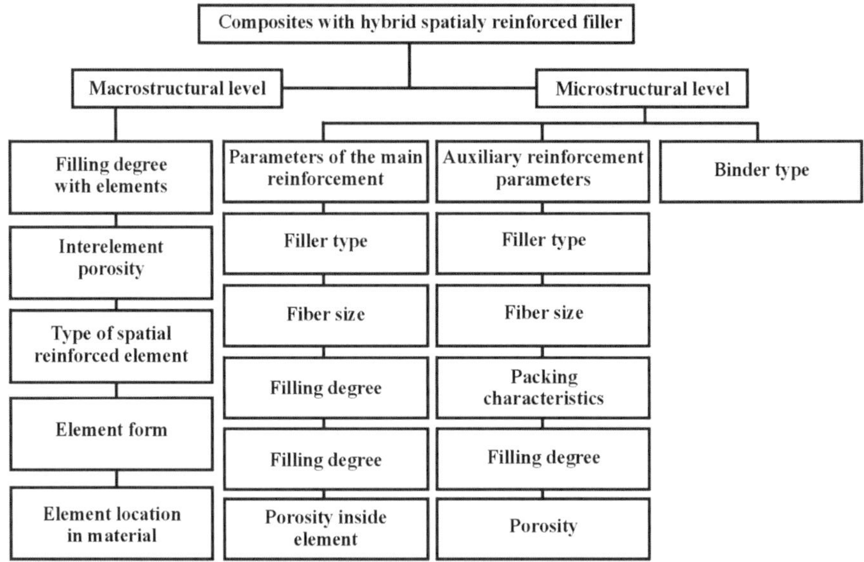

Fig. 2.5. Schematic diagram of the main structural parameters of the material

tension forces of the main and auxiliary reinforcement, and the type of the main reinforcing filler. Macrostructural parameters are governed by the technological procedure of material shaping from the elements, including forming peculiarities, laying spacing, tension force, molding pressure, winding speed, curing process and its regime.

A close interrelation is observed between the structural parameters of the material at each level, the technological process and all stages of producing composites with spirally reinforced filler. This interrelation is, in the long run, the determining factor for the composite material structure and its characteristics.

2.2 Filling Degree of the Material with Spirally Reinforced Filler

One of the most important characteristics of reinforced materials is the filling degree, which specifies their physical and mechanical properties. Unidirectional reinforced composites are known to have a certain limiting filling degree, which is dependent on the disposition of the reinforcing fibers within the material bulk.

For tetragonal laying it constitutes 0.785, for hexagonal −0.907. Because of the small diameter of the elementary fibers and their great quantity in the filler, limiting degrees of filling are not attained as a rule during the technological procedure.

Since it is hard to adhere to an even arrangement of fibers in the material, pores appear with increasing fiber content that lead to difficulties in reaching the needed saturation quality. In this connection, a specific optimum filling degree has been developed for each type of composite amounting to 50–70%, which excess results in impaired elastic and strength characteristics.

The filling degree of the composites with spirally reinforced fillers is determined by the sum of filling degrees of the main φ_0 and auxiliary φ_a reinforcements

$$\varphi_0 = \varphi_0 + \varphi_a \,. \tag{2.1}$$

Correspondingly, most important for material properties in the direction of the main reinforcement is the filling degree with the main reinforcement. Its value is calculated as follows

$$\varphi_0 = \varphi_1 \cdot \varphi_2 \,, \tag{2.2}$$

where φ_1 is the filling degree of the element core with fibrous filler, which maximum value $\varphi_1 = 0.907$ corresponds to the hexagonal packing of fibers; φ_2 – filling degree of material with element cores.

As indicated earlier, deformation of the original spirally reinforced filler leads to variation of its initial circular shape. The element acquires the form of an ellipse or, in the limiting case, of a polygon and its filling degree with the main reinforcing fibers increases:

$$\varphi_1 = \varphi_{in} \cdot q \,, \tag{2.3}$$

where φ_{in} – initial filling degree of the element core.

Assuming that the core of the spirally reinforced element acquires as a result of deformation the form of an ellipse with semiaxes a and b, and the element core perimeter remains intact, it is possible to derive a relation for q factor with allowance for the element deformation

$$q = \frac{4E^2(\varepsilon^2)}{\pi^2\sqrt{1-\varepsilon^2}} \,, \tag{2.4}$$

where ε – eccentricity of the ellipse; $E(\varepsilon^2)$ – is the complete second order elliptical integral.

Table 2.1 shows the interrelation between q values, ellipse eccentricity value ε and its semiaxes ratio $m = a/b$, using parameters for the complete second order elliptical integral $E(\varepsilon^2)$. The dependence of $q(\varepsilon)$ on $q(m)$ is shown in Fig. 2.6. the case of insignificant deformation of the circular section of the element $(\varepsilon \to 0)$, one has

$$E(\varepsilon^2) \sim \frac{\pi}{2}\left[1 - 2\left(\frac{\varepsilon^2}{8}\right) - 3\left(\frac{\varepsilon^2}{8}\right) - 10\left(\frac{\varepsilon^2}{8}\right) - \frac{175}{4}\left(\frac{\varepsilon^2}{8}\right)^4 - \cdots\right] \,.$$

Table 2.1. The Dependence of q on Geometrical Parameters of the Spirally Reinforced Element Core

m	ε	ε^2	$E(\varepsilon^2)$	$\sqrt{1-\varepsilon^2}$	q
1.0	0	0	1.5708	1	1
1.5	0.745	0.556	1.3219	0.667	1.0617
2.0	0.866	0.75	1.2110	0.5	1.1890
2.5	0.917	0.84	1.1507	0.4	1.3415
3.0	0.943	0.889	1.1135	0.333	1.5078
4.0	0.968	0.938	1.0716	0.25	1.8621
5.0	0.98	0.96	1.0505	0.2	2.2371
6.0	0.986	0.972	1.0377	0.167	2.6178
7.0	0.99	0.98	1.0286	0.143	2.9985
8.0	0.992	0.984	1.02355	0.125	3.3967
9.0	0.994	0.988	1.0185	0.111	3.835
10.0	0.995	0.99	1.01599	0.10	4.3252

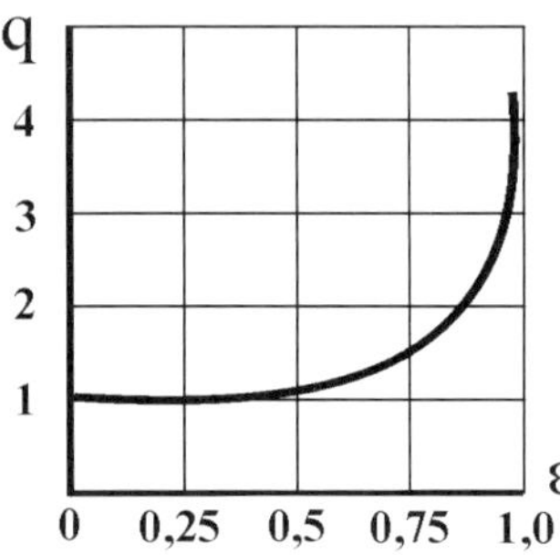
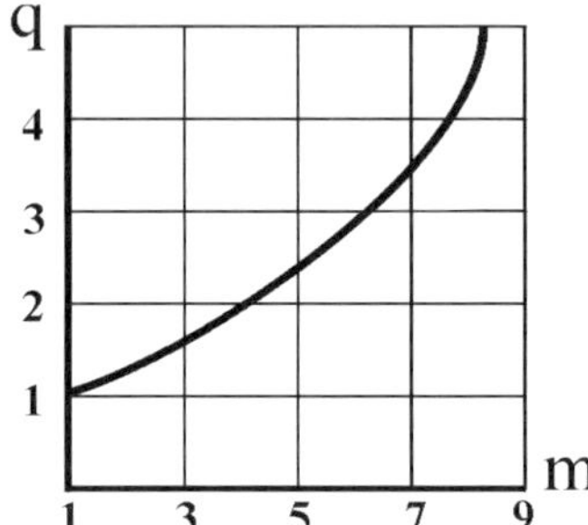

Fig. 2.6. Variation of q factor depending on geometrical parameters of the spirally reinforced element core

$$q \sim \frac{\left[1 - \dfrac{\varepsilon^2}{4} - \dfrac{3\varepsilon^4}{64} - \cdots\right]}{\sqrt{1-\varepsilon^2}} \approx \frac{1}{1 - \dfrac{3}{32}\varepsilon^4}. \tag{2.5}$$

Relation (2.5) is true for $\varepsilon < 0.5$ or $m < 1.2$.

Under higher $m \gg 1$ the asymptotic formula is true.

$$E(\varepsilon^2) \sim 1 + \frac{1}{2}\left(\Lambda - \frac{1}{2}\right)(1-\varepsilon^2) + \frac{3}{16}\left(\Lambda - \frac{3}{12}\right)(1-\varepsilon^2)$$
$$+ \frac{15}{128}\left(\Lambda - \frac{6}{5}\right)(1-\varepsilon^2)^3 + \ldots,$$

where $\Lambda = \ln \dfrac{4}{1-\varepsilon^2}$.

It is easy to show that asymptotically the next equality is true:

$$q = \frac{4\left(m + \dfrac{\ln 2m}{m}\right)}{\pi^2}. \tag{2.6}$$

Hence, within the active values of m or ε one must use the accurate formula (2.4) or data from Table 2.1 for the dependence $q(\varepsilon)$. For limiting cases when $\varepsilon \to 0$, $m \to 1.0$ or $\varepsilon \to 1.0$, one can use for q asymptotic formulas (2.5) and (2.6) yielding an error in the third sign within the indicated limits. Moreover, if one uses an approximate formula for the ellipse perimeter through its semiaxes a and b

$$L = \pi[1,5(a + b) - \sqrt{ab}],$$

it is possible to obtain a simpler relation for q:

$$q = \frac{\left[\dfrac{3}{2}(1 + m) - \sqrt{m}\right]}{4m}. \tag{2.7}$$

The formula thus obtained is sufficiently accurate. With growing $m \to \infty$, its relative error rises, which is also true for m presented in Table 2.1, i.e. for $m \to 0$ the error reaches the order of 0.01.

The degree of the composite filling with cores of spirally reinforced elements φ_2 is related to geometrical parameters and the element shape as well as the manner of joining the layers of the auxiliary reinforcement.

To find φ_2 it would be expedient to consider random packing of a sectional plane with elliptical spirally reinforced elements (Fig. 2.7a). It is seen that the major semi-axis coincides with axis $0_1 U$. Let us make the affine transformation of the plane, that stretches perpendicularly to axis $0_1 U$ with coefficient $m = a/b$. Then, the ellipses will transfer to tangent circles of $R^* = a$ radius and triangle $O_1' O_2' O_3'$ will be equilateral with sides $2a$ (Fig. 2.7b).

Since abscissas remained invariable at affine transformation and the ordinates are incremented at tension factor m, then

$$\operatorname{tg} \alpha' = m \operatorname{tg} \alpha; \quad \operatorname{tg} \beta' = m \operatorname{tg} \beta; \quad \alpha' + \beta' = \pi/3.$$

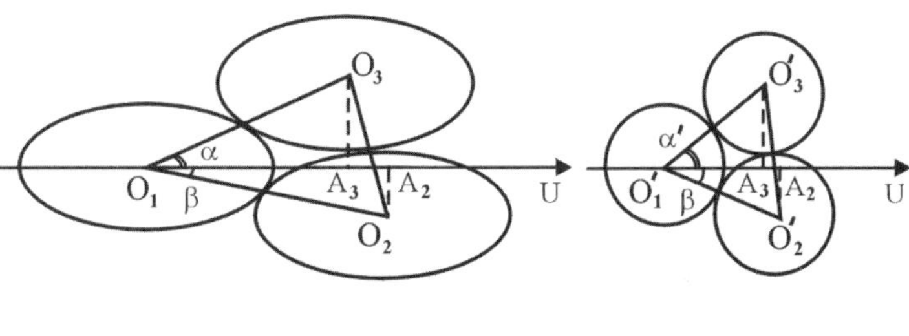

Fig. 2.7a,b. Design model for calculation of filling degree φ_2

Thus, at a given value of angle β it is possible to calculate angle α:

$$\operatorname{tg}\alpha = \frac{\sqrt{3} - mtg\beta}{m(1 + \sqrt{3}mtg\beta)}\;.$$

From the latter it follows that contact angles of ellipse β and α can be random and interrelated, and their relation is determined by the semiaxes ratio. Taking $\beta = 0$ or $\alpha = \beta$, particular cases of the packing can be obtained when, correspondingly $m > 1.0$ or $m < 1.0$.

Insofar as the ratio of figure areas remains invariable before and after affine transformation, then the area of triangle $0_1 0_2 0_3$, specifying the elementary cell, can be calculated thus

$$F_1 = \frac{1}{m}[a^2\sqrt{3}(1 + k)^2]\,,$$

and φ_2 is found from

$$\varphi_2 = \frac{\pi}{4t \cdot S(1 + k)(1 + km)}\,, \tag{2.8}$$

where $k = \delta/a$ specifies the auxiliary reinforcement layer thickness and coefficient t – conditions of layer joining. In the case of unbroken winding $t = 1.0$, and at mutual cohesion of turns

$$t = \frac{(2 + k)(2 + km)}{4(1 + k)(1 + km)}\,. \tag{2.9}$$

Factor S takes account of the packing type of elements in the material. At tetragonal it is $S = 1.0$ and at hexagonal $S = \sqrt{3}/2$.

From (2.8) and (2.9) it follows that the filling degree φ_2 depends not only on the auxiliary reinforcement layer thickness but also on the ratio of ellipse semiaxes. However, with reducing layer thickness, i.e. when $k \to 0$, φ_2 grows to a limiting value that is, i.e. at hexagonal packing, equal to $\varphi_2 = \pi/2\sqrt{3} = 0.906$.

The dependence of filling degree φ_2 on k for round elements ($m = 1$) at various values of S and t is illustrated in Fig. 2.8. From the given data it follows that with incrementing thickness of the winding layer, the filling degree diminishes significantly. However, when coils are engaged the variation is less evident and has a region where the tetragonal packing allows for a higher filling degree than the hexagonal one with a continuous winding layer. The filling degree by the auxiliary reinforcement can be presented as

$$\varphi_f = \varphi_3 \cdot \varphi_4\,, \tag{2.10}$$

where φ_3 is the filling degree by fibers inside the winding reinforcement whose maximum value at hexagonal packing is $\varphi_3 = -0.907$; φ_4 – filling degree layerwise.

When the layer of auxiliary reinforcement is not thick, value φ_4 is calculated by

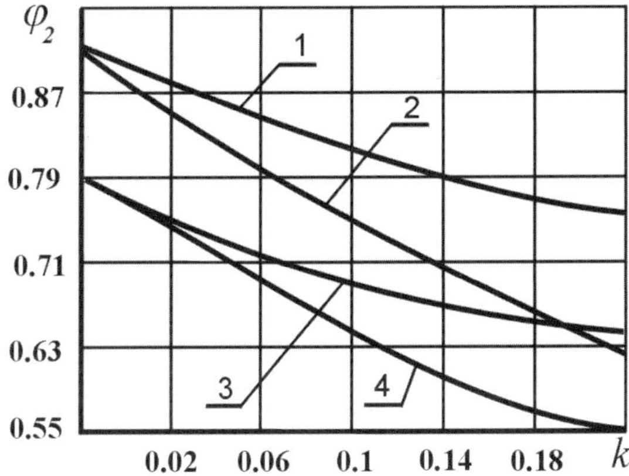

Fig. 2.8. The effect of packing with round elements on filling degree φ_2 is dependent on the interlayer thickness. *1, 2* – hexagonal; *3, 4* – tetragonal; *1, 3* – with coil engagement; *2, 4* – continuous winding

$$\varphi_4 = \frac{\pi n}{4S(1+k)(1+km)}\,, \tag{2.11}$$

where factor n is dependent on the winding type. Thus, for a continuous layer

$$n = (1+k)(1+km) - 1\,, \tag{2.12}$$

and for a stepwise winding with engagement of the coils:

$$n = \frac{k}{t}\sqrt{\frac{1}{8}[m^2(2+k)^2 + (2+mk)^2]}\,. \tag{2.13}$$

The analysis performed has shown that along with the main reinforcement content the filling degree of the composite material with spirally reinforced filler is also conditioned by the size of the spirally reinforced element, thickness and kind of bonding of the auxiliary reinforcement layers. Hence, to attain the highest filling parameters in the main direction it is expedient to use fine filaments as the auxiliary reinforcement, since they can be wound on the main reinforcement fibers at a pitch that provides for mutual engagement of the coils of the neighboring spirally reinforced elements.

2.3 Structural Analysis of Composites with Spirally Reinforced Filler

It is possible to change the shape of the spirally reinforced element during processing. This is conditioned by the filling degree of the element core with fibers of the main reinforcement. When the filling degree is sufficiently high, the element displays significant rigidity, and is negligibly deformed. In the case of a little fiber content, the element is, vice versa, subject to deformation to a stronger extent. It is, therefore, evident that the type of material structure much influences its physico-mechanical characteristics. Therefore, it is necessary to define the parameters of the thus obtained structure and the shape of the spirally reinforced element.

2.3.1 Material Structure at High Filling Degree of the Element Core

The original material structure where each layer is packed with spirally reinforced elements so as to form a clearance h between them and the subsequent layer is laid in the hollows of the previous one, is depicted in Fig. 2.9. When the structure is compressed in a vertical direction the shape of the element core and the structure as a whole change. Since the deformability of the element is rather small, the emerging spots of flattening of ρ length are conjugated as is presented in Fig. 2.10. So far, by introducing the notion of the material percentage reduction $\gamma = 0_3 0_4 / 0'_3 0'_4$ (Fig. 2.10b), one can express the sought parameters of the structure as follows,

$$r = \frac{1}{2}(b - p)\sin\beta\,;$$

$$\operatorname{tg}\frac{\alpha}{2} = \frac{\gamma\sqrt{4R^2 - (R + h/2)^2}}{R + h/2}\,;$$

$$\omega_2 = \sqrt{4\gamma^2 R^2 + (1 - \gamma^2)(R + h/2)^2}\,;$$

$$\omega_1 = 2R + h\,. \tag{2.14}$$

To determine the flattening p let us denote

$$x = p\cos\frac{\beta}{2}\,;\ y = p\sin\frac{\beta}{2}\,;\ m_1 = \omega_1\cos\frac{\alpha}{2}\,;\ n_1 = \omega_1\sin\frac{\alpha}{2}$$

and, proceeding from conditions of the contact between neighboring elements (see Fig. 2.11), we obtain an equation that interrelates structural parameters and element dimensions:

$$m_1 x - x^2 = n_1 y - y^2\,. \tag{2.15}$$

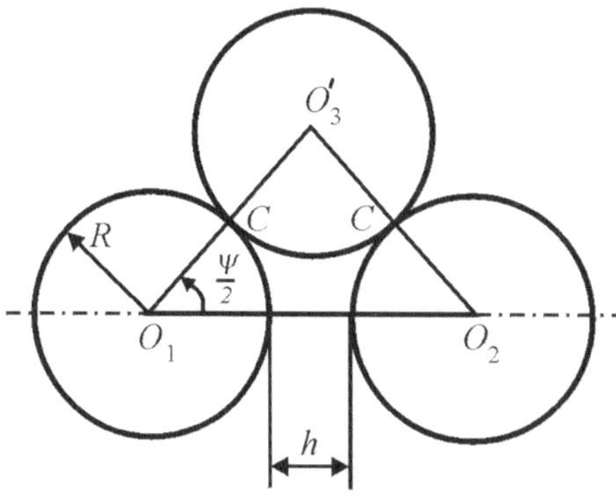

Fig. 2.9. The original material structure

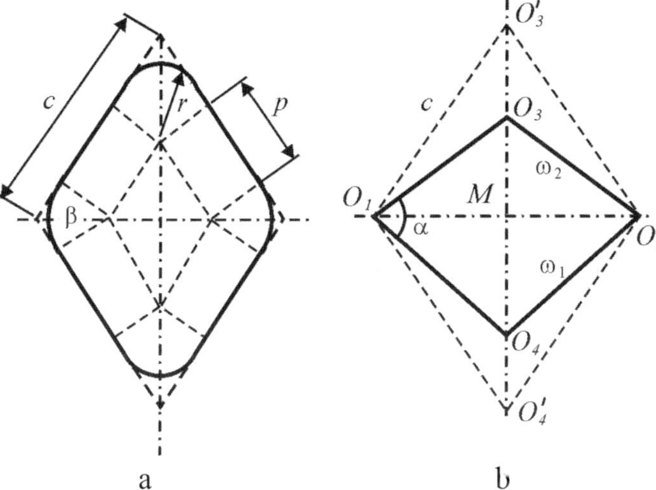

Fig. 2.10. Variations of material structure and core shape of the spirally reinforced element under vertical squeezing

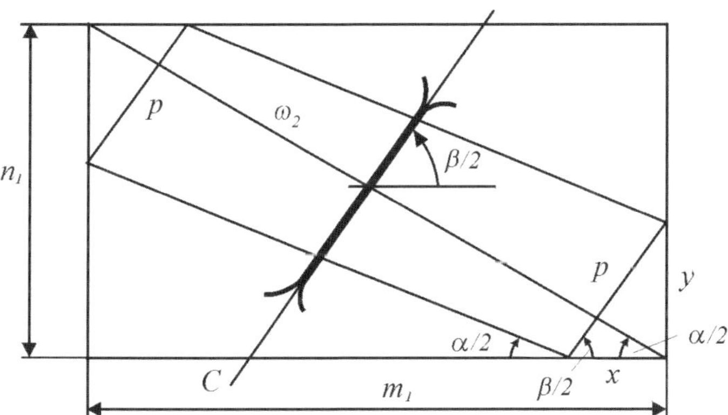

Fig. 2.11. To calculate the flattening value of the spirally reinforced element

Together with (2.15) invariance conditions of the element section perimeter define the system of equations about x and y, which is equivalent to an eighth-power equation.

$$m_1 x - x^2 = n_1 y - y^2 \, ;$$

$$4\sqrt{x^2 + y^2} + \pi \left[\frac{1}{2} \frac{\sqrt{x^2 + y^2}(my + nx)}{xy} - \sqrt{x^2 + y^2}\,\frac{2xy}{x^2 + y^2} \right] = 2\pi R \, ; \quad (2.16)$$

Let us solve the stated problem, taking the smallness of parameter p and neglecting x^2 and y^2.

The solution will be sought as

$$\frac{\beta}{2} = \frac{\pi}{2} - \frac{\alpha}{2} + \xi.$$ (2.17)

Then, with accuracy to the 1st order of smallness, the second equation of system (2.16) will be of the form

$$p^2 \cos\alpha + 2p^2\xi \sin\alpha = \omega_2 p\xi.$$ (2.18)

Wherefrom, by neglecting $2p^2\xi \sin\alpha$, one gets:

$$\xi = \frac{p \cos\alpha}{\omega_2}.$$ (2.19)

Therefore, the relation between b and ω_2 is determined as follows:

$$b = \frac{m_1}{2\cos\dfrac{\beta}{2}} + \frac{n_1}{2\sin\dfrac{\beta}{2}} = \omega_2\frac{\cos\xi}{\sin\beta}.$$ (2.20)

Substituting the defined values into (2.18) and neglecting the member of 3rd order smallness, we have

$$p = \frac{\pi(2R - \omega_2)}{4\sin\alpha}.$$ (2.21)

It follows that the shape of the spirally reinforced element and structural parameters can be determined by relations (2.14), (2.17), (2.20) and (2.21).

As an example, structural parameters and dimensions of the spirally reinforced element under various percentage reduction values γ, clearance h and $R = 0.5$ are presented in Table 2.2.

Table 2.2. Structural Parameters of Materials with a Highly Filled Element Core

h	γ	α^0	ω_2	p	β^0	C	r	F_{EL}	$\dfrac{F_0^*}{F_{EL}}$
0.25R	1.0	102.64	1.0	0	77.36	1.025	0.5	0.7854	1.0
	0.95	99.75	0.9698	0.024	79.77	0.986	0.473	0.7492	1.0483
	0.9	96.69	0.9403	0.047	82.64	0.447	0.714	0.7136	1.1007
	0.8	89.95	0.8835	0.092	90.06	0.884	0.396	0.6460	1.2158
0.2R	1.0	106.26	1.0	0	73.74	1,042	0,.5	0.7854	1.0
	0.95	103.42	0.9683	0.023	75.88	0.999	0.472	0.7480	1.05
	0.9	100.39	0.9372	0.050	78.51	0.956	0.711	0.7109	1.0511
	0.8	93.70	0.8773	0.096	85.49	0.88	0.391	0.6392	1.2288
0.15R	1.0	109.8	1.0	0	70.2	1.063	0.5	0.7854	1.0
	0.95	107.01	0.9668	0.027	72.04	1.016	0.470	0.7472	1.0511
	0.8	97.40	0.8712	0.102	80.87	0.882	0.385	0.6336	1.2396
0.1R	1.0	113.27	1.0	0	66.73	1.089	0.5	0.7854	1.0
	0.95	110.54	0.9654	0.029	68.25	1.039	0.469	0.7468	1.0516
	0.8	101.08	0.8654	0.108	76.18	0.891	0.380	0.6294	1.2478

* F_0 – the original element area; F_{EL} – the deformed element area

2.3.2 Structural Features of Material with a Slightly Filled Spirally Reinforced Element

Under deformation of a slightly filled core with the chief reinforcement material in a vertical direction ($\gamma < 1$), the structure and shape of the element core change as shown in Fig. 2.12. Structural parameters ω_1, ω_2 and α can be determined in this case by (2.14) whereas R_1 and R_2 are found as

$$R_1 = \frac{1}{2}\left(\omega_2 - p\,\mathrm{tg}\,\frac{\alpha}{2}\right)$$
$$R_2 = \frac{1}{2}\left(\omega_1 - p\,\mathrm{ctg}\,\frac{\alpha}{2}\right) \qquad (2.22)$$

where flattening value p is found from the invariance condition of the element perimeter

$$p = \frac{\pi(2R - \omega_2)}{4 - \pi\,\mathrm{tg}\,\dfrac{\alpha}{2} - 2\alpha\,\mathrm{ctg}\,\alpha} . \qquad (2.23)$$

Shown in Fig. 2.12, a core form of a spirally reinforced element is easily approximated by an ellipse whose semi-axis can be readily expressed through the initial values and h:

$$a = \frac{1}{2}\left[\omega_2 + \frac{p}{\cos\dfrac{\alpha}{2}}\left(1 - \sin\frac{\alpha}{2}\right)\right] ;$$
$$b = \frac{1}{2}\left[\omega_2 + \frac{p}{\sin\dfrac{\alpha}{2}}\left(1 - \cos\frac{\alpha}{2}\right)\right] . \qquad (2.24)$$

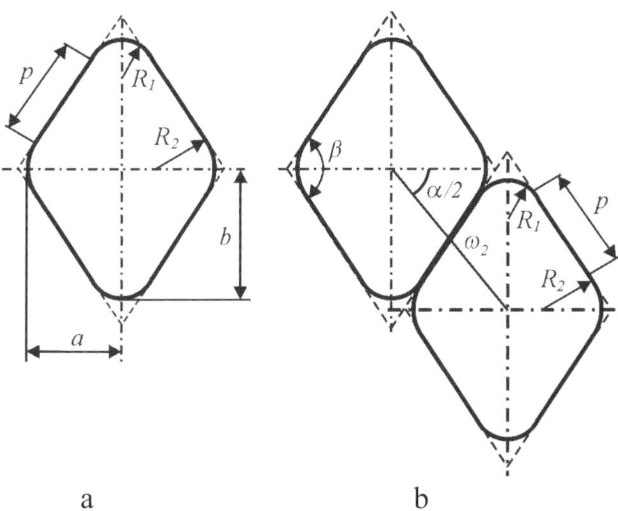

Fig. 2.12. Variation of material structure and element shape under compression in a vertical direction for the case of slight filling with the main reinforcement

Structural parameters of the material with the initial radius of the element $R = 0.5$, various h and percentage reduction γ obtained by the above relations are given in Table 2.3. Limiting γ can be determined under a condition when the contact between polar ellipses in either vertical or horizontal direction is absent:

$$b \leq R + \frac{h}{2}; \quad a \leq \left(R + \frac{h}{2}\right) \operatorname{tg} \frac{\alpha}{2}.$$

For instance, for $h = 0.25$ given in Table 2.3, this value will be $\gamma > 0.7$.

When the elements are originally in a hexagonal packing, i.e. at $h = 0$, the structural parameters can be found from

$$\begin{aligned}
\alpha &= \operatorname{arctg} \gamma\sqrt{3}\,; \\
\omega_2 &= R\sqrt{3\gamma^2 + 1}\,; \\
\omega_1 &= 2R\,.
\end{aligned}$$
(2.25)

Hence, the material deformation scheme changes because each element contacts six neighboring ones. Consequently, six flattening areas arise at deformation, where curvature radii R_1 and R_2 differ generally. The microstructure of the described material is shown in Fig. 2.13 and the element design model is illustrated in Fig. 2.14.

Dimensions of the elements and the approximating ellipse semiaxes are calculated as follows

$$\begin{aligned}
R_1 &= R - \frac{p}{2} \operatorname{ctg} \frac{\alpha}{2}\,; \\
R_2 &= \frac{1}{2}\left(R\sqrt{3\gamma^2 + 1} - p\gamma\sqrt{3}\right); \\
p &= R\frac{\pi - 2\alpha - \left(\dfrac{\pi}{2} - \alpha\right)\sqrt{3\gamma^2 + 1}}{3 - \operatorname{ctg}\alpha - \left(\dfrac{\pi}{2} - \alpha\right)\gamma\sqrt{3}}\,; \\
a &= \frac{p}{2\cos\alpha} + \frac{1}{2}\left(R\sqrt{3\gamma^2 + 1} - p\gamma\sqrt{3}\right); \quad b = R\,.
\end{aligned}$$
(2.26)

Table 2.3. Structural Parameters of the Material versus Molding Conditions with a Slight Filling of the Main Reinforcement

h	γ	α^0	ω_2	C	p	R_1	R_2	b	a	F_{EL}	$\dfrac{F_0}{F_{EL}}$
0.25	1.0	102.6	1.0	1.025	0	0.5	0.5	0.5	0.5	0.785	1.0
	0.9	96.69	0.940	0.947	0.217	0.348	0.377	0.512	0.519	0.777	1.011
	0.85	93.43	0.911	0.913	0.323	0.284	0.304	0,.52	0.526	0.758	1.036
	0.8	89.95	0.883	0.884	0.426	0.229	0.229	0.53	0.53	0.736	1.068
	0.75	86.26	0.856	0.858	0.524	0.183	0.148	0.542	0.532	0.712	1.104
	0.7	82.3	0.830	0.838	0.616	0.146	0.063	0.555	0.531	0.687	1.144
0.1	1.0	113.3	1.0	1.089	0	0.5	0.5	0.5	0.5	0.785	1.0
	0.85	104.5	0.888	0.927	0.962	0.216	0.309	0.511	0.538	0.776	1.012
	0.8	101.1	0.865	0.882	0.484	0.139	0.234	0.519	0.547	0.736	1.068

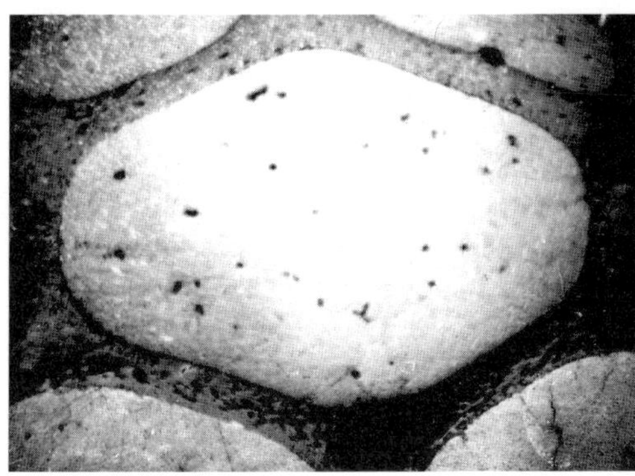

Fig. 2.13. Shape of a spirally reinforced element with slight filling degree of the main reinforcement. Magnification ×120, carbon plastics

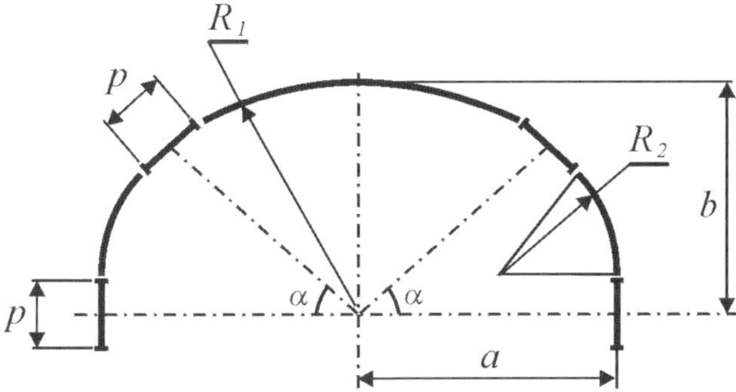

Fig. 2.14. Design model to define parameters of a deformed element with a low filling degree

2.3.3 Calculation of Size Limits of Spirally Reinforced Element

When the pliability of the original element is sufficient, i.e. at a corresponding filling degree with the main reinforcement, an ultimate strain can be reached. As a result, separate elements start to close up across the entire perimeter. The filling degree of the core of such deformed elements can be the maximum possible. In the case of hexagonal packing of the main reinforcing fibers it reaches 0.906. When the filling of each closing element turns out to be less than the limiting one, deformation of the material may continue and the winding will be violated. So far, the initial filling degree of the spirally reinforced element can be determined by judging the limiting filling and maintaining strain-free winding.

To determine the dimensions of this element it is worthwhile examining the structural packing of the material by the original round elements of outer radius R and winding layer thickness δ. Each subsequent layer of the material is packed at a certain pitch d in relation to the previous one (Fig. 2.15).

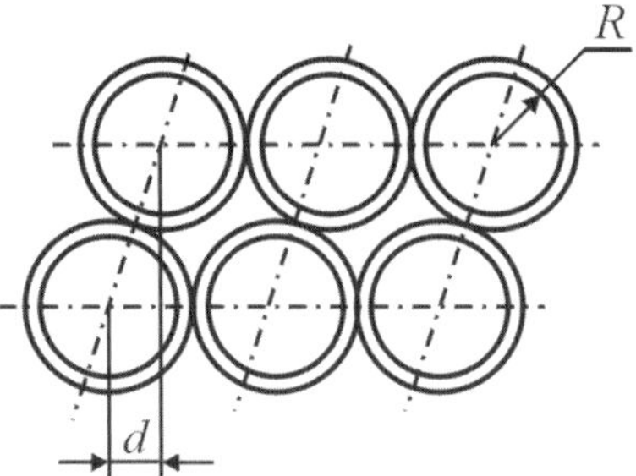

Fig. 2.15. The original material structure

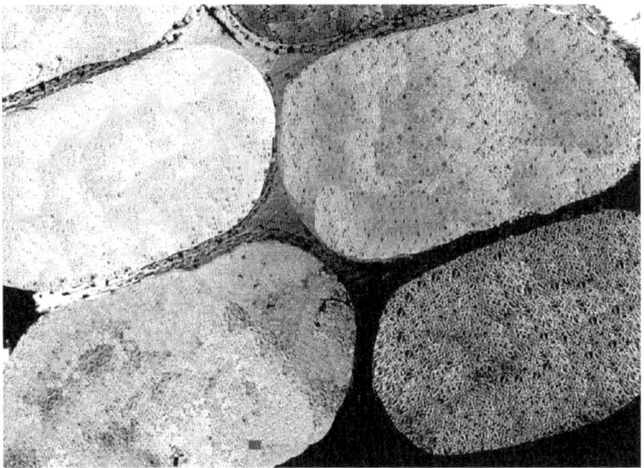

Fig. 2.16. Structure of carbon-glass plastics at a stepwise packing of material layers and ultimate strain

During the first straining stage, the elements shift vertically and flattening areas appear at the contact spots. Further straining brings diagonally placed elements into contact and leads to formation of the corresponding flattening site perpendicular to a line connecting their centers. The final view of the described structure is illustrated in Fig. 2.16.

In accordance with the design model presented in Fig. 2.17, the main dimensions of the element can be determined

$$KQ = \left(\frac{t_1}{\cos \varphi} - \omega_1 \right) \operatorname{ctg} \varphi \,;$$

$$KS = t_1 \cdot \operatorname{tg} \varphi - \frac{1}{\sin \varphi} \left(\frac{t_1}{\cos \varphi} - \omega_1 \right) \,;$$

$$DS = \frac{1}{\sin \alpha} \left(\omega_1 - KS \sin \varphi \right) \,; \tag{2.27}$$

where $t_1 = \sqrt{\omega_1^2 + \omega_2^2 - 2\omega_1\omega_2 \cos \alpha}$; $\varphi = \operatorname{arctg} \dfrac{\omega_2 \sin \alpha}{\omega_1 - \omega_2 \cos \alpha}$.

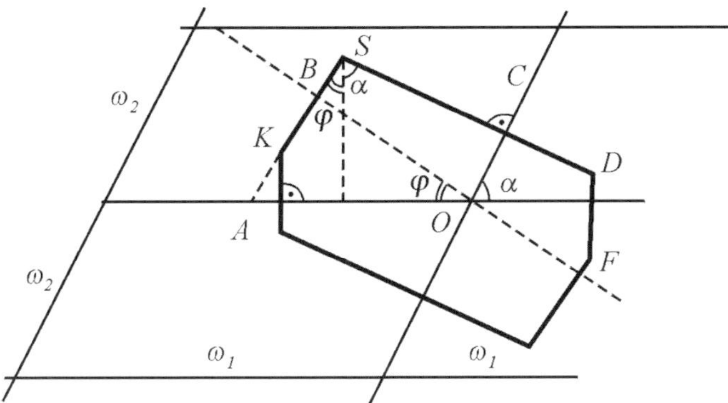

Fig. 2.17. The design model for determining dimensions of a spirally reinforced element under ultimate strain

The parameters of the structure are found by the following relations

$$\omega_2 = \sqrt{4R^2 + (4R^2 - d^2)(\gamma^2 - 1)}; \quad \omega_1 = 2R;$$

$$\alpha = \arccos \frac{d}{\sqrt{4R^2 + (4R^2 - d^2)(\gamma^2 - 1)}}. \tag{2.28}$$

The strained element under consideration can be presented by three possible regular figures shown in Fig. 2.18. At $\alpha = 90^0$ the element looks like a rectangle with sides ω_1 and ω_2. For this condition the ultimate strain value is found from the relation

$$\omega_1 + \omega_2 = \pi R.$$

The second symmetrical form of the element is possible at $d = R$. The derived polygon turns out to be oriented in a horizontal direction, so sizes a and b are determined by

$$a = \omega_1;$$

$$b = \frac{\omega_2 \cos(\alpha - \varphi)}{\sin \varphi}. \tag{2.29}$$

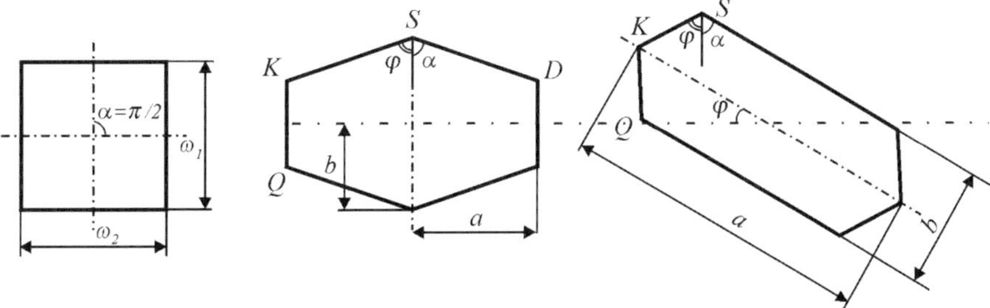

Fig. 2.18. Possible symmetrical forms of the spirally reinforced element

The third symmetrical form of the element is probable when equality $KQ = KS$ is met and dimensions a, b and the slope angle of element β axis are found as

$$a = \frac{1}{4\sin^2\varphi}\left[(2\omega_2\sin\alpha - \omega_1\sin 2\varphi)^2 + 4\omega_1^2\sin^4\varphi\right]^{1/2};$$

$$b = \omega_2;$$

$$\beta = \operatorname{arctg}\frac{2\omega_2\sin\alpha - \omega_1\sin 2\varphi}{2\omega_1\sin^2\varphi}. \tag{2.30}$$

Proceeding from the invariance condition of the perimeter of the spirally reinforced element at a given packing pitch, deformation values $\gamma = \gamma_{PR}$ will be defined from condition

$$2(KQ + KS + SD) = 2\pi R. \tag{2.31}$$

Parameters of the structure, dimensions of the element and limiting reduction values are illustrated in Table 2.4 versus the packing pitch. When $d = R$ one has hexagonal packing of the elements. Hence, limiting reduction values γ can be found from the condition of the equality of the element areas and material cell ($\gamma_U \leq 0.8$), and condition $KS = R$ ($\gamma_L \geq 0.66$).

In the case of packing the spirally reinforced elements (Fig. 2.9) a cellular structure may be formed under ultimate strain (Fig. 2.19). The upper limit of reduction γ_U, leading to the entire filling of the structure, can be found from the equality condition of the element areas and the material cell

$$\gamma_u = \frac{\omega_2\sin\alpha}{t_2}, \tag{2.32}$$

where $t_2 = \sqrt{12R^2 - 2Rh - h^2}$, and structural parameters are calculated by

$$\omega_2 = \left[(R + h/2)^2 + (\gamma t_2)^2\right]^{1/2};$$

$$\alpha = \operatorname{arctg}\frac{\gamma \cdot t_2}{R + h/2}. \tag{2.33}$$

Table 2.4. Limiting Reduction Values and the Main Structural Parameters of the Material Dependent on the Lay-out Pitch of the Spirally Reinforced Elements

d	γ_{PR}	ω_2	α	KQ	DS	F_{EL}	$\dfrac{F_0}{F_{EL}}$	a	β
0	0.57	0.57	90	0.57	0	1.0	0.57	1.378	1.15	29.79
0.2	0.63	0.63	80.78	0.48	0.17	0.91	0.619	1.269	1.11	25.78
0.4	0.67	0.69	73.33	0.41	0.32	0.84	0.657	1.195	1.08	22.34
0.6	0.71	0.74	65.89	0.37	0.43	0.77	0.679	1.156	1.07	20.05
0.8	0.75	0.80	59.58	0.34	0.53	0.69	0.690	1.138	1.06	18.90
1.0	0.80	0.85	54.18	0.33	0.62	0.62	0.691	1.136	1.05	18.33

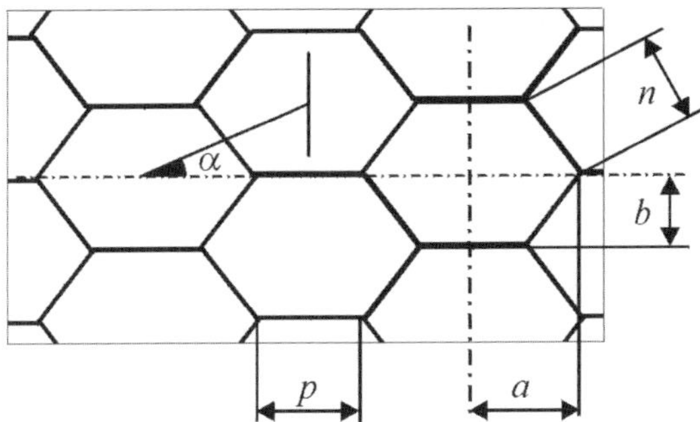

Fig. 2.19. Structural diagram of the material with initial packing of spirally reinforced elements with spacing

Thus, the dimensions of the elements are found from

$$h = \left[(R - p + h/2)^2 + (\gamma t_2)^2\right]^{1/2} ;$$

$$a = R + \frac{h - p}{2} ;$$

$$b = \gamma t_2 . \tag{2.34}$$

To determine flattening p, it is convenient to use the invariance condition of the element perimeter

$$2n + R = \pi R . \tag{2.35}$$

Theoretically, the elements might continue straining with increasing percentage reduction due to changes in their main dimensions. Thus, the lower γ_L can be found from the relation

$$\gamma_L = \frac{\sqrt{\pi^2 R^2 - (2R + h)^2}}{2t_2} . \tag{2.36}$$

The main parameters of the structure depending on percentage reduction γ and clearance h at $R = 0.5$ are presented in Table 2.5.

2.3.4 Applicability Limits of Obtained Dependencies

Taking (2.4), the final filling degree being equal to the limit possible for the case of hexagonal disposition of the main reinforcing fibers inside the core of a spirally reinforced element, a dependence can be drawn $q = f(\varphi_1)$, shown in Fig. 2.20.

Thus, having the initial filling degree, one can find the value of factor q under which the structure experiences the ultimate strain. Moreover, the above-derived dependencies make it possible to find current values of factor q for a given percentage reduction γ under the noted limitations. In Fig. 2.21 the dependence of factor q on percentage reduction is shown for the case

Table 2.5. Variation of Structural Parameters and Element Form Depending on Percentage Reduction and Initial Clearance Values

h	γ	α^0	ω_2	p	a	b	n	F_{EL}	$\dfrac{F_0}{F_{EL}}$
1.5R	0.8	23.89	0.967	0.520	0.615	0.387	0.402	0.678	1.159
	0.7	21.17	0.938	0.394	0.678	0.339	0.525	0.593	1.324
	0.6	18.36	0.922	0.323	0.714	0.290	0.588	0.508	1.545
	0.5	15.46	0.908	0.273	0.739	0.242	0.604	0.424	1.854
	0.4	12.48	0.896	0.237	0.757	0.194	0.649	0.339	2.318
1.25R	0.75	28.29	0.913	0.534	0.546	0.437	0.514	0.710	1.105
	0.7	26.67	0.909	0.377	0.624	0.408	0.597	0.663	1.184
	0.6	23.28	0.885	0.256	0.684	0.350	0.658	0.568	1.382
	0.5	19.74	0.863	0.183	0.721	0.292	0.694	0.474	1.658
	0.4	16.01	0.845	0.132	0.747	0.233	0.719	0.379	2.073
1.00R	0.7	31.69	0.881	0.360	0.570	0.463	0.606	0.695	1.1309
	0.6	27.89	0.849	0.176	0.662	0.397	0.697	0.595	1.3193
	0.5	23.80	0.820	0.085	0.708	0.331	0.743	0.496	1.5832
	0.4	19.43	0.795	0.022	0.739	0.265	0.774	0.397	1.9790
0.5R	0.65	39.07	0.805	0.543	0.354	0.507	0.514	0.634	1.2383
0.25R	0.66	45.82	0.807	0.298	0.414	0.578	0.636	0.651	1.2063

of packing with a clearance $h = 0.25$. As follows from the above, formulas (2.14) and (2.21) can be employed to determine structural parameters at $1.0 \geq \gamma \geq \gamma_1 = 0.95$. The limiting percentage reduction γ_1 is found under the least difference between R_1 and R_2, calculated by (2.22). With further deformation of the elements the percentage reduction value is found to be $\gamma_2 = 0.7$ under conditions of no contact between them. This means that if $\gamma_2 < \gamma \leq \gamma_1$, then the structural parameters and dimensions of the element are calculated using formulas (2.22), (2.23), and (2.24).

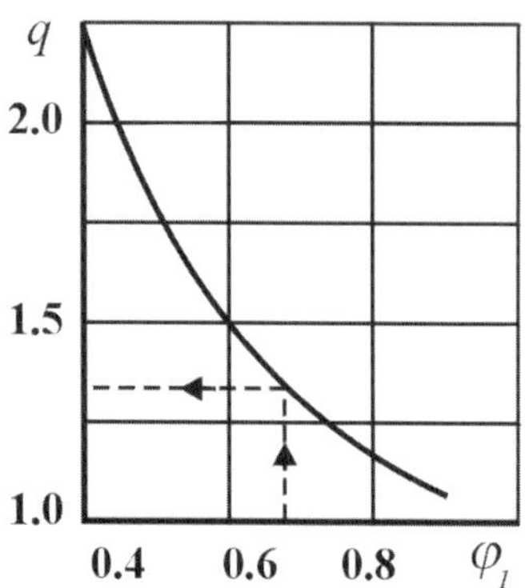

Fig. 2.20. Variations of factor q limiting value

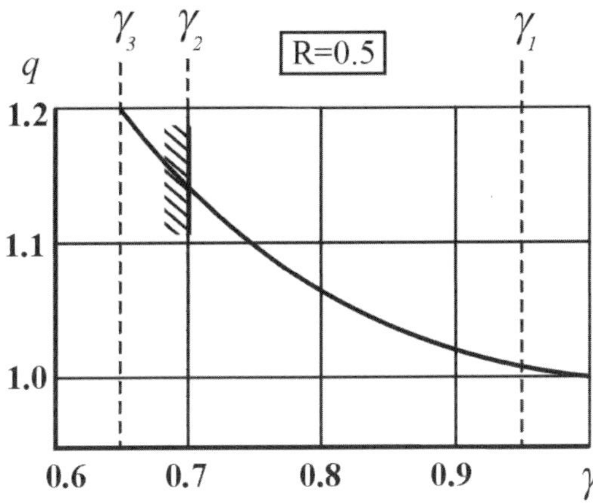

Fig. 2.21. Factor q dependence on percentage reduction at packing with a clearance ($h = 0.25$) upon the initial filling degree of the spirally reinforced element

With increasing percentage reduction the limiting value of $\gamma = \gamma_3$ can be found, which transforms the element into a rhombus. For the considered case $h = 0.25$, this value is $\gamma_3 = 0.66$. Hence, at $\gamma_3 \leq \gamma \leq \gamma_2$ the element form and structural parameters are described by relations (2.33) and (2.34).

Similar data were derived under other h values. For instance, in Fig. 2.22 dependencies $q = f(h)$ are depicted at $h = 0.05;\ 0.15;\ 0.25$ and $R = 0.5$ obtained from (2.33) and (2.34). It is quite apparent that with diminishing h limiting γ_2 values reduce.

In Fig. 2.23 the dependence of the q factor on percentage reduction is shown for a hexagonal packing of the spirally reinforced elements without a clearance. Similarly to the previous case, it is expedient to use formulas (2.25) and (2.26) to calculate the structural parameters and dimensions of the elements under small percentage reduction values ($0.9 \leq \gamma_1 \leq 1.0$). The limiting value $\gamma_2 = 0.8$ is calculated at further deformation under a condition of equality between the material cell areas and the element. The element dimensions and structural parameters are found from (2.27) and (2.28).

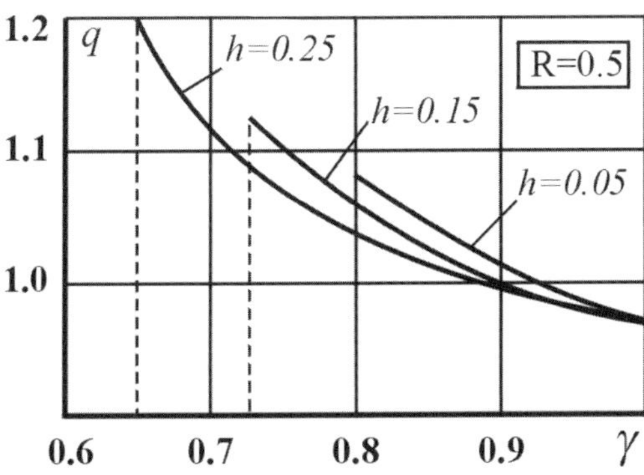

Fig. 2.22. Variation of q factor depending on percentage reduction for different h values

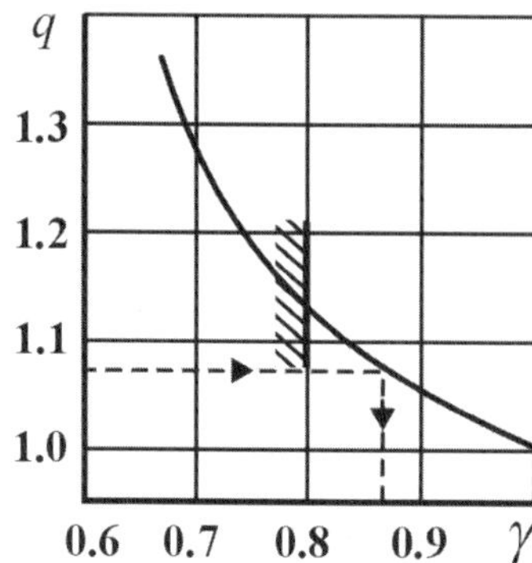

Fig. 2.23. Variation of q factor as dependent upon percentage reduction in case of the original hexagonal packing of spirally reinforced elements

For a layer-by-layer packing of the elements, the limiting percentage reduction and q are found to depend on the packing pitch. Data obtained from (2.27) and (2.28) are shown in Fig. 2.24. It is evident that with extending packing pitch the limiting percentage reduction increases while factor q reduces.

Thus, having the initial filling degree φ_1, one can determine the limiting value of the variation ratio of the filling degree q. Then, depending on the type of packing, percentage reduction γ, the main parameters of the structure and the element formed are found. For example, at $\varphi_1 = 0.85$ we have from Fig. 2.20 that $q = 1.065$. Then, for the case of initial hexagonal packing, we find by Fig. 2.23 that $\gamma_{PR} = 0.87$, so the size of the element can be calculated by (2.25) and (2.26).

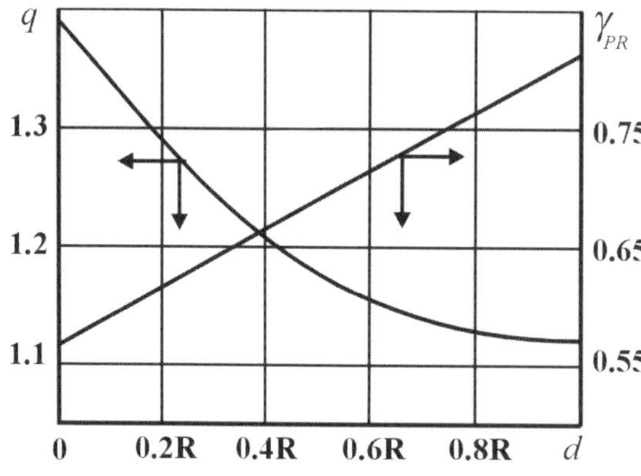

Fig. 2.24. Variation of limiting percentage reduction and q factor depending on packing pitch

2.4 Inner Stress in Composites with Spirally Reinforced Filler

The initial technological stresses arise in the material during its manufacture and thermal treatment. The reasons for their appearance and methods of calculation are discussed at length in [1, 120, 123, 125, 200, 201, 299, 301, 302]. Along with technological stresses additional stresses and strains might appear in the materials with spatially-reinforced filler. This is attributed to interactions between structural elements with different coefficients of linear thermal expansion in various directions.

Let us examine thermal deformation of a spirally reinforced element typical of a spatially-reinforced material. When the temperature changes by $\mathrm{d}T$ due to differences in the linear thermal expansion coefficient of the layer and the element core, there arise stresses in the element, which bring additional deformation. If one neglects transversal stresses and strains in the fine winding filament, then the relative lengthwise deformation is

$$\varepsilon_F = \frac{P}{F_F E_F} + \int_{T_1}^{T_2} \alpha_F \mathrm{d}F , \qquad (2.37)$$

where T_1 and T_2 are the initial and final temperatures of thermal treatment; E_F, F_F – are elasticity modulus and cross-sectional area of the winding filament; P – the force appearing in the winding filament upon cooling down; α_F – coefficient of linear thermal expansion of the layer material.

Elastic strain in the winding filament, whose length is taken to be equal to one turn, is calculated by the relation:

$$A_F = \frac{P^2 L_F}{2 E_F F_F} + \frac{1}{2} \int_{T_1}^{T_2} P L_F \alpha_F \mathrm{d}T , \qquad (2.38)$$

where $L_F = L_0/\sin\alpha$ – is the winding filament length; L_0 – core perimeter of the spirally reinforced element; α – winding angle.

The core of the spirally reinforced element is found in a complex strained state being under the force of the winding filament. Taking the smallness of winding pitch t_0, it is considered that the spirally reinforced element experiences in a hardened state a load evenly distributed over its length and corresponding to the packing pitch of the winding filament. Believing that the core of the spirally reinforced element consists of fibers packed in the direction of the main reinforcement, it can be considered to be a transverse isotropic rod that bears an evenly distributed load on its side surface, which is conditioned by force P, axial force P_0 and applied on the girth torque M_{tr} (Fig. 2.25).

To define stress and strain in the element core one must introduce rectangular coordinates with the origin point in the gravity center of the element's cross-section and the z axis passing along the longitudinal axis of the spirally

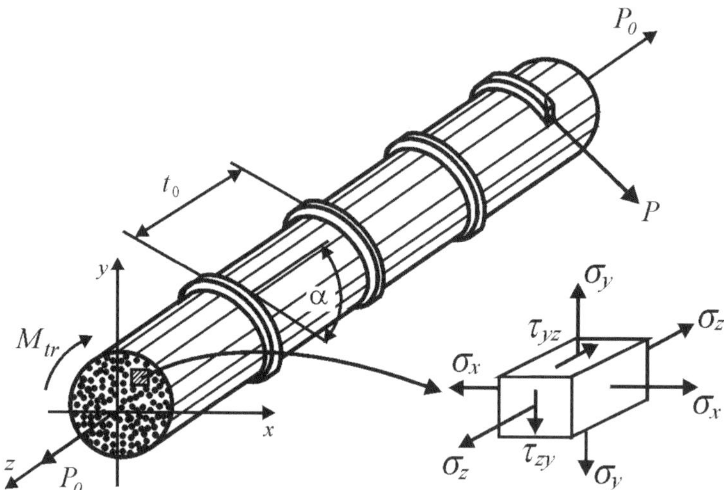

Fig. 2.25. Analysis of thermal stresses in spirally reinforced filler

reinforced element. The force acting on the winding filament P forms the following angles with the axes of the chosen coordinates

$$\alpha'_Z = \alpha; \quad \alpha'_X = \arccos\frac{\mathrm{d}x}{\mathrm{d}L}; \quad \alpha'_Y = \arccos\frac{\mathrm{d}y}{\mathrm{d}L}, \tag{2.39}$$

where $\mathrm{d}L$ is the differential of arc of the element's cross-sectional profile.

By denoting

$$P_X = -\frac{P\cos\alpha'_X}{t_0}, \quad P_Y = -\frac{P\cos\alpha'_Y}{t_0},$$

it is possible to calculate stresses σ_{XK} and σ_{YK} on the element profile

$$\sigma_{XK} = \frac{\mathrm{d}P_X}{\mathrm{d}L} = -\frac{P}{t_0}\frac{\mathrm{d}^2x}{\mathrm{d}L^2};$$

$$\sigma_{YK} = \frac{\mathrm{d}P_Y}{\mathrm{d}L} = -\frac{P}{t_0}\frac{\mathrm{d}^2y}{\mathrm{d}L^2}. \tag{2.40}$$

To determine stress at any point of the element core let us solve an additional generalized plane problem for a transverse isotropic rod with boundary conditions (2.40). Introducing function F [202] meeting the relation

$$\frac{\partial^4 F}{\partial x^4} + 2\frac{\partial^4 F}{\partial x^2\partial y^2} + \frac{\partial^4 F}{\partial y^4} = 0, \tag{2.41}$$

stresses in the element core are calculated as

$$\sigma_X = \frac{\partial^2 F}{\partial y^2}; \quad \sigma_Y = \frac{\partial^2 F}{\partial x^2}; \quad \tau_{XY} = \frac{\partial^2 F}{\partial x\partial y}; \quad \sigma_Z = \nu_{XZ}(\sigma_X + \sigma_Y). \tag{2.42}$$

Since the characteristic differential equation (2.40) has equal pairs of roots $(\mu_1 = \mu_2 = i,\ \overline{\mu}_1 = \overline{\mu}_2 = -i)$, so function F is sought in the form [202]:

$$F = \mathrm{Re}\left\{\overline{\xi}F_1(\xi) + F_2(\xi)\right\},\tag{2.43}$$

where $\xi = x + iy$, and functions F_1 and F_2 can be presented as [203]:

$$F_1 = \sum_{k=1}^{\infty} a_k \xi^k;\quad F_2 = \sum_{k=1}^{\infty} b_k \xi^k.\tag{2.44}$$

Therefore, stresses in the element core are derived from the equations

$$\sigma_x = -\frac{P\sin\alpha}{t_0}\sum_{k=1}^{\infty}\mathrm{Re}\left\{a_k[2k\xi^{k-1} + k(k-1)\overline{\xi}\xi^{k-2}] + b_k k(k-1)\xi^{k-2}\right\};$$

$$\sigma_y = -\frac{P\sin\alpha}{t_0}\sum_{k=1}^{\infty}\mathrm{Re}\left\{a_k[2k\xi^{k-1} - k(k-1)\overline{\xi}\xi^{k-2}] - b_k k(k-1)\xi^{k-2}\right\};$$

$$\tau_{xy} = -\frac{P\sin\alpha}{t_0}\sum_{k=1}^{\infty}\mathrm{Im}\left\{k(k-1)[a_k\overline{\xi}\xi^{k-2} + b_k\xi^{k-2}]\right\}.\tag{2.45}$$

Bounding the series (2.45) by some number n and using boundary conditions (2.40), by the collocation method, a system is obtained of $2n - 1$ linear equations relative to unknown $a_1, \ldots, a_n, b_2, \ldots, b_n$.

By using the longitudinal constituent of force P one can calculate stress σ_Z operating along the rod axis

$$\sigma_Z = \nu_{XZ}(\sigma_X + \sigma_Y) - \frac{P\cos\alpha}{F_0},\tag{2.46}$$

where F_0 is the rod's cross-sectional area.

It is possible to calculate stresses τ_{xz} and τ_{yz} using known dependencies [59], where the torque value is found from

$$M_{tr} = \int_{L_0} (P_Y x - P_X y)\mathrm{d}L = -\frac{P\sin\alpha}{t_0}\int_{L_0} (y\mathrm{d}x - x\mathrm{d}y).\tag{2.47}$$

Taking account of the obtained stress values, the work spent on deformation of the element core is determined by [130]:

$$A_E = \int_V \sigma_{ij}\cdot\varepsilon_{ij}V;\tag{2.48}$$

where $\varepsilon_i = \sum_{j=1}^{6} a_{ij}\sigma_{ij} + \int_T \alpha_i\mathrm{d}T$.

Since the work of the elastic forces acting in the winding filament (2.38) is spent in deforming the element core, the following equality is true

$$A_F = -A_E,\tag{2.49}$$

wherefrom, the force P appearing on thermal shrinkage in the winding filament is determined, then the stress in the rod and the twisting angle with its length equal to the winding pitch

$$\varphi = \frac{M_{tr}t_0}{G_{XZ}J_P}\;; \tag{2.50}$$

where G_{XZ} – shear modulus; J_P – polar inertia moment of the section.

In case the core of the spirally reinforced element remains round, an equation for determining the force in the winding filament will be

$$P = -\int_{T_1}^{T_2} \frac{\left[(1+\mathrm{ctg}^2\,\alpha)\alpha_H - \alpha_X - \alpha_Z\,\mathrm{ctg}^2\,\alpha\right]\mathrm{d}T}{2\sin\alpha\left[\dfrac{(1-\nu_{XY})\,\mathrm{tg}\,\alpha}{\pi D^2 E_X} + \dfrac{4\,\mathrm{ctg}\,\alpha}{\pi D^2 G_{XZ}} + \dfrac{(1+\mathrm{ctg}^2\,\alpha)^{3/2}}{2E_H F_H}\right]} \,. \tag{2.51}$$

Thermal stress and strain characteristics have been calculated using the proposed procedure for different materials as well as correlation of the element core dimensions of the spirally reinforced filler based on carbon fiber and various auxiliary reinforcements, such as glass filaments (Fig. 2.27) and organic SHP (synthetic high-polymer materials, analogous to Kevlar-49) threads (Fig. 2.26). The semiaxes ratio is in this case $a/b = 1.0$. Analysis of the obtained data has shown that the type of material used exerts a considerable effect on thestress distribution in the element. Thus, at $\alpha = 35$–$40°$, which corresponds to a packing pitch of the auxiliary reinforcement of 2.5–3.4 mm, substitution of organic fibers by glass ones raises the tangential stress almost 3-fold and changes the torque direction.

Analogous stress dependencies have been obtained under other correlations of the main and auxiliary reinforcements. Fig. 2.28 illustrates their variation for organic plastic and Fig. 2.29 the glass plastic. A prominent feature of the organic-glass-reinforced plastic is a perceptible reduction of force factors, stresses and strains at winding angle $\alpha = 40$–$50°$. This feature makes

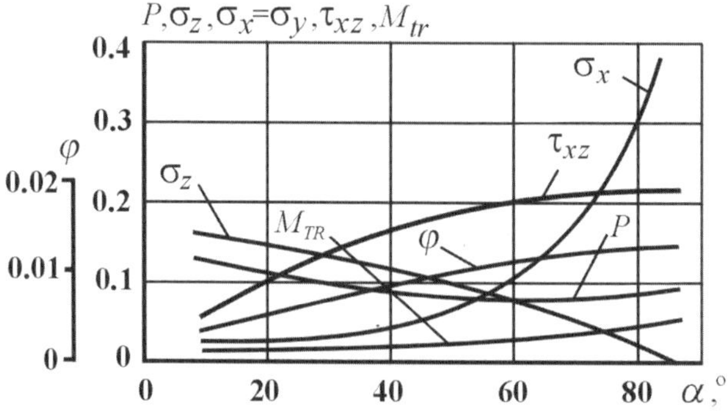

Fig. 2.26. Variation of thermal stresses and strains of carbon-glass-reinforced plastic depending on winding angle

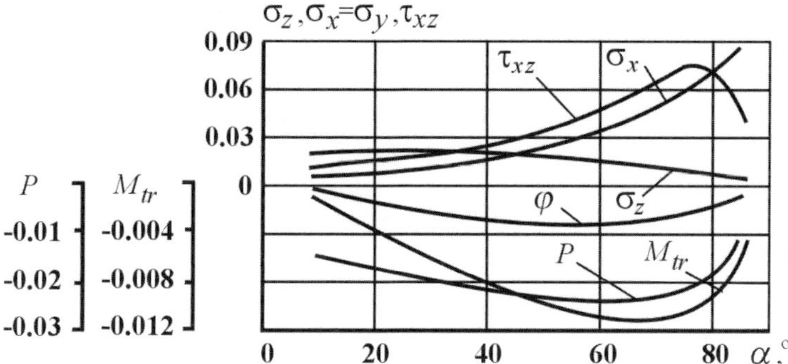

Fig. 2.27. Thermal stress and strain dependence on winding angle for carbon-organoplastic

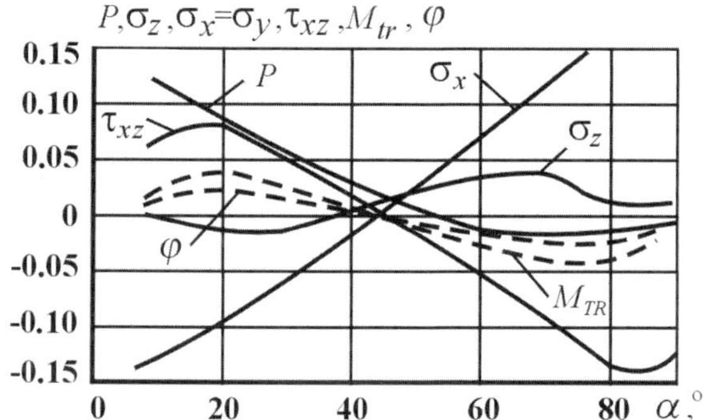

Fig. 2.28. Variation of thermal stresses and strains of organic-glass-reinforced plastic depending on winding angle

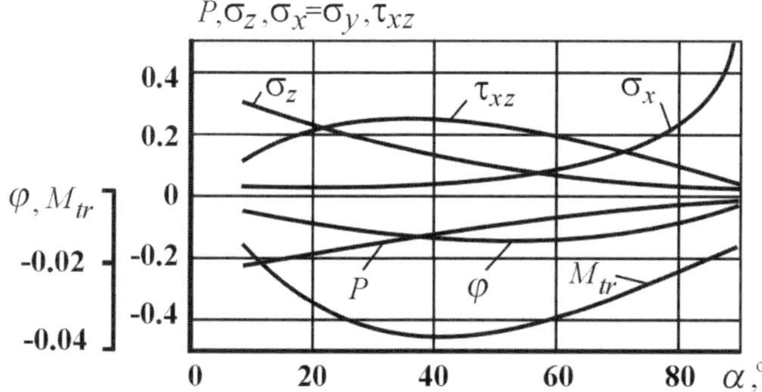

Fig. 2.29. Dependence of thermal stress and strain on winding angle for glass-reinforced organoplastic

it possible to choose the winding pitch of the auxiliary reinforcement so as to reduce to a minimum thermal stresses in the material.

Studies of tension in the winding filament, torque and twist angle dependence on characteristic dimensions of the spirally reinforced element have shown that the named parameters remain practically invariable within the $a/b = 1.0$–2.0 range. As, e.g. at $\alpha = 80°$ within the named range of a/b variation, the tension value increases for the carbon-glass-reinforced plastic by 7% and for carbon-organoplastic by 5.6%.

It is evident that tension in the winding filament can be calculated by a simplified formula (2.51) and the order of the stress values by using the dependencies:

$$\sigma_X = \sigma_Y = \frac{2P}{Dt_0}\sin\alpha\,;$$

$$\sigma_Z = \frac{4P}{\pi D^2}\cos\alpha\,;$$

$$\tau_{XZ} = \frac{16P\sin\alpha}{\pi D^3}\rho\,, \qquad (2.52)$$

where ρ is current radius.

From (2.51) it follows that the direction of tension is a function of relative values of linear thermal expansion factors and forces of the temperature field upon heating and cooling on binder polymerization. For instance, at material cooling when the condition $\alpha_X > \alpha_F$ is met compressive stress occurs in the auxiliary reinforcement layer, whereas in the core there is tensile stress. Thus, the spirally reinforced element turns by an angle

$$d\varphi = \frac{16P\sin\alpha}{G_{XZ}\pi D^3}dz\,. \qquad (2.53)$$

Insofar as the material consists of a great number of spirally reinforced elements evenly distributed across its section, then the full moment in the section can be found as the sum of the torques of all the elements.

$$M_{tr\Sigma} = \sum_{i=1}^{n} M_{tr}^i = nM\,. \qquad (2.54)$$

2.5 Design of Material Structure with Spirally Reinforced Filler

Analysis of the equations derived in Sect. 2.2 for the filling degree of materials with spirally reinforced filler has proved that the degree is always lower than the limiting one of a unidirectional material. This is especially true for filling with a main reinforcement that conditions the characteristics of the whole composite in the reinforcing direction. Therefore, the design of materials with spirally reinforced fillers must be carried out so as to minimize worsening of the chief characteristics in the reinforcing direction whilst attaining the maximum positive effect on filling with the spiral winding.

It follows from (2.8) that from the above standpoint it is expedient to wind the main reinforcement with fine filaments allowing one to reach $k \leq 0.2$ at a pitch when each coil is laid in the spaces between the coils of neighboring elements.

If we use a unidirectional winding of all spirally reinforced elements, then the coils of the winding filaments of neighboring elements intersect. Thus, to make them engage, it is necessary to lay an auxiliary reinforcement with an extremely stable winding pitch. As a result, some areas are left between coils of the auxiliary reinforcement remaining unfilled with the reinforcement, which impairs the rated filling degree. Moreover, it follows from (2.47), in the cross-section of the material with the spirally reinforced filler there is a torque found, which is induced by the difference between the coefficients of the linear thermal expansion of the main and auxiliary reinforcements along and across the direction of the reinforcing fibers. This circumstance may lead to lengthwise curling of the finished article, i.e. warping.

To eliminate this drawback, a structure of the composite material is proposed that diminishes the winding pitch to $t_0 = 2b_0/\sin\alpha$ (b_0 is the width of the winding filament) and augments filling by both main and auxiliary reinforcements. In this case the product is to be formed by winding or laying a strip or veneer sheet containing an even number of spirally reinforced elements, which should be used where the directions of the even and odd elements are to be different. The succeeding reinforcement layers are laid in sequence over the material thickness with a shift equal to the element diameter or half-diameter on hexagonal packing. The directions of the winding spiral coils are alternated in staggered rows, while those of the winding filaments of the contacting elements are identical (Fig. 2.30). In turn, the winding coils of each element are located in the interwinding spaces between

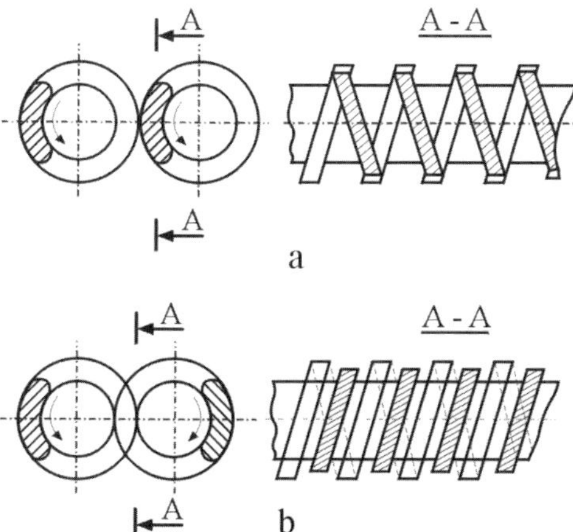

Fig. 2.30. Disposition of winding filament coils for (**a**) similar, and (**b**) alternating directions of winding of neighboring elements

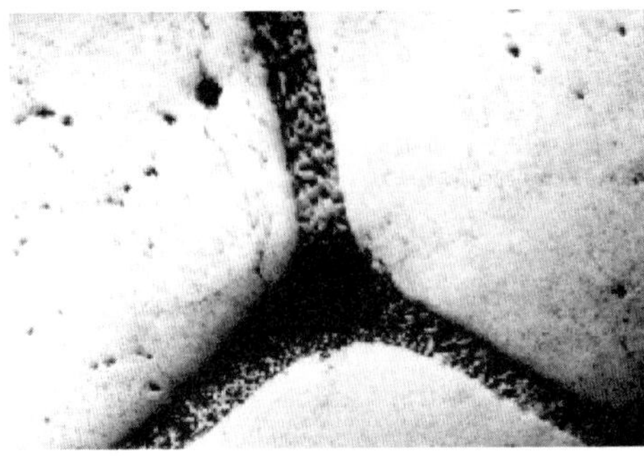

Fig. 2.31. Auxiliary reinforcement distribution between cores of the elements. Magnification $\times 250^*$, Carbon-glass-reinforced plastic

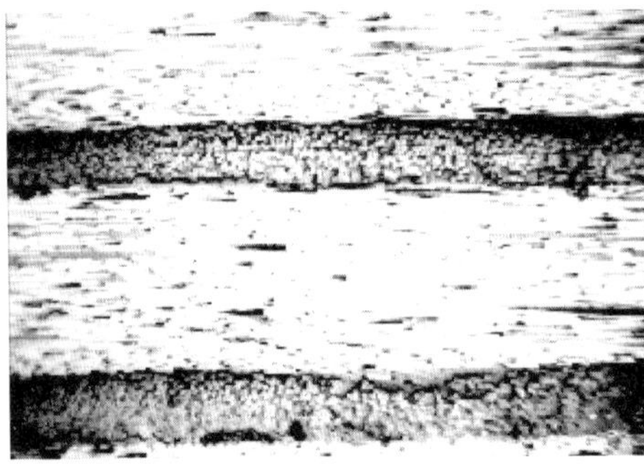

Fig. 2.32. Distribution of auxiliary reinforcement along the filler length. Magnification $\times 120^*$, Carbon-glass-reinforced plastic

neighboring elements, thus contributing to the formation of the unbroken interlayer just as between element cores (Fig. 2.31), with lengthwise spirally reinforced filler (Fig. 2.32). Assuming that during material processing under pressure the winding filaments of neighboring elements tend to shift and occupy a more favorable, in respect of the energy position in the interfibrous space, the demands on stability of the winding pitch may be lowered a little. Furthermore, with such an arrangement of elements with oppositely-directed windings, the torques of each pair of neighboring elements acquire opposite charges and become mutually balanced, whereas the total moment in the material section turns out to be equal to zero.

An analogous effect is reached when two-layer coaxial reinforced elements are fabricated with opposite direction of inner and outer winding layers. The packing angle of the spiral windings is chosen so as to make the torques, which appear due to temperature deformation in the sections, mutually compensate one another.

2.6 The Choice of a Representative Element of the Material Structure

It was shown earlier that the chief structural parameters of the composite and dimensions of the spirally reinforced element are highly dependent on the initial filling degree and filler diameter. To find parameters of such a semifinished product, various components were examined. For reinforcement in the main direction, threads and plaits based on carbon, organic and glass fibers were used. For the winding reinforcement mainly filaments on organic and glass fiber bases were employed. Epoxide and epoxyphenolic binders were also used. As investigations have proved, variations in technological regimes of manufacturing the spirally reinforced filler and the material in its base result in corresponding changes in filling degree inside the element. Note that the realized filling degree of the element φ_k is lower than the maximum one that corresponds to a hexagonal packing of the fibers. Hence, limiting values of factor q, indicating variations in filling degree with the main reinforcement on material deforming, can be determined as $q^* = \varphi_k/\varphi_f$.

The results obtained are presented in Fig. 2.6, where $\varphi_f^{\max}$ and $\varphi_f^{\min}$ are the maximum and minimum filling degrees of the original spirally reinforced element, correspondingly; D – element diameter. For carbon plastic a carbon plait from VMN-4 (carbon braid) was used for the main reinforcement, for the organoplastic SHP filament (58.8 tex), and for glass-reinforced plastic-glass filament BS-6. For the auxiliary reinforcement in all cases an NSK-150/2 glass fiber (150 fibers, 2 twists) was used. It should be noted that substitution of glass fiber for an SHP filament (14.3 tex) does not much change the dimensions of the resultant elements. It follows from the achieved data, limiting q* for carbon plastics is within 1.25–1.5 limits, for organoplastics -1.2–2.1, and glass-reinforced plastics -1.12–2.1.

During packing of subsequent layers of the material, a sufficiently rigid plait of the spirally reinforced filler tends to move apart previously laid plaits. Therefore, it is necessary to take measures to *avoid* the effect and not to let the filler emerge through the spaces during firming. Investigations of the material microstructure have shown that in cases where the first layer is laid without a clearance between elements, a clearance $h = (0.1 - 0.3)R$ appears when laying the next layer. Thus, a limiting value of $q' = 0.8 - 1.15$ can be defined for this type of structure, as shown above. Comparison of the data obtained to those from Table 2.6 has proved that independently of the type of components the following forms of the spirally reinforced elements are feasible: at $q = 1.0$ the round form, at $q' > q > 1.0$ – ellipsoidal, and at $q' = q$ – polygonal. Along with this, some spreading of the layers at a certain pitch is probable on compaction, which value is, as microstructural analysis has shown, about $(0.7 - 1.0)R$. It turns out that $q' = 1.145$, i.e. all other intermediate forms of the elements are also probable.

By way of example let us examine the microstructure of a pressed carbon-glass-reinforced plastic produced under the following initial param-

Table 2.6. Chief Parameters of Original Spirally Reinforced Fillers

Parameters	Carbon-glass-reinforced plastic			Organo-glass-reinforced plastic	Glass-fiber-reinforced plastic
	2 layers	3 layers	5 layers		
φ_f^{max}	0.60	0.57	0.53	0.70	0.75
φ_f^{min}	0.50	0.50	0.50	0.40	0.40
D, mm	1.00	1.20	1.70	1.00	1.00
φ_k	0.75	0.75	0.75	0.84	0.84
q_{min}^*	1.25	1.32	1.42	1.20	1.12
q_{max}^*	1.50	1.50	1.50	2.10	2.10

eters: $D = 1.0\,mm$, $h = 0.17\,mm$, $\varphi_f = 0.56$, $\gamma = 0.77$. The type of structure is depicted in Fig. 2.33. In Table 2.7 the results of theoretical calculations and statistical data on the main structural parameters are given. The microstructure of carbon-organoplastic obtained by winding under the following initial parameters $D = 1.0\,mm$, $h = 0.1\,mm$, $\varphi_f = 0.58$, $\gamma = 0.97$ as shown in Fig. 2.34. As in the previous case, deviations of the theoretical and experimental data are negligible.

Proceeding from the above limitations on the packing pitch, element diameter, initial filling degree and its limiting value, and considering agreement between theoretical and experimental data, it is possible to determine the

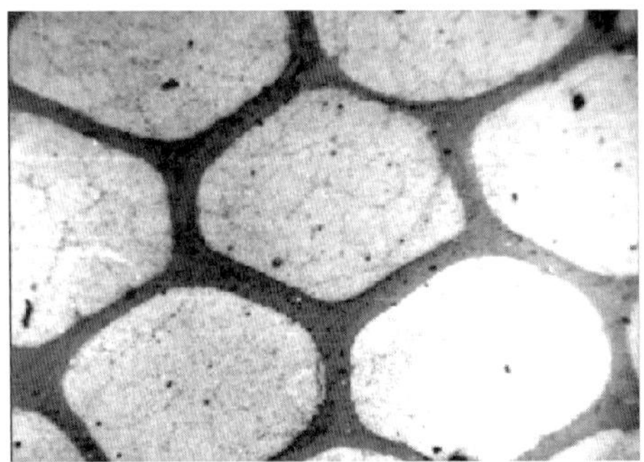

Fig. 2.33. Microstructure of carbon-glass-reinforced plastic ($\gamma = 0.77$). Magnification $\times100^*$

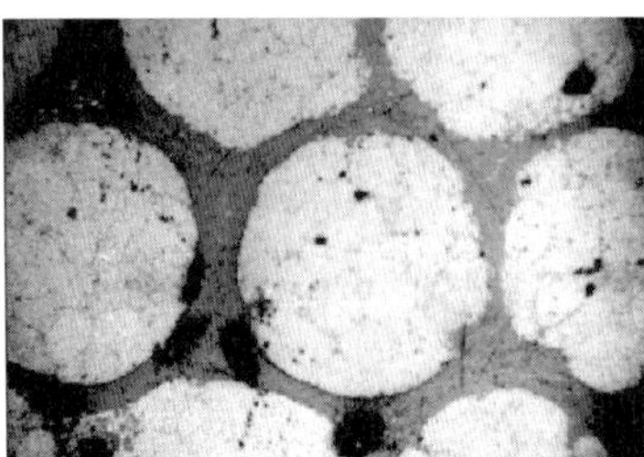

Fig. 2.34. Microstructure of carbon-organoplastic ($\gamma = 0.97$). Magnification $\times100^*$

Table 2.7. Results of Statistical Processing of Carbon-glass-reinforced Plastic Microstructure

	α	ω_2	a	b	p	R_1	R_2	h	δ
m	53^006	0.8456	0.4304	0.5625	0.4862	0.1471	0.1949	0.1691	0.0275
σ^2	0.0007	0.0017	0.0004	0.0005	0.0005	0.0001	0.0001	0.0001	0.27×10^{-5}
σ	0.0258	0.0410	0.0212	0.0223	0.0231	0.0104	0.0071	0.0072	0.0016
A	-0.2427	-0.1719	-0.091	-0.113	-0.104	-0.009	-0.0101	-0.0098	-0.0005
ε	-0.0413	0.0839	0.0311	0.0254	0.062	0.0031	0.0057	0.0003	0.002
Δm	0.0047	0.0074	0.0039	0.0041	0.0042	0.0019	0.0013	0.0015	0.0003
x^T	47.64	0.8553	0.455	0.5460	0.5044	0.1451	0.1884	0.17	–

Table 2.8. Results of Statistical Processing of Carbon-organoplastic Microstructure

	α	ω_2	ω_1	p	r	h	δ
m	59^059	1.089	1.115	0.028	0.490	0.1154	0.032
σ^2	0.0024	0.0038	0.002	1.6×10^{-6}	0.00086	3.3×10^{-5}	1.2×10^{-6}
σ	0.4870	0.062	0.044	0.0013	0.0294	0.0057	0.0015
A	-0.4224	-0.341	-2.251	-0.0087	0.1128	0.1012	-0.0101
ε	0.0709	0.063	0.085	0.0033	0.0097	0.0064	0.0019
Δm	0.0089	0.013	0.008	0.00024	0.0054	0.0010	0.00028
x^T	56	0.973	1.10	0.024	0.48	0.1	–

Table 2.9. Variation Range of the Main Macrostructural Parameters of the Composite Based on Spirally Reinforced Filler

Parameters	$\alpha,^\circ$	$\dfrac{\omega_2}{\omega_1}$	$\dfrac{a}{b}$	$\dfrac{\delta}{a}$
Minimum	54	0.55	0.66	0.02
Maximum	63	1.00	1.50	0.20

type of the representative element and probable bounds of the main varying macrostructural parameters (Table 2.9)

2.7 Conclusions

1. The developed material structure on the basis of a hybrid spirally reinforced filler makes provision for efficient separation of fiber bunches of the main reinforcing material and their regular arrangement in the composite.
2. To produce spirally reinforced fillers it is recommended to use fine filaments as an auxiliary reinforcement, which are wound on the main rein-

forcing fibers at a pitch that allows for mutual engagement of the coils of neighboring spirally reinforced elements.

3. Investigations conducted have established the relation between the structural and technological parameters. Based on the investigation results a calculation method of determining the chief structural characteristics of the material is proposed proceeding from the given original filling degree of the spirally reinforced filler and the percentage reduction.

4. Based on analysis of the auxiliary reinforcement effect on the stress-strain state of the spirally reinforced element of the material under thermal treatment, the relations for determining internal thermal stresses and strains have been derived, which gave grounds for the elaboration of a new material structure, capable of avoiding any warping of the product.

5. The fulfilled theoretical and experimental analyses of composite structures based on spirally reinforced fillers have made it possible to define the type of preliminary element and variation ranges of the main structural parameters. This serves as a base for investigations of the stress-strain state of materials and the optimization of their structure under different types of loading.

3 Stress-Strain State in Composites Based on Hybrid Spirally Reinforced Fillers Under Transverse Loading

The stress-strain state of composites has been examined from the standpoint of the linear theory of elasticity in a great number of studies [123, 160, 176, 197, 204]. This interest is related to, firstly, exhaustive description of composite behavior by the elasticity theory and, secondly, the relative simplicity and maturity of the theory itself. Most developed is considered to be the part of the theory dealing with 2D problems. The problems are commonly studied in terms of the theory of the complex variable function. They have been studied by a great many scientists, including Muskhelishvily [205] (for isotropic materials) and Lekhnitsky [60] (for anisotropic matter). The considered problems are solved through a combination of stress functions, or complex potentials with some arbitrary parameters, so that differential equations of the state are decided identically. The arbitrariness of the chosen parameters is discarded by observing boundary conditions.

Boundary conditions of the stress functions (complex potentials) are fulfilled in the considered problems in discrete contour points (collocation method). The method was first proposed by Kantorovich [206] for differential equations with partial derivatives. Convergence of the method depends to some extent on the position of collocation points, although their optimum location for a random contour is difficult to foresee. Nevertheless, the collocation method turns out to be highly effective for a series of elasticity theory problems.

A number of studies report on tress evaluation in the microstructure of composites, including hybrid ones [40, 119, 128, 132, 190, 207–212]. A major part of the work dealing with the determination of efficient elasticity module is based on the theory of elasticity or some other theoretical procedures (e.g. rule of mixtures, etc.) and experimental approaches [213].

The approximate problem solution proceeding from the theory of complex variable functions mainly takes one of two paths, that is, reduction to an integral equation [214] or the collocation method [20, 212, 215]. The complex potentials introduced here may either automatically meet the twofold stress periodicity [210, 215] or the latter condition will be met approximately during solution [20, 212]. In this case, a representative member is specified and boundary conditions are determined on the fiber-to-matrix mating contour, along with conditions for twofold periodicity at the representative member bounds. The described procedure seems bulky compared to [215], since the number of resolving equations increases perceptibly. The collocation method is much simpler than that of integral equations, which is why the majority of quantitative results have been obtained by the collocation method.

Recently, a series of studies has appeared dealing with laminates, in which the elementary layer is studied as an infinite strip with periodically disposed wafers or holes in it [216, 217]. These studies are, however, far from complete, especially in contrast to problems of the doubly periodic system of inclusions.

It has been shown that it is possible to separate the fiber bunches of the main reinforcement by winding them with an auxiliary reinforcement. As a result, a specific interlayer or separating layer forms on the bunch surfaces. It is easy to determine both geometrical parameters and properties of the layer theoretically or experimentally using model samples of the microplastic. It should be noted that the so-called interfacial layer also forms in usual composite materials as a result of physico-chemical phenomena occurring at the interface between components but it is difficult to determine its physico-mechanical properties. The absence of these layers in the material structure should, evidently, lead to variations in its characteristics, which make the stress-strain problem and analysis of the layers extremely actual. However, very few reports are devoted to this problem at present [20, 55, 132, 218] and the properties of the interlayer are expressed in an inexplicit form [20, 132]. The effect of the layer properties on efficient elastic moduli was studied in [20], where the layer was taken to be very thin and isotropic. In this work, a conclusion was made that the layer only insignificantly influences the properties of the composite as a whole. In contrast, in [55] it was stated that the layer strongly affects the stress values of the composite and efficient elastic moduli of the macromodel. It was also proved that the layer constituting only 1.5% of the total composite volume influences the reduced elasticity modulus, which varies within 5–6% depending on the elasticity modulus of the layer [55]. On transition through the layer there appears stress leap, so it is proposed [55] to form a few layers of varying rigidity to smooth the stresses and to avoid this.

A specific feature of composites based on spirally reinforced fillers consists of a wide range of properties acquired by the intermediate layers formed on the surface of the fiber bunches of the main reinforcement. Furthermore, the layers can be very thin and isotropic or, vice versa, thick and orthotropic. Notice that such questions as the effect of the orthotropic layers on the composite stress state and reduced elastic moduli values have not yet been posed.

3.1 Statement of the Problem

We will now consider the transverse loading of composites based on spirally reinforced filler. The material structure is in this case ordered, since spirally reinforced elements are located in the nodes of the double-periodic lattice and the spaces between them are filled by the binder-matrix, which is considered to be isotropic. We will assume that the cores of the spirally reinforced elements represent long transtropic cylinders of random cross-section, on whose side surfaces layers of the auxiliary reinforcement are located. We will take the layers as infinitely long hollow cylinders and cylindrically orthotropic in a general case. Depending on the impregnation, the properties of the auxiliary reinforcing layer may change across the thickness. When the impregnation is made by the binder found inside the spirally reinforced element, layer portions located near the element core turn out to be the best impregnated. On impregnation by the binder located in spaces between the elements, the most saturated turn out to be remote from the core layer. Preliminary impregnation of the winding reinforcement helps to attain constancy of the layer properties across the thickness. Analogous variations in the interfacial layer properties are common for conventional composites when physical or chemical treatment is applied during fabrication.

Insofar as stresses and displacements remain invariable along the z-axis at transverse loading, investigation of the composite under study is reduced to solution of the plane problem of the elasticity theory for an anisotropic in a general case body. In this connection, it is sufficient to consider only the cross-section perpendicular to the reinforcement direction (Fig. 3.1). Since the structure is ordered, geometrical centers of inclusions form a double-periodic system of nodes with periods ω_1 and ω_2. It is proposed, for simplification, to admit that composite elements display geometrical symmetry relative to the OX and OY axes (Fig. 3.2).

Proceeding from the Kolosov–Muskhelishvily theory of complex potentials [205] plane problem stresses and displacements at the isotropic binder and layer boundary can be written as

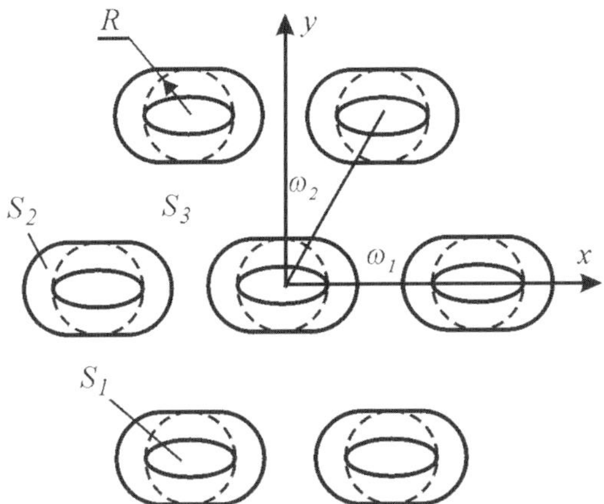

Fig. 3.1. Model of the composite based on spirally reinforced filler

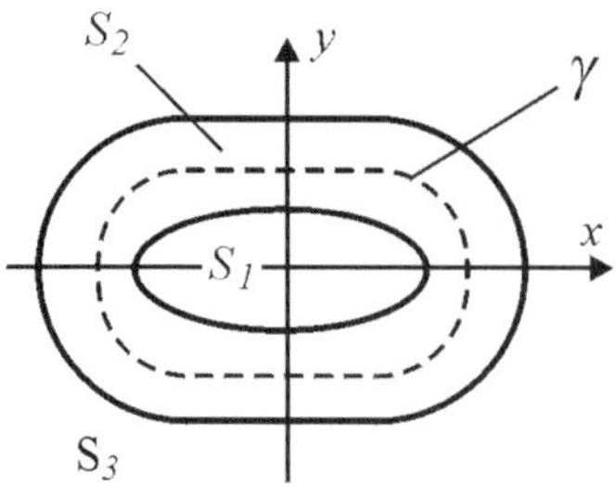

Fig. 3.2. Representative element of the composite model

$$\sigma_x^{(3)} = 2\operatorname{Re}\{\varphi_3'(t)\} - \operatorname{Re}\{\bar{t}\varphi_3''(t) + \psi_3'(t)\}\,;$$
$$\sigma_y^{(3)} = 2\operatorname{Re}\{\varphi_3'(t)\} + \operatorname{Re}\{\bar{t}\varphi_3''(t) + \psi_3'(t)\}\,;$$
$$\tau_{xy}^{(3)} = \operatorname{Im}\{\bar{t}\varphi_3''(t) + \psi_3'(t)\}\,; \tag{3.1}$$

$$\begin{matrix} u^{(3)} \\ \nu^{(3)} \end{matrix} = \frac{1}{2G_3}\begin{matrix} \operatorname{Re} \\ \operatorname{Im} \end{matrix}\{\kappa_3\varphi_3(t) - t\varphi_3'(t) - \psi_3'(t)\}\,, \tag{3.2}$$

where $\varphi_3(t), \psi_3(t)$ – limiting values of complex potentials; t – affix of boundary points between layer and matrix; $\kappa_3 = 3 - 4\nu_3$; ν_3 – matrix Poisson's ratio; G_3 – matrix shear modulus.

Let us assume that the mean stresses in the composite (or infinite stresses) are reduced to the main stresses P_1 and P_2, and P_1 forms angle α with axis OX. From this statement it follows that stresses in the composite are double-periodic. This condition can be met by taking complex potentials $\Phi(z) = \varphi_3'(z)$ and $\Psi(z) = \psi_3'(z)$ as analytic outside some radius R, like [210]

$$\frac{1}{2G_3}\Phi(z) = \Gamma + \alpha_0 + \sum_{k=0}^{\infty}\alpha_{2k}\frac{R^{2k+2}}{z^{2k+2}} + \sum_{k=0}^{\infty}\sum_{j=0}^{\infty}\alpha_{2k+2}R^{2k+2}r_{j,k}z^{2j}\,;$$

$$\frac{1}{2G_3}\Psi(z) = \Gamma' + \beta_0 + \sum_{k=0}^{\infty}\beta_{2k+2}\frac{R^{2k+2}}{z^{2k+2}} + \sum_{k=0}^{\infty}\sum_{j=0}^{\infty}\beta_{2k+2}R^{2k+2}r_{j,k}z^{2j}$$

$$- \sum_{k=0}^{\infty}\sum_{j=0}^{\infty}(2k+2)\alpha_{2k+2}R^{2k+2}s_{j,k}z^{2j}\,. \tag{3.3}$$

Here:

$$r_{j,k} = \frac{(2j+2k+1)!}{(2j)!(2k+1)!}\frac{q_{j+k+1}}{2^{2j+2k+2}}\,;\quad r_{0,0} = 0\,;$$

$$s_{j,k} = \frac{(2j+2k+2)!}{(2j)!(2k+2)!}\frac{\rho_{j+k+1}}{2^{2j+2k+2}}\,;\quad s_{0,0} = 0\,;$$

$$\Gamma = \frac{P_1+P_2}{8G_3}\,;\quad \Gamma' = \frac{P_2-P_1}{4G_3}e^{-2i\alpha}\,, \tag{3.4}$$

where R – radius of circumferences inscribed into element contour (Fig. 3.2); constants q_{j+k+1}, ρ_{j+k+1} for different packing parameters are presented elsewhere [210].

The parameters α_{2k+2}, β_{2k+2} in (3.3) are true on composite tension-compression in the direction $\alpha = 0$ or imaginary at shear $\left(\alpha = \dfrac{\pi}{4};\ P_1 = -P_2\right)$. In addition, a condition is imposed on factors $\alpha_0, \beta_0, \alpha_2$ and β_2

$$\alpha_0 = (k_0\alpha_2 + k_1\beta_2)R^2; \quad \beta_0 = (k_2\alpha_2 + k_3\beta_2)R^2 . \tag{3.5}$$

The values of constants k_0, k_1, k_2, k_3 are also given in [210].

From relations (3.1) and (3.2) using (3.3) stresses $\sigma_x^{(3)}, \sigma_y^{(3)}, \tau_{xy}^{(3)}$ and shifts $u^{(3)}, v^{(3)}$ are determined in the matrix on the layer and matrix bonding contour.

$$\frac{\sigma_x^{(3)}}{2G_3} = 2\Gamma - \mathrm{Re}\{\Gamma'\} + 2\alpha_0 - \beta_0 + 2\,\mathrm{Re}\sum_{k=0}^{\infty}\alpha_{2k+2}$$

$$\times\left\langle \frac{R^{2k+2}}{t_{2k+2}} + \sum_{j=0}^{\infty}\{[r_{j,k} + (k+1)S_{j,k}]R^{2k+2}t^{2j}\right.$$

$$\left. -jr_{j,k}\bar{t}t^{2j-1}R^{2k+2}\} + (k+1)\bar{t}\frac{R^{2k+2}}{t^{2k+3}}\right\rangle$$

$$- \mathrm{Re}\sum_{k=0}^{\infty}\beta_{2k+2}\left[\frac{R^{2k+2}}{t_{2k+2}} + \sum_{k=0}^{\infty}r_{j,k}R^{2k+2}t^{2j}\right];$$

$$\frac{\sigma_y^{(3)}}{2G_3} = 2\Gamma + \mathrm{Re}\{\Gamma'\} + 2\alpha_0 + \beta_0 + 2\,\mathrm{Re}\sum_{k=0}^{\infty}\alpha_{2k+2}$$

$$\times\left\langle \frac{R^{2k+2}}{t_{2k+2}} + \sum_{j=0}^{\infty}\{[r_{j,k} - (k+1)s_{j,k}]R^{2k+2}t^{2j}\right.$$

$$\left. +jr_{j,k}R^{2k+2}t^{2j-1}\bar{t}\} - (k+1)\frac{R^{2k+2}}{t^{2k+3}}\bar{t}\right\rangle$$

$$+ \mathrm{Re}\sum_{k=0}^{\infty}\beta_{2k+2}\left[\frac{R^{2k+2}}{t_{2k+2}} + \sum_{j=0}^{\infty}r_{j,k}R^{2k+2}t^{2j}\right];$$

$$\frac{\tau_{xy}^{(3)}}{2G_3} = Jm\{\Gamma'\} + 2Jm\sum_{k=0}^{\infty}\alpha_{2k+2}$$

$$\times\left\{ -(k+1)\bar{t}\frac{R^{2k+2}}{t^{2k+3}} + \sum_{j=0}^{\infty}[jr_{j,k}\bar{t}R^{2k+2}t^{2j-1}\right.$$

$$\left. -(k+1)S_{j,k}R^{2k+2}t^{2j}]\right\}$$

$$+ Jm\sum_{k=0}^{\infty}\beta_{2k+2}\left\{\frac{R^{2k+2}}{t^{2k+2}} + \sum_{j=0}^{\infty}r_{j,k}R^{2k+2}t^{2j}\right\};$$

$$
\begin{aligned}
u^{(3)}_{} = {} & (\kappa_3 - 1)(\Gamma + \alpha_0)\frac{\mathrm{Re}}{Jm}\{t\} - \frac{\mathrm{Re}}{Jm}\{t(\overline{\Gamma}' + \beta_0)\} \\
v^{(3)} & \\
& + \sum_{j=0}^{\infty}\left[r_{j,k}\left(\frac{\kappa_3}{2j+1}R^{2k+2}t^{2j+1} - t\bar{t}^{2j}R^{2k+2} \right)\right. \\
& + \frac{\mathrm{Re}}{Jm}\sum_{k=0}^{\infty}\alpha_{2k+2}\left\{ -\frac{\kappa_3}{2k+1}\frac{R^{2k+2}}{t^{2k+1}} - t\frac{R^{2k+2}}{t^{2k+2}} \right. \\
& \left.\left. + (2k+2)S_{j,k}\frac{R^{2k+2}\bar{t}^{2j+1}}{2j+1} \right]\right\} + \frac{\mathrm{Re}}{Jm}\sum_{k=0}^{\infty}\beta_{2k+2} \\
& \times \left[\frac{R^{2k+2}}{\bar{t}^{2k+1}(2k+1)} - \sum_{j=0}^{\infty}r_{j,k}\frac{R^{2k+2}\bar{t}^{2j+1}}{2j+1} \right].
\end{aligned}
$$

In order to define the stress state of the rest of the anisotropic elements, it is expedient to use a sufficiently elaborated plane theory of elasticity for anisotropic bodies [202]. Since use of complex potentials (3.3) automatically meets the requirement of the double-periodicity of tresses, we will further consider only one representative structural element (Fig. 3.2).

3.2 Determination of Reduced Elastic Characteristics

Along with the stress state study and analysis of strength properties it is necessary to find the rigidity of the material – assumed so-called reduced elastic characteristics. For this purpose, the composite is to be a homogenous medium displaying, in a general case, orthotropic properties. The homogeneous medium and the composite material are considered to be equivalent [168] if mean stresses or displacements coincide in them.

Let us use the method for finding displacements in a composite material. We will assume that factors $\alpha_0, \beta_0, \alpha_2, \beta_2$ from expansions (3.3) can be found for some loads at infinity.

Let the stress state of a homogeneous equivalent medium be induced by the main tensions at infinity, that is $P_1 = \tau$ and $P_2 = -\tau$, and P_1 forms angle $\pi/4$ with axis OX. Then, strain in the medium will be of the kind

$$
\frac{\partial u}{\partial y} + \frac{\partial v}{\partial x} = \frac{i\tau}{G_{PR}}, \tag{3.6}
$$

where G_{PR} – reduced shear modulus.

Under such loading the displacement field in a continuum should be homogeneous, and it then follows from (3.6)

$$
u = \frac{\tau y}{2G_{PR}}; \quad v = \frac{\tau x}{2G_{PR}};
$$

$$
q^*(z, \bar{z}) = u + iv = \frac{i\tau\bar{z}}{2G_{PR}}. \tag{3.7}
$$

For the shear stress state the considered, complex potentials should correspond to (3.3) and look like

$$\varphi_3(z) = \varphi_0(z); \quad \psi_3(z) = i\tau z + \psi_0(z),$$ (3.8)

where $\varphi_0(z)$, $\psi_0(z)$ – quasiperiodic function.

It is easy to prove that factors α_k, β_k in (3.3) are purely imaginary under the above-specified loading conditions,

$$\alpha_k = i\alpha_k^*; \quad \beta_k = i\beta_k^*$$

α_k^*, β_k^* are true factors.

Using dependencies (2.2) one can find displacement vector $q = u^{(3)} + iv^{(3)}$ in the composite:

$$2G_3 q(z,\bar{z}) = i\tau\bar{z} + \kappa_3\varphi_0(z) - z\overline{\varphi_0'(z)} - \overline{\psi_0(z)}.$$ (3.9)

From (3.7) and (3.9) taking account of (3.3) we obtain the following relations

$$\begin{aligned}
\Omega_1 &= q(z + \omega_1, \bar{z} + \overline{\omega_1}) - q(z,\bar{z}) \\
&= \frac{i\tau\overline{\omega_1}}{2G_3} + i\left[\alpha_2^* R^2(\gamma_1 - \kappa_3\delta_1) + \beta_2^* R^2\bar{\delta} + (\kappa_3 - 1)\alpha_0^*\omega_1 - \beta_0^*\overline{\omega_1}\right];
\end{aligned}$$

$$\Omega_1 = q^*(z + \omega_1, \bar{z} + \overline{\omega_1}) = \frac{i\tau\overline{\omega_1}}{2GT_{PR}},$$ (3.10)

where γ, δ – some constants whose values for different periods ω_1 are presented in [210].

By comparing relations (3.10) for equivalent materials we have (at $\tau = 2G_3$)

$$\begin{aligned}
G_{xy}^* &= \frac{G_{PR}}{G} \\
&= \left[1 + \alpha_2^* R^2(\gamma_1 - \kappa_3\delta_1) + \beta_2^* R^2\bar{\delta}_1 + (\kappa_3 - 1)\alpha_0^*\omega_1 - \beta_0^*\overline{\omega_1}\right].
\end{aligned}$$ (3.11)

The above-cited shear modulus G_{xy}^* of the equivalent material agrees with that from [210].

For a special case of hexagonal disposition of inclusion centers formula (3.11) is of the following type:

$$G_{xy}^* = \frac{\sqrt{3}}{\sqrt{3} + 2(1 - v_3)\pi(\beta_2^* - 2\alpha_2^*)}.$$ (3.12)

The rest elastic characteristics of the equivalent medium at hexagonal locations of structural elements are

$$E_x^* = \frac{E_x^{PR}}{E_3} = \frac{2}{A + B}; \quad v_1^* = \frac{A - B}{A + B}; \quad v_2^* = \frac{v_1^* E_y^*}{E_x^*};$$

$$E_y^* = \frac{E_y^{PR}}{E_3} = \frac{\sqrt{3}}{\sqrt{3} + 8\alpha_2 E_x^*\pi},$$ (3.13)

where

$$A = 1 + v_3 + \frac{2\pi}{\sqrt{3}}(\beta_2 - 2\alpha_2);$$

$$B = 1 - v_3 + \frac{2\pi}{\sqrt{3}}(\beta_2 - 2\alpha_2).$$

Parameters α_2, β_2 are calculated at a uniaxial tension of the composite: $P_1 = P$; $P_2 = 0$ [210].

3.3 Study of the Material Model with Orthotropic and Variable Across Thickness Layer Properties

Let us denote the inner and outer contours of the layer by, correspondingly L_1 and L_2, occupying zone S_2 in the representative element section. The zone occupied by the core of the spirally reinforced element will be denoted through S_1. We will suppose that the element core is transversely isotropic and the layer shows a cylindrical orthotropy.

Conditions of a rigid contact on contours L_1 and L_2 (Fig. 3.2) are written as [20]

$$u^{(j)} = u^{(j+1)}; \quad v^{(j)} = v^{(j+1)};$$

$$\sigma_n^{(j)} = \sigma_n^{(j+1)}; \quad \tau^{(j)} = \tau^{(j+1)}; \quad (j = 1, 2), \tag{3.14}$$

where $u^{(j)}, v^{(j)}$ are components of the shear vector; $\sigma_n^{(j)}$; $\tau^{(j)}$ are normal and tangential components of a stress tensor related to zones S_j, and:

$$\sigma_n^{(j)} = \sigma_x^{(j)} \cos^2 \alpha_1 + \sigma_y^{(j)} \sin^2 \alpha_1 + \tau_{xy}^{(j)} \sin 2\alpha_1;$$

$$\tau^{(j)} = -\frac{\sigma_x^{(j)} + \sigma_y^{(j)}}{2} \sin 2\alpha_1 \tau_{xy}^{(j)} + \tau_{xy}^{(j)} \cos 2\alpha_1 \tag{3.15}$$

$$\sigma_n^{(2)} = \sigma_r^{(2)} \cos^2 \beta_1 + \sigma_\theta^{(2)} \sin^2 \beta_1 + \tau_{r\theta}^{(2)} \sin 2\beta_1;$$

$$\tau^{(2)} = -\frac{\sigma_r^{(2)} + \sigma_\theta^{(2)}}{2} \sin 2\beta_1 \tau_{r\theta}^{(2)} + \tau_{r\theta}^{(2)} \cos 2\beta_1 \tag{3.16}$$

α_1, β_1 are here, correspondingly, angles between directions of axes X or r, with a normally contoured L_1 or L_2 (depending on what contour $\sigma_n^{(j)}$; $\tau^{(j)}$ are calculated to be).

Let us introduce stress function F into zones S_1 and stresses will be written as follows [202, 203]:

$$\sigma_x^{(1)} = \frac{\partial^2 F}{\partial y^2}; \quad \sigma_y^{(1)} = \frac{\partial^2 F}{\partial x^2}; \quad \tau_{xy}^{(1)} = \frac{\partial^2 F}{\partial x \partial y}.$$

For a transtropic body parameters $\beta_{ij}^{(1)}$ [202] are equal to:

$$\beta_{11}^{(1)} = \beta_{22}^{(1)} = \beta_{12}^{(1)} + \frac{\beta_{66}^{(1)}}{2} = \frac{1 - v'v''}{E_1},$$

the rest parameters $\beta_{ij} = 0$; v – Poisson's ratio characterizing reduction in plane OXY at tension in this plane; v' – the same at tension in the reinforcing direction; v'' – characterizes reduction in the reinforcement direction at tension in plane OXY; E_1 – elasticity modulus of the element core in the isotropic plane.

Hence, from equilibrium equations we get $\Delta\Delta F = 0$, which allows one to utilize the theory of representing stresses and displacements in zone S_1 developed in [203]. Then, the factor used in [203] must be of some other type for the problem under consideration. Thus, for zone S_1 [202]:

$$\kappa_1 = \frac{3 - 4v'v'' - v}{1 + v}; \quad G_1 = \frac{E_1}{2(1 + v)}.$$

Therefore, for the components of displacements $u^{(1)}$ and $v^{(1)}$ of stresses $\sigma_x^{(1)}, \sigma_y^{(1)}, \tau_{xy}^{(1)}$ we have

$$2G_1(u^{(1)} + iv^{(1)}) = \kappa_1\varphi_1(z) - z\overline{\varphi'(z)} - \overline{\psi'_1(z)}; \tag{3.17}$$

$$\sigma_x^{(1)} + \sigma_y^{(1)} = 4\,\mathrm{Re}\{\varphi'_1(z)\};$$
$$\sigma_y^{(1)} + i\tau_{xy}^{(1)} = 2\,\mathrm{Re}\{\varphi'_1(z)\} + \overline{z}\varphi''_1(z) + \psi'_1(z). \tag{3.18}$$

Bearing in mind the symmetry of the problem, it is possible to prove that the analytic in zone S_1 functions $\varphi_1(z)$ and $\psi_1(z)$ can be roughly presented as polynoms

$$\varphi_1(z) = 2G_1\sum_{k=1}^{N_1}\Phi_{1k}z^{2k-1}; \quad \psi_1(z) = 2G_1\sum_{k=1}^{N_1}\Phi_{2k}z^{2k-1}. \tag{3.19}$$

Then, from (3.17) and (3.18) we obtain

$$\left\{\begin{array}{c}u^{(1)}\\v^{(1)}\end{array}\right\} = R\sum_{k=1}^{N_1}\left[\Phi_{1k}\begin{array}{c}\mathrm{Re}\\\mathrm{Im}\end{array}\{\kappa_1 t^{2k-1} - (2k-1)t\overline{t}^{2k-2}\} - \Phi_{2k}\begin{array}{c}\mathrm{Re}\\\mathrm{Im}\end{array}\{\overline{t}^{2k-1}\}\right];$$

$$\sigma_x^{(1)} = 2G_1 R\sum_{k=1}^{N_1}(2k-1)\,\mathrm{Re}\{\Phi_{1k}[2t^{2k-2}-(2k-2)\overline{t}t^{2k-3}]-\Phi_{2k}t^{2k-2}\};$$

$$\sigma_y^{(1)} = 2G_1 R\sum_{k=1}^{N_1}(2k-1)\,\mathrm{Re}\{\Phi_{1k}[2t^{2k-2}+(2k-2)\overline{t}t^{2k-3}]+\Phi_{2k}t^{2k-2}\};$$

$$\tau_{xy}^{(1)} = 2G_1 R\sum_{k=1}^{N_1}(2k-1)\,\mathrm{Im}\{\Phi_{1k}(2k-2)\overline{t}t^{2k-3} + \Phi_{2k}t^{2k-2}\}. \tag{3.20}$$

Functions $\sigma_n^{(1)}$ and $\tau^{(1)}$ included in boundary conditions (3.14) are calculated through unknown yet true parameters Φ_{jk} by formulas (3.15) and (3.16).

To calculate stresses in zone S_2 of the layer let us write equilibrium equations for the plane strain as [202]:

$$\frac{\partial \sigma_r^{(2)}}{\partial r} + \frac{1}{r}\frac{\partial \tau_{r\theta}^{(2)}}{\partial \theta} + \frac{\sigma_r^{(2)} - \sigma_\theta^{(2)}}{r} = 0;$$

$$\frac{\partial \tau_{r\theta}^{(2)}}{\partial r} + \frac{1}{r}\frac{\partial \sigma_\theta^{(2)}}{\partial \theta} + \frac{2}{r}\tau_{r\theta}^{(2)} = 0, \tag{3.21}$$

where r, θ – polar coordinates.

Hook's law with allowance for orthotropic properties is as follows

$$\varepsilon_r^{(2)} = \frac{\partial u_r^{(2)}}{\partial r} = \beta_{11}^{(2)}(r)\sigma_r^{(2)} + \beta_{12}^{(2)}(r)\sigma_\theta^{(2)};$$

$$\varepsilon_\theta^{(2)} = \frac{1}{r}\frac{\partial u_\theta^{(2)}}{\partial \theta} + \frac{u_r^{(2)}}{r} = \beta_{12}^{(2)}(r)\sigma_r^{(2)} + \beta_{22}^{(2)}(r)\sigma_\theta^{(2)};$$

$$\gamma_{r\theta}^{(2)} = \frac{1}{r}\frac{\partial u_r^{(2)}}{\partial \theta} + \frac{\partial u_\theta^{(2)}}{\partial r} - \frac{u_\theta^{(2)}}{r} = \beta_{66}^{(2)}(r)\tau_{r\theta}^{(2)}, \tag{3.22}$$

where

$$\beta_{11}^{(2)}(r) = \frac{1 - v_{rz}v_{zr}}{E_r}; \quad \beta_{12}^{(2)}(r) = -\frac{v_{r\theta} + v_{rz}v_{z\theta}}{E_r};$$

$$\beta_{22}^{(2)}(r) = \frac{1 - v_{\theta z}v_{z\theta}}{E_\theta}; \quad \beta_{66}^{(2)}(r) = \frac{1}{G_{r\theta}}.$$

E_r, E_θ – variables and Young's moduli in a general case in the main elasticity directions (radial r and tangential θ); $G_{r\theta}$ – shear modulus characterizing variation in the angle between r and θ directions; and $v_{zr} = v_{rz}$, $v_{\theta z} = v_{\theta r}$, $v_{z\theta} = v_{r\theta}$ – also variable Poisson's ratio characterizing transverse compression in the direction of the axis marked by the first index under tension in the direction of the axis marked by the second index.

By solving jointly (3.21) and (3.22) relative to $u_r^{(2)}, v_\theta^{(2)}$, we have:

$$\frac{\beta_{22}^{(2)}}{\gamma}\frac{\partial^2 u_r^{(2)}}{\partial r^2} + \frac{\partial u_r^{(2)}}{\partial r}\left[\frac{\dot{\beta}_{22}^{(2)}}{\gamma} - \frac{\dot{\gamma}\beta_{22}^{(2)}}{\gamma^2} + \frac{\beta_{22}^{(2)}}{\gamma r}\right] + u_r\left[\frac{\dot{\gamma}\beta_{12}^{(2)}}{r\gamma^2} - \frac{\dot{\beta}_{12}^{(2)}}{r\gamma} - \frac{\beta_{11}^{(2)}}{\gamma r^2}\right]$$

$$+ \frac{1}{\beta_{66}^{(2)}}\frac{\partial^2 u_r^{(2)}}{\partial \theta^2} + \frac{\partial^2 u_\theta^{(2)}}{\partial r \partial \theta} \times \left[\frac{1}{r\beta_{66}^{(2)}} - \frac{\beta_{12}^{(2)}}{r\gamma}\right]$$

$$+ \frac{\partial u_\theta^{(2)}}{\partial \theta}\left[\frac{\dot{\gamma}\beta_{12}^{(2)}}{r\gamma^2} - \frac{\dot{\beta}_{12}^{(2)}}{r\gamma} - \frac{\beta_{11}^{(2)}}{\gamma r^2} - \frac{1}{r^2 \beta_{66}^{(2)}}\right] = 0;$$

$$\frac{\partial^2 u_r^{(2)}}{\partial r \partial \theta}\left[\frac{1}{r\beta_{66}^{(2)}} - \frac{\beta_{12}^{(2)}}{r\gamma}\right] + \frac{\partial u_r^{(2)}}{\partial \theta}\left\{\frac{\beta_{11}^{(2)}}{\gamma r^2} - \frac{\beta_{66}^{(2)}}{r[\beta_{66}^{(2)}]^2} - \frac{1}{r^2 \beta_{66}^{(2)}} + \frac{2}{r^2 \beta_{66}^{(2)}}\right\}$$

$$+ \frac{1}{\beta_{66}^{(2)}}\frac{\partial^2 u_\theta^{(2)}}{\partial r^2} + \frac{\partial u_\theta^{(2)}}{\partial r}\left\{\frac{2}{r\beta_{66}^{(2)}} - \frac{\beta_{66}^{(2)}}{[\beta_{66}^{(2)}]^2} - \frac{1}{r\beta_{66}^{(2)}}\right\}$$

$$+ u_\theta^{(2)}\left\{\frac{\beta_{66}^{(2)}}{r[\beta_{66}^{(2)}]^2} + \frac{1}{r^2 \beta_{66}^{(2)}} - \frac{2}{r^2 \beta_{66}^{(2)}}\right\} + \frac{\beta_{11}^{(2)}}{\gamma r^2}\frac{\partial^2 u_\theta^{(2)}}{\partial \theta^2} = 0, \tag{3.23}$$

where the dot above the function denotes **r** derivative and

$$\gamma = \beta_{11}^{(2)}\beta_{22}^{(2)} - [\beta_{12}^{(2)}]^2 \, .$$

To solve equations (3.23) let us present functions $u_r^{(2)}, v_\theta^{(2)}$ as a series with respect to cosines and sines written with regard to the problem symmetry as [219]:

$$u_r^{(2)} = \sum_{k=0}^{N_2} u_{kr}(r) \cos 2k\theta \, ;$$

$$u_\theta^{(2)} = \sum_{k=1}^{N_2} u_{k\theta}(r) \sin 2k\theta \, . \tag{3.24}$$

u_{kr} and $v_{k\theta}$ are in (3.24) some known functions, to determine which ones we will substitute (3.24) in (3.23) and equate coefficients of trigonometric functions of similar arguments. Thus, we obtain

$$k = 0: \quad \frac{\beta_{22}^{(2)}}{\gamma}\ddot{u}_{0r} + \left[\frac{\dot{\beta}_{22}^{(2)}}{\gamma} - \frac{\dot{\gamma}\beta_{22}^{(2)}}{\gamma^2} + \frac{\beta_{22}^{(2)}}{\gamma r}\right]\dot{u}_{0r}$$

$$+ \left[\frac{\dot{\gamma}\beta_{12}^{(2)}}{\gamma^2 r} - \frac{\dot{\beta}_{12}^{(2)}}{\gamma} - \frac{\beta_{11}^{(2)}}{\gamma r^2}\right]u_{0r} = 0 \, ; \tag{3.25}$$

$$k \geq 1: \quad \frac{\beta_{22}^{(2)}}{\gamma}\ddot{u}_{kr} + \left[\frac{\dot{\beta}_{22}^{(2)}}{\gamma} - \frac{\dot{\gamma}\beta_{22}^{(2)}}{\gamma^2} + \frac{\beta_{22}^{(2)}}{\gamma r}\right]\dot{u}_{kr}$$

$$+ \left[\frac{\dot{\gamma}\beta_{12}^{(2)}}{\gamma^2 r} - \frac{\dot{\beta}_{12}^{(2)}}{\gamma r} - \frac{\beta_{11}^{(2)}}{\gamma r^2} - \frac{4k^2}{\beta_{66}^{(2)} r^2}\right]u_{kr} + 2k\left[\frac{1}{r\beta_{66}^{(2)}} - \frac{\beta_{12}^{(2)}}{\gamma r}\right]\dot{u}_{k\theta}$$

$$+ \left[\frac{\dot{\gamma}\beta_{12}^{(2)}}{\gamma^2 r} - \frac{\dot{\beta}_{12}^{(2)}}{\gamma r} - \frac{\beta_{11}^{(2)}}{\gamma r^2} - \frac{4k^2}{\beta_{66}^{(2)} r^2}\right]2k^2 u_{k\theta} = 0 \, ;$$

$$-2k\left[\frac{1}{r\beta_{66}^{(2)}} - \frac{\beta_{12}^{(2)}}{\gamma r}\right]\dot{u}_{kr} - 2k\left\{\frac{\beta_{11}^{(2)}}{\gamma r^2} - \frac{\dot{\beta}_{66}^{(2)}}{r[\beta_{66}^{(2)}]} + \frac{1}{r^2\beta_{66}^{(2)}}\right\}u_{kr}$$

$$+ \frac{\ddot{u}_{k\theta}}{\beta_{66}^{(2)}} + \left\{\frac{1}{r\beta_{66}^{(2)}} - \frac{\dot{\beta}_{66}^{(2)}}{[\beta_{66}^{(2)}]^2}\right\}\dot{u}_{k\theta} + \left\{\frac{\dot{\beta}_{66}^{(2)}}{r[\beta_{66}^{(2)}]^2} - \frac{1}{r^2\beta_{66}^{(2)}}\right\}u_{k\theta}$$

$$- \frac{\beta_{11}^{(2)}}{\gamma r^2}4k u_{k\theta} = 0 \, . \tag{3.26}$$

Differential equations (3.25) and (3.26) cannot be solved analytically for random functions of the type $\beta_{jk}^{(2)}(r)$. However, if they are presented approximately like [202]:

$$\beta_{11}^{(2)} = b_{11}r^a; \quad \beta_{12}^{(2)} = b_{12}r^a; \quad b_{22}^{(2)} = b_{22}r^a \, ;$$

$$\beta_{66}^{(2)} = b_{66}r^a; \quad \gamma = \gamma_1 r^{2a}; \quad \gamma_1 = b_{11}b_{22} - b_{12}^2 \, , \tag{3.27}$$

then, (3.25) and (3.26) result in the known Euler's relations [203] whose solutions are the functions:

$$k = 0: \quad u_{0r}(r) = \sum_{j=1}^{2} C_{j0} r^{\lambda j0} \, ;$$

$$k = 1: \quad u_{kr}(r) = 2 \sum_{j=1}^{4} \Phi'_{jk} C_{jk} r^{\lambda jk} \, ;$$

$$u_{k\theta}(r) = \sum_{j=1}^{4} \Psi'_{jk} C_{jk} r^{\lambda jk} \, , \tag{3.28}$$

where C_{jk} – constants of integration,

$$\Phi'_{jk} = \frac{b_{11} - ab_{12}}{\gamma_1} + \frac{1}{b_{66}} - \left(\frac{1}{b_{66}} - \frac{b_{12}}{\gamma_1} \right) \lambda_{jk} \, ;$$

$$\Psi'_{jk} = \frac{b_{22}}{\gamma_1} \lambda_{jk}^2 - \frac{ab_{22}}{\gamma_1} \lambda_{jk} + \frac{ab_{12} - b_{11}}{\gamma_1} - \frac{4k^2}{b_{66}} \, ,$$

λ_{jk} – characteristic equation roots:

$$A_4 \lambda_k^4 + A_3 \lambda_k^3 + A_2 \lambda_k^2 + A_1 \lambda_k + A_0 = 0 \, ;$$

$$A_0 = (4k^2 - 1)[(1 - a)(ab_{12} - b_{11}) + 4k^2 b_{11}] \, ;$$
$$A_1 = a[b_{22}(1 - a) + b_{11} - ab_{12} + 4k^2(b_{66} + 2b_{12})] \, ;$$
$$A_2 = ab_{12} - b_{11} + b_{22}(a^2 + a - 1) - 4k^2(b_{66} + 2b_{12}) \, ;$$
$$A_3 = -2ab_{22} \, ;$$
$$A_4 = b_{22} \, .$$

Components $u_r^{(2)}$ and $u_\theta^{(2)}$ of the displacement vector and those of the stress tensor $\sigma_r^{(2)}, \sigma_\theta^{(2)}, \tau_{r\theta}^{(2)}$ are found from (3.28).

Functions $u^{(2)}$ and $v^{(2)}$ incorporated in boundary conditions (3.14) on curvilinear contours L_1 and L_2 are expressed through $u_r^{(2)}$ and $u_\theta^{(2)}$ using dependencies [203]:

$$u^{(2)} = u_r^{(2)} \cos\theta - u_\theta^{(2)} \sin\theta \, ;$$
$$v^{(2)} = u_r^{(2)} \sin\theta + u_\theta^{(2)} \cos\theta \, . \tag{3.29}$$

Functions $\sigma_P^{(2)}$ and $\tau^{(2)}$ are found in terms of $\sigma_r^{(2)}, \sigma_\theta^{(2)}, \tau_{r\theta}^{(2)}$ from dependencies (3.16). Thus, we finally have:

$$\frac{\sigma_n^{(2)}}{E_r} = \sum_{k=0}^{N_2} \sum_{j=1}^{4} E_{kn}^{(j)} C_{jk} \, ; \qquad \frac{\tau^{(2)}}{E_r} = \sum_{k=0}^{N_2} \sum_{j=1}^{4} E_{k\tau}^{(j)} C_{jk} \, ; \tag{3.30}$$

$$u^{(2)} = \sum_{k=0}^{N_2} \sum_{j=1}^{4} E_{ku}^{(j)} C_{jk} \, ; \qquad v^{(2)} = \sum_{k=0}^{N_2} \sum_{j=1}^{4} E_{kv}^{(j)} C_{jk} \, . \tag{3.31}$$

It is evident that in (2.30) and (2.31) at $j = 3, 4$

$$E_{on}^{(j)} = E_{o\tau}^{(j)} = E_{ou}^{(j)} = E_{ov}^{(j)} = 0.$$

Stresses on the curvilinear contour L_2 in matrix $\sigma_n^{(3)}$ and $\tau^{(3)}$ entering boundary conditions (3.14), and displacements $u^{(3)}, v^{(3)}$ are determined by (3.15) and relations presented earlier.

Hence, all values incorporated in (3.14) are determined through so far unknown constant parameters

$$\Phi_{jk}(j = 1, 2; \ k = 1, \ldots, N_1); \quad C_{10}, C_{20}, C_{jk} \ (j = 1, \ldots, 4; k = 1, \ldots, N_2);$$
$$\alpha_{2k+2}, \beta_{2k+2} \ (k = 0, \ldots, N_3). \tag{3.32}$$

The parameters of (3.32) are found by the collocation method using equations with boundary conditions (3.14) of the kind

$$\sum_{k=1}^{N_1}(2k - 1)(A_{kn}^{(1)}\Phi_{1k} + A_{kn}^{(2)}\Phi_{2k}) = \delta_1 \sum_{k=0}^{N_2}\sum_{j=1}^{4} E_{kn}^{(j)}C_{jk};$$

$$\sum_{k=1}^{N_1}(2k - 1)(A_{k\tau}^{(1)}\Phi_{1k} + A_{k\tau}^{(2)}\Phi_{2k}) = \delta_1 \sum_{k=0}^{N_2}\sum_{j=1}^{4} E_{k\tau}^{(j)}C_{jk};$$

$$\sum_{k=1}^{N_1}(A_{ku}^{(1)}\Phi_{1k} + A_{ku}^{(2)}\Phi_{2k}) = \sum_{k=0}^{N_2}\sum_{j=1}^{4} E_{ku}^{(j)}C_{jk};$$

$$\sum_{k=1}^{N_1}(A_{kv}^{(1)}\Phi_{1k} + A_{kv}^{(2)}\Phi_{2k}) = \sum_{k=0}^{N_2}\sum_{j=1}^{4} E_{kv}^{(j)}C_{jk};$$

$$(x, y, r \in L_1);$$

$$\delta_2 \sum_{k=0}^{N_2}\sum_{j=1}^{4} E_{kn}^{(j)}C_{jk} = \sum_{k=0}^{N_3}(\alpha_{2k+2}B_{kn}^{(1)} + \beta_{2k+2}B_{kn}^{(2)}) + G_n;$$

$$\delta_2 \sum_{k=0}^{N_2}\sum_{j=1}^{4} E_{k\tau}^{(j)}C_{jk} = \sum_{k=0}^{N_3}(\alpha_{2k+2}B_{k\tau}^{(1)} + \beta_{2k+2}B_{k\tau}^{(2)}) + G_\tau;$$

$$\sum_{k=0}^{N_2}\sum_{j=1}^{4} E_{ku}^{(j)}C_{jk} = \sum_{k=0}^{N_3}(\alpha_{2k+2}B_{ku}^{(1)} + \beta_{2k+2}B_{ku}^{(2)}) + G_u;$$

$$\sum_{k=0}^{N_2}\sum_{j=1}^{4} E_{kv}^{(j)}C_{jk} = \sum_{k=0}^{N_3}(\alpha_{2k+2}B_{kv}^{(1)} + \beta_{2k+2}B_{kv}^{(2)}) + G_v; \tag{3.33}$$

$$(x, y, r \in L_2)$$

$$\delta_1 = \frac{E_r(S_1)}{E_1}; \quad \delta_2 = \frac{E_r(S_2)}{E_3}; \quad S_1 \in L_1, \ S_2 \in L_2.$$

Relations for coefficients of the system (3.33) are the following

$$A_{kn}^{(j)} = T_x^j \cos^2 \alpha_1 + T_y^j \sin^2 \alpha_1 + T_{xy}^j \sin 2\alpha_1 \, ;$$

$$A_{k\tau}^{(j)} = -0,5(T_x^j + T_y^j) \sin 2\alpha_1 + T_{xy}^j \cos 2\alpha_1 \, ;$$

$$\left\{ \begin{array}{c} A_{ku}^{(1)} \\ A_{kv}^{(1)} \end{array} \right\} = R \left\{ \begin{array}{c} \mathrm{Re} \\ Jm \end{array} \right\} \{ \kappa_1 t^{2k-1} - (2k-1)t\bar{t}^{2k-2} \} \, ;$$

$$\left\{ \begin{array}{c} A_{ku}^{(2)} \\ A_{kv}^{(2)} \end{array} \right\} = R \left\{ \begin{array}{c} \mathrm{Re} \\ Jm \end{array} \right\} \{ \bar{t}^{2k-1} \} \, ;$$

$$T_x^1 = 2R \operatorname{Re}\{ t^{2k-2} - (k-1)\bar{t}t^{2k-3} \} \, ;$$
$$T_y^1 = 2R \operatorname{Re}\{ t^{2k-2} + (k-1)\bar{t}t^{2k-3} \} \, ;$$

$$T_x^2 = T_y^2 = R \operatorname{Re}\{ t^{2k-2} \} \, ;$$
$$T_{xy}^1 = Jm\{ \bar{t}t^{2k-3} \}(2k-2) \, ;$$
$$T_{xy}^2 = Jm\{ t^{2k-2} \} \, ;$$

$$E_{0\tau}^{(1)} = \frac{B_{22}^{(2)} \lambda_0 - B_{12}^{(2)} + B_{11}^{(2)} \lambda_0 - B_{12}^{(2)} \lambda_0}{\gamma B_{66}^{(2)}} r^{\lambda_0-1} \sin 2\beta \, ;$$

$$E_{0n}^{(1)} = \frac{B_{22}^{(2)} \lambda_0 - B_{12}^{(2)}}{\gamma B_{66}^{(2)}} r^{\lambda_0-1} \cos^2 \beta + \frac{B_{11}^{(2)} + B_{12}^{(2)} \lambda_0}{\gamma B_{66}^{(2)}} r^{\lambda_0-1} \sin^2 \beta \, ;$$

$$E_{0\tau}^{(2)} = \frac{B_{22}^{(2)} \lambda_0 + B_{12}^{(2)} - B_{11}^{(2)} - \lambda_0 B_{12}^{(2)}}{\gamma B_{66}^{(2)}} r^{-\lambda_0-1} \sin 2\beta \, ;$$

$$E_{0n}^{(2)} = -\frac{B_{22}^{(2)} \lambda_0 + B_{12}^{(2)}}{\gamma B_{66}^{(2)}} r^{-\lambda_0-1} \cos^2 \beta + \frac{B_{11}^{(2)} - B_{12}^{(2)} \lambda_0}{\gamma B_{66}^{(2)}} r^{-\lambda_0-1} \sin^2 \beta \, ;$$

$$E_{0u}^{(2)} = r^{-\lambda_0} \cos \Theta; \quad E_{0u}^{(1)} = r^{\lambda_0} \cos \Theta \, ;$$

$$E_{0v}^{(2)} = r^{-\lambda_0} \sin \Theta; \quad E_{0v}^{(1)} = r^{\lambda_0} \sin \Theta \, ;$$

$$E_{kn}^{(j)} = \frac{2k \cos 2k\Theta \, r^{\lambda_{jk}-1}}{\gamma B_{66}^{(2)}} [(B_{22}^{(2)} \lambda_{jk} - B_{12}^{(2)})\Phi_{jk} - B_{12}^{(2)} \Psi_{jk}] \cos^2 \alpha$$

$$- 2k r^{\lambda_{jk}-1} \cos 2k\Theta \frac{B_{12}^{(2)} \lambda_{jk} - B_{11}^{(2)} \Phi_{jk} - B_{11}^{(2)} \Psi_{jk}}{\gamma B_{66}^{(2)}} \sin^2 \alpha$$

$$+ r^{\lambda_{jk}-1} \sin 2k\Theta \frac{(\lambda_{jk}-1)\Phi jk - 4k^2 \Psi_{jk}}{B_{66}^{(2)}} \sin 2\alpha \, ;$$

$$E_{k\tau}^{(j)} = \frac{k}{\gamma B_{66}^{(2)}} r^{\lambda_{jk}-1} \cos 2k\Theta [(B_{12}^{(2)} - B_{22}^{(2)}\lambda_{jk} + B_{12}^{(2)}\lambda_{jk} - B_{11}^{(2)})\Phi_{jk}$$

$$+ (B_{12}^{(2)} - B_{11}^{(2)})\Psi_{jk}]\sin 2\alpha$$

$$+ \frac{1}{B_{66}^{(2)}} r^{\lambda_{jk}-1}[(\lambda_{jk}-1)\Phi_{jk} - 4k^2\Psi_{jk}]\sin 2k\Theta \cos 2\alpha \, ;$$

$$\frac{E_{ku}^{(j)}}{E_r} = 2k\Phi_{jk}r^{\lambda_{jk}}\cos 2k\Theta \cos\Theta - \Psi_{jk}r^{\lambda_{jk}}\sin 2k\Theta \sin\Theta \, ;$$

$$\frac{E_{kv}^{(j)}}{E_r} = 2k\Phi_{jk}r^{\lambda_{jk}}\cos 2k\Theta \sin\Theta - \Psi_{jk}r^{\lambda_{jk}}\sin 2k\Theta \cos\Theta \, ;$$

$$B_{kn}^{(1)} = \left\langle -\frac{\pi}{2\sqrt{3}}\varepsilon(k) + 2\,\mathrm{Re}\left\{\frac{R^{2k+2}}{\bar{t}^{2k+2}} + \sum_{j=0}^{\infty} R^{2k+2}[(r_{j,k} + (k+1)S_{j,k})t^{2j}\right.\right.$$

$$\left.-j_0 r_{j,k}t^{2j-1}\bar{t}] + (k+1)\bar{t}\frac{R^{2k+2}}{\bar{t}^{2k+3}}\right\}\bigg\rangle\cos^2\alpha + \left\langle\frac{\pi}{2\sqrt{3}}\varepsilon(k)\right.$$

$$+ 2\,\mathrm{Re}\left\{\frac{R^{2k+2}}{t^{2k+2}} + \sum_{j=0}^{\infty} R^{2k+2}[(r_{j,k} - (k+1)S_{j,k})t^{2j}\right.$$

$$\left.+ jr_{j,k}t^{2j-1}\bar{t}] - (k+1)\bar{t}\frac{R^{2k+2}}{t^{2k+3}}\right\}\bigg\rangle\sin^2\alpha + 2Jm\left\{-\frac{k+1}{t^{2k+3}}\bar{t}R^{2k+2}\right.$$

$$\left.+ \sum_{j=0}^{\infty} R^{2k+2}[jr_{j,k}\bar{t}t^{2j-1} - (k+1)S_{j,k}t^{2j}]\right\}\sin 2\alpha \, ;$$

$$B_{kn}^{(2)} = \left[\frac{\pi}{2\sqrt{3}}\varepsilon(k) - \mathrm{Re}\left\{\frac{R^{2k+2}}{t^{2k+2}} + \sum_{j=0}^{\infty} r_{j,k}t^{2j}R^{2k+2}\right\}\right]\cos^2\alpha$$

$$+ \left[\frac{\pi}{2\sqrt{3}}\varepsilon(k) + R^{2k+2}\,\mathrm{Re}\left\{\frac{1}{t^{2k+2}} + \sum_{j=0}^{\infty} r_{j,k}t^{2j}\right\}\right]\sin^2\alpha$$

$$+ Jm\left\{\frac{R^{2k+2}}{t^{2k+3}} + R^{2k+2}\sum_{j=0}^{\infty} r_{j,k}t^{2j}\right\}\sin 2\alpha \, ;$$

$$B_{k\tau}^{(1)} = -\left[-\frac{\pi}{2\sqrt{3}}\varepsilon(k) + 4\,\mathrm{Re}\left\{\frac{R^{2k+2}}{t^{2k+2}}\sum_{j=0}^{\infty} r_{j,k}t^{2j}\right\}\right]\sin 2\alpha$$

$$+ 2Jm\left\{-(k+1)\bar{t}\frac{R^{2k+2}}{t^{2k+3}} + R^{2k+2}\sum_{j=0}^{\infty} jr_{j,k}t^{2j-1}\bar{t} - (k+1)S_{j,k}t^{2j}\right\}\cos 2\alpha ;$$

$$B_{k\tau}^{(2)} = -\frac{\pi}{2\sqrt{3}}\varepsilon(k)\sin 2\alpha + Jm\left\{\frac{R^{2k+2}}{t^{2k+2}} + R^{2k+2}\sum_{j=0}^{\infty} r_{j,k}t^{2j}\right\}\cos 2\alpha\,;$$

$$\frac{G_u}{G_v} = (\kappa_3 - 1)\Gamma - \Gamma'\frac{\mathrm{Re}}{Jm}\{t\};\quad G_n = \frac{P_1\cos^2\alpha + P_2\sin^2\alpha}{2G_3}\,;$$

$$G_v = -2\Gamma\sin 2\alpha;\quad \varepsilon(k) = \left\{\begin{array}{l}0, k\neq 0\\ 1, k = 0\end{array}\right\}\,;$$

$$\frac{B_{ku}^{(1)}}{B_{ku}^{(2)}} = -\frac{\pi}{2\sqrt{3}}\varepsilon(k)\frac{\mathrm{Re}}{\mathrm{Im}}\{t\} + \frac{\mathrm{Re}}{\mathrm{Im}}\left\{-\frac{\kappa_3}{2k+1}\frac{R^{2k+1}}{t^{2k+1}} - t\frac{R^{2k+2}}{t^{2k+2}}\right.$$

$$\left. +R^{2k+2}\sum_{j=0}^{\infty} r_{j,k}\left(\frac{\kappa_3}{2k+1}t^{2j+1} - t\bar{t}^{2j}\right) + (2k+2)S_{j,k}\frac{\bar{t}^{2j+1}}{2j+1}\right\}\,;$$

$$\frac{B_{kv}^{(1)}}{B_{kv}^{(2)}} = (\kappa_3-1)\frac{\pi}{4\sqrt{3}}\varepsilon(k)\frac{\mathrm{Re}}{\mathrm{Im}}\{t\} + \frac{\mathrm{Re}}{\mathrm{Im}}\left\{\frac{R^{2k+1}}{(2k+1)\bar{t}^{2k+1}} - R^{2k+2}\sum_{j=0}^{\infty} r_{j,k}\frac{\bar{t}^{2j+1}}{2j+1}\right\}.$$

The total number of parameters included in (3.33) is

$$N = 2N_1 + 4N_2 + 2N_3 + 4\,.$$

The number of collocation points in each equation is N divisible by 8, and $N/8+1$ at N aliquant to 8. In the case of circular contours, the system (3.33) is simpler, since the radial direction coincides with that normal to contours and L_1 and L_2.

This suggests that the problem of plane strain of a composite based on a spirally reinforced filler results in the determination of the coefficients α_k, β_k of expansions (3.3) and complex potentials φ_1 and ψ_1. Thus, it is necessary to make use of the rigid contact conditions of zones S_1, S_2 and S_3 for specific conditions of material processing considered below.

For a general form of contours L_1 and L_2 the problem is not solved by its accurate statement, nevertheless, an approximate solution can be found using numerical methods. One of them is the collocation method, which is practically independent of the form of contours L_1 and L_2 and presents a relatively simple computer algorithm. The method yielded satisfactory results for the problems with both circular and non-circular forms of contours.

3.3.1 Investigation of Interlayer Inhomogeneity Effect on the Composite Stress State

The effect of inhomogeneous properties of the layer of auxiliary reinforcement upon the stress distribution in the composite with the spirally reinforced filler will be studied for the case of low filling. If one admits that the layer section is bounded by concentric circles, then the solution of the problem of a uniaxial loading of the composite having an inhomogeneous orthotropic layer by a transverse element core and isotropic matrix is reduced to a finite number (12) of linear algebraic equations. This makes the solution easy and

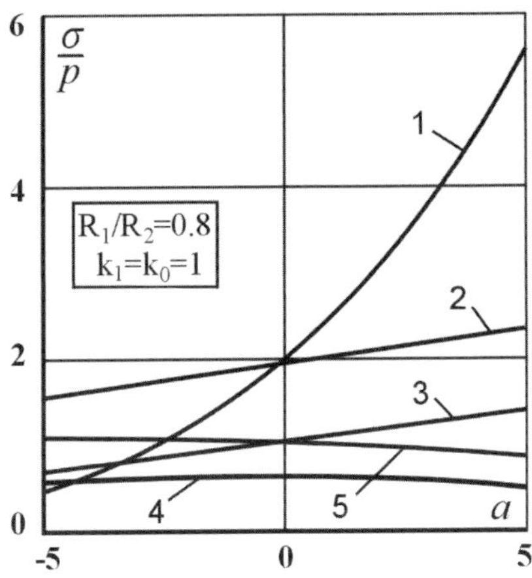

Fig. 3.3. Stress dependence in material ingredients upon the "a" parameter of the layer inhomogeneity at biaxial loading. $1 - \sigma_{\theta 21}$; $2 - \sigma_{\theta 23}$; $3 - \sigma_{r3}$; $4 - \sigma_{r1}, \sigma_{\theta 1}$; $5 - \sigma_{\theta 3}$

with some accuracy. In the case of the all-sided tension-compression of the material, the problem is reduced to the solution of just four linear algebraic equations, whose roots can be obtained in a finite form. Let us examine the dependence of the composite stress state on the parameter of inhomogeneity of the auxiliary reinforcement layer "a" employed in (3.27), and elastic characteristics of the element core and the layer.

In Fig. 3.3 one can see the dependencies of stress tensor components appearing in the matrix at a biaxial loading and given ratio of elastic characteristics of the element core and the layer. It is evident that with an increasing "a" parameter, the stresses in the layer neighboring the core rise most strongly. Other stresses change, but negligibly. Thus, with increasing "a" the probability of the auxiliary reinforcing layer failure rises, this is why inhomogeneity of the layer properties is to be minimized at composite production.

Variations of elastic characteristics of the element core transversely expressed by factor k_0 lead to changes of stress magnitude in the material components (Fig. 3.4). Note that factor k_0 denotes deviation of the elastic moduli of the element core and matrix from some mean typical of unidirectional composites value. Elevated elastic characteristics of the core ($k_0 > 1.0$) result in considerable stress reduction in the layer and its insignificant augmentation in all other components. Consequently, improved elastic properties of the element core lessen the danger of auxiliary reinforcement layer breakage under a biaxial tension.

The data illustrated in Figs. 3.4a and b for stresses in the layer at different "a" show that if the parameter diminishes, stresses diminish too independently of the factor k_0 magnitude. For instance, at $k_0 = 1.0$ stress concentration $\sigma_{\theta 21}$ in the layer close to the core reduces from 5.7 to 1.8 with an "a" reduction to zero.

When various winding filaments are used as an auxiliary reinforcement, the properties of the layer change as a function of the material filling degree.

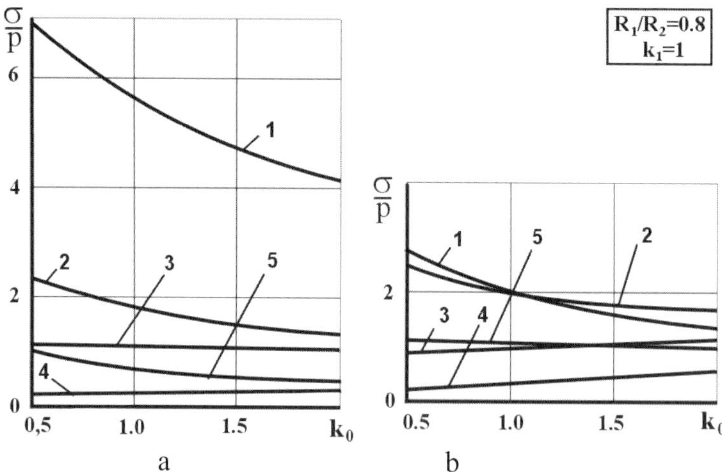

Fig. 3.4. Stress variations in material ingredients in transversal direction depending on the element core properties: (**a**) "a" = 4.92; (**b**) "a" = 0; notations correspond to Fig. 3.3

Let us examine the dependence of stresses emerging at interfaces between components upon the properties of the layer itself. Variations of the properties will be characterized by factor k_1, reflecting deviation of the ratio of the radial elasticity modulus of the layer to the elasticity modulus of the matrix from some optimum calculated for a common unidirectional composite. As seen from Fig. 3.5, improved elasticity parameters of the layer ($k_0 > 1.0$) augment only insignificantly the tangential stresses in it. Highest are stresses

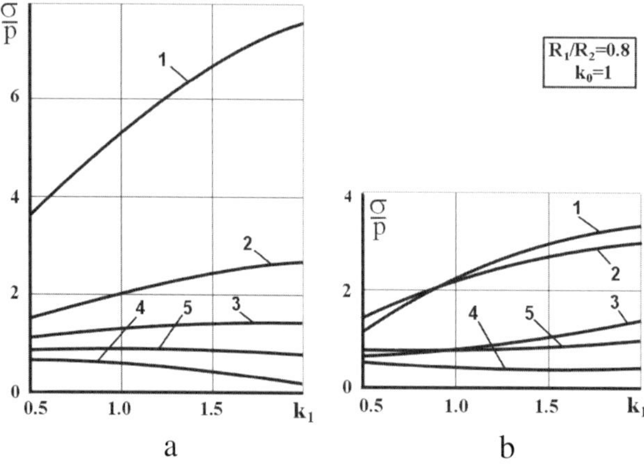

Fig. 3.5. Stress dependence at contact interfaces of material ingredients on elastic characteristics of auxiliary reinforcement layer under biaxial loading (**a**) "a" = 4.92; (**b**) "a" = 0. Notations are similar to Fig. 3.3

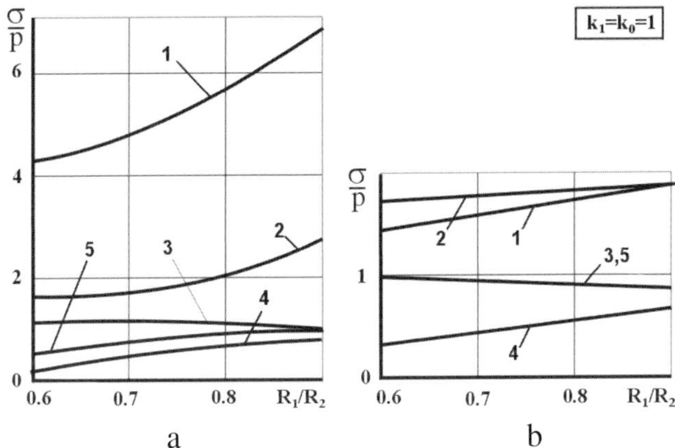

Fig. 3.6. Stress concentration variations at contact boundaries of ingredients depending on layer thickness under biaxial loading (**a**) "a" > 0; (**b**) "a" = 0. Notations are similar to Fig. 3.3

at the core boundary with the layer. It is, however, known that an increase in elastic properties is, in a number of cases, accompanied by improving strength characteristics that is why stress growth is not so critical. At the same time, by reducing the "a" value it is possible to lower stresses more than threefold.

Dimensions of the auxiliary reinforcement layer also essentially affect stress distribution in the material bulk. In Fig. 3.6, one can observe stress variations as a function of the element inner to outer radius relation. It follows that diminishing layer thickness leads to stress growth in the components. As in previous cases, the highest stresses occur at the interface with the element core. To reduce the stresses one must lessen the parameter "a" value. Thus, at $a = 0$, stress $\sigma_{\theta 21}$ reduces 2.5 times compared to those at $a = 4.92$.

In Fig. 3.7, stress variation in the element core and matrix at the interface with the layer is shown at a uniaxial tension dependent on angle θ. Stresses were calculated under two "a" values, namely 0 and 4.92, and $R/R_2 = 0.8$; $k_1 = k_0 = 1.0$. It is obvious that reduction of "a" does not essentially affect stress values. If one considers stresses appearing in the layer of the auxiliary

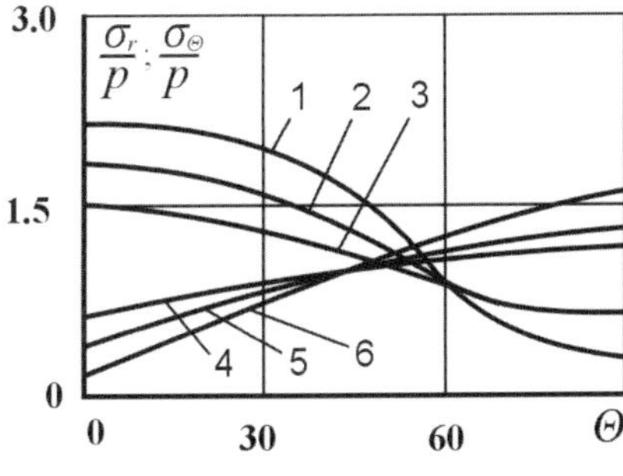

Fig. 3.7. Stress variations in the matrix and element core at interfaces with the layer under uniaxial tension: 1 – $\sigma_{r1}(a = -4.92)$; 2 – $\sigma_{r1}(a = 0)$; 3 – $\sigma_{r3}(a = 0; \ a = -4.92)$; 4 – $\sigma_{\theta 3}(a = 0)$; 5 – $\sigma_{\theta 3}(a = -4.92)$; 6 – $\sigma_{\theta 3}(a = 0; \ a = -4.92)$

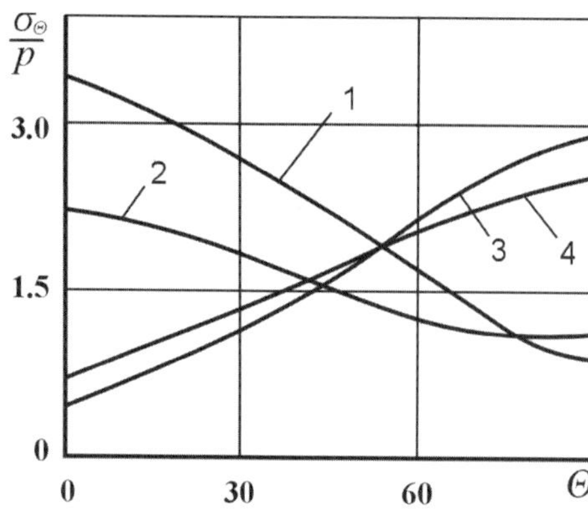

Fig. 3.8. Changes in stresses on contact boundaries of components in the layer at uniaxial tension: *1* – $\sigma_{\theta 21}(a = 0)$; *2* – $\sigma_{\theta 21}(a = -4.92)$; *3* – $\sigma_{\theta 23}(a = -4.92)$; *4* – $\sigma_{\theta 23}(a = 0)$

reinforcement (Fig. 3.8), they are seen to vary only $\sigma_{\theta 21}$ from 3.2 down to 2.4 at "a" variation from 0 to 4.92. Consequently, negative values of the parameter do not perceptibly reduce stress concentration in the structure.

Hence, to diminish stresses within the material structure one has to make the auxiliary reinforcement layer more homogeneous, in which case $a = 9$. It is necessary to promote high-grade impregnation of the auxiliary reinforcement layer by using prepregs of the main reinforcing material to ensure the needed quantity of the binder in the interelement spaces, and employ presaturated filaments when producing the layer.

3.3.2 Numerical Analysis

A series of known problems on the tension of a plane with a double-periodic system of circular isotropic elastic inclusions or holes was used to estimate the accuracy of the obtained data [210]. Furthermore, the accuracy of the solution was estimated by obeying boundary conditions between collocation points and stress σ_θ values not included in the boundary conditions of the following test problems:

a) tension of a model with the element core and a layer made of the same material ($\sigma_{\theta 1} = \sigma_{\theta 21}$) (Fig. 3.9);

b) tension of a model with the layer and matrix made of the same material ($\sigma_{\theta 3} = \sigma_{\theta 23}$).

These problems for models a and b, consisting of isotropic materials, were studied similarly to those for composites that did not contain the layer. In this case, values of expansion coefficients (3.3) were compared to those cited in [210]. It has been established that calculation error is here below 0.5%, in contrast to that from [210].

In Fig. 3.10, stress dependencies of material ingredients having the following characteristics: $E_1/E_3 = 18.8$; $v = v' = v'' = 0.25$; $E_r/E_3 = 3.9$; $E_\theta/E_3 = 15$; $G_{r\theta}/E_3 = 1.4$; $v_{rz} = v_{\theta z} = v_{z\theta}$; $v_3 = 0.35$ are presented ver-

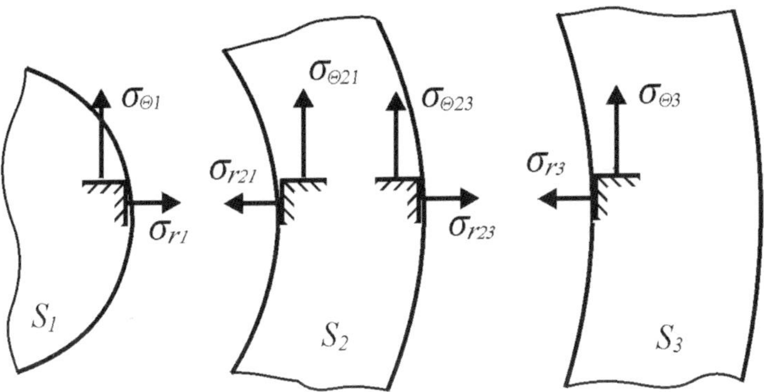

Fig. 3.9. Diagram of stresses acting in ingredients of the composite based on spirally reinforced filler

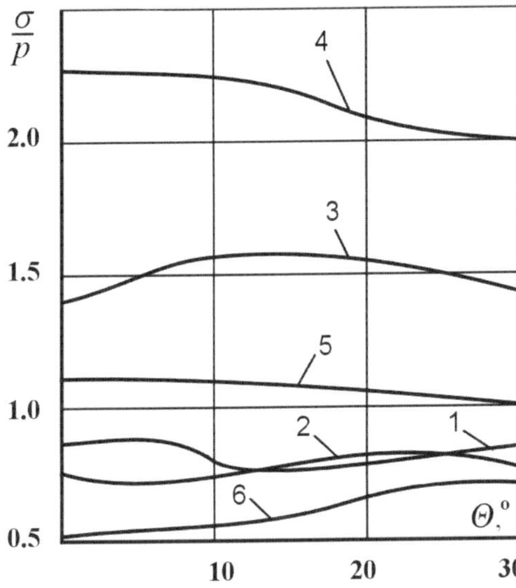

Fig. 3.10. Stress dependence in material ingredients versus angle θ: $1 - \sigma_{\theta 1}$; $2 - \sigma_{r1}$; $3 - \sigma_{\theta 21}$; $4 - \sigma_{\theta 23}$; $5 - \sigma_{r3}$; $6 - \sigma_{\theta 3}$

sus polar angle θ. Note that the element core and the layer are taken to be circular, where the ratio of the layer thickness to outer radius of the element is $\delta/R_1 = 0.2$ and the filling is made with spirally reinforced elements $\varphi_2 = 0.58$. For comparison, the same stresses are shown in Fig. 3.11, for the case when $\delta/R_1 = 0$, i.e. the layer is absent and the filling degree is analogous to $\varphi_2 = 0.58$.

It follows that introduction of an intermediate layer into the material structure perceptibly effects stress values. Thus, stress σ_θ in the element core augments on introduction of the layer (curve 1), while stresses in the binder go down by more than 65% (curve 6). Similar variations take place with stresses σ_r.

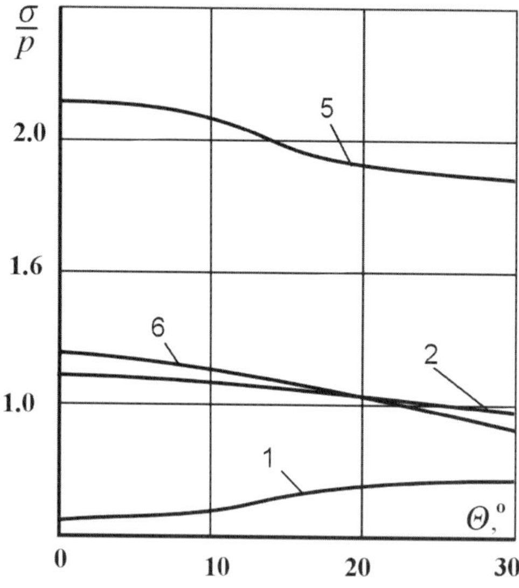

Fig. 3.11. Stress variations in material ingredients without the layer at $\varphi_2 = 0.58$. Notations are similar to 3.10

As the filling degree of the material with spirally reinforced elements grows, stresses initiations are also greatly redistributed. In Fig. 3.12, one can just see how the stresses change in the material without the layer ($\delta/R_1 = 0$) at filling degree $\varphi_2 = 0.73$. It is also evident that the highest stress values increase with growing filling degree.

When elastic characteristics of the components change even insignificantly stresses also redistribute greatly. In Fig. 3.13, stress values are presented over material contours under the following parameters of its ingredients $E_1/E_3 =$

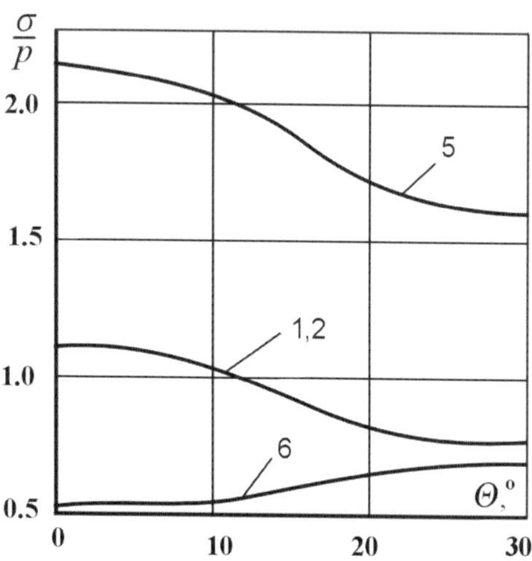

Fig. 3.12. Stress to angle θ dependence without the layer at $\varphi_2 = 0.73$. Notations correspond to Fig. 3.10

Table 3.1. Variation of Radial and Tangential Stress Tensor Components on Composite Contours at Uniaxial and Biaxial Tension

P_x, P_y	φ	θ^0	$\sigma_{r1} = \sigma_{r21}$	$\sigma_{r3} = \sigma_{r23}$	$\sigma_{\theta21}$	$\sigma_{\theta23}$	$\sigma_{\theta3}$	$\sigma_{\theta1}$
$P_x = P$	0.2	0	0.74	1.00	1.42	2.36	0.73	0.93
		10	0.70	0.98	1.44	2.33	0.78	0.95
		20	0.65	0.95	1.45	2.30	0.86	0.97
		30	0.62	0.94	1.48	2.28	0.93	1.00
$P_y = P$	0.4	0	0.84	1.10	1.30	2.15	0.45	0.80
		10	0.82	1.08	1.35	2.10	0.55	0.83
		20	0.76	1.05	1.38	2.05	0.65	0.87
		30	0.69	1.00	1.43	2.00	0.70	0.90
$P_x = P$	0.5	0	1.01	1.08	-0.20	-0.81	-0.22	-0.33
$P_y = 0$		15	1.42	1.26	-0.28	-1.00	-0.28	-0.44
		30	0.83	0.80	-0.53	1.12	0.29	0.32
		45	0.32	0.33	1.02	1.32	0.39	0.72
		60	0.30	0.23	1.41	1.74	0.64	0.74
		75	-0.28	-0.18	1.53	1.82	0.72	0.82
		90	-0.33	-0.25	1.61	2.04	0.78	0.79

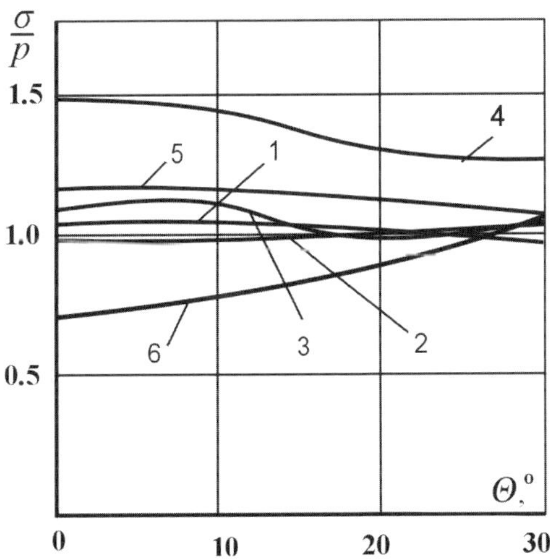

Fig. 3.13. Stress variations in material ingredients under lower material and element core rigidity. Notations are similar to Fig. 3.10

15; $E_r/E_3 = 3.9$; $G_{r\theta}/E_3 = 1.4$; $v = v_{rz} = 0.4$; $v' = 0.07$; $v'' = v_{\theta z} = 0.28$; $v_3 = 0.35$; $\delta/R_1 = 0.2$; $\varphi_2 = 0.58$.

From the above data it follows that the interlayer exerts a significant effect on stress redistribution in the composite (Fig. 3.14). This is also true for materials with a thinner interlayer.

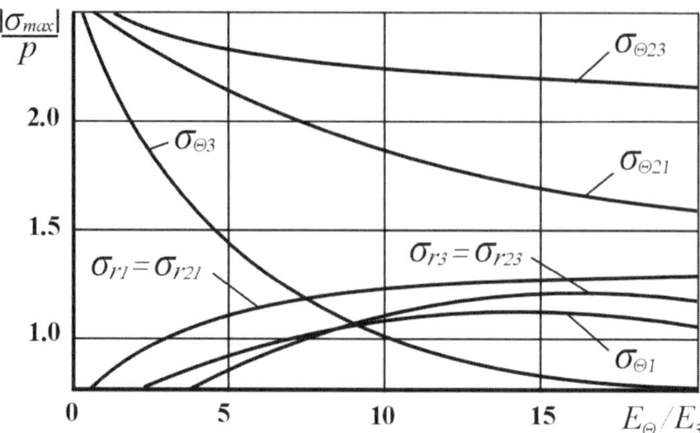

Fig. 3.14. Maximum stress dependence on composite contours upon the properties of the auxiliary reinforcement layer

In Table 3.1, values of radial $\sigma_{r1}, \sigma_{r21}, \sigma_{r23}$ and tangential $\sigma_{\theta1}, \sigma_{\theta21}, \sigma_{\theta23}, \sigma_{\theta3}$ stress tensor components are presented on material contours at biaxial and uniaxial tensions. The parameters were calculated for a carbon plastic based on a spirally reinforced filler with the following elastic constants: $E_1 = 7000\,\mathrm{MPa}$, $E_3 = 3300\,\mathrm{MPa}$, $E_r = 14000\,\mathrm{MPa}$, $E_\theta = 64800\,\mathrm{MPa}$, $G_{r\theta} = 6100\,\mathrm{MPa}$, $v = 0.35$, $v_{\theta r} = 0.28$, $v_3 = 0.35$ and $\delta/R_1 = 0.05$. Notice that under the uniaxial tension most critical turn out to be two points $\theta = 15°$ where radial stresses σ_{r3} reach the highest magnitude under which the matrix might collapse. Moreover, at $\theta = 90°$ stresses $\sigma_{\theta23}$ are maximal and the interlayer, most probably, fails. Under the biaxial loading the critical point is $\theta = 0°$, at which both stresses approach their most hazardous values.

Variations of radial and tangential stress tensor components on composite contours at uniaxial and biaxial tension and variations of the elasticity modulus dependent on filling degree φ_2 are described in Fig. 3.15, for three types of composites, carbon-organoplastic (curves 1), carbon-glass-reinforced plastic (curves 2) and glass-glass-fiber-reinforced plastic (curves 3) under $\delta/R_1 = 0.059$ and 0.1. From the above-cited data it follows that use of materials with higher modulus value for the interlayer considerably increases the transversal elasticity modulus. However, with reducing layer thickness, the elasticity modulus decreases.

The element core shape also affects the stress distribution in the material. In Fig. 3.16, one can see the dependencies of stress tensor components on the angle θ under uniaxial tension. Elastic characteristics of the components are given for organic-carbon plastic whose filling degree is $\varphi_2 = 0.73$, layer thickness to major semiaxis ratio, $\delta/a = 0.2$ and semiaxes ratio $a/b = 1.5$.

Proceeding from the above, a conclusion can be made that internal stresses are strongly interrelated with the element's cross-sectional form.

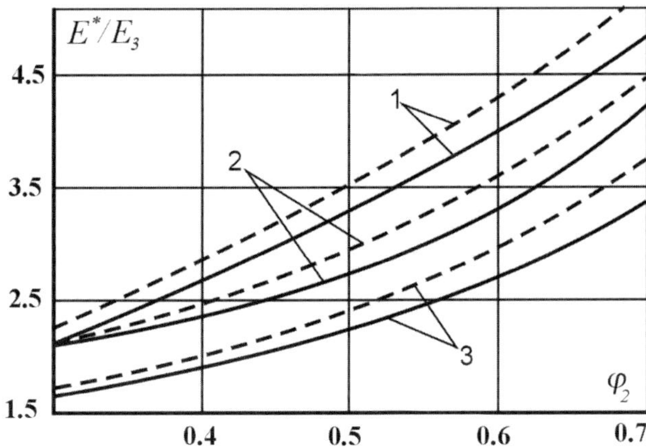

Fig. 3.15. Dependence of the transversal elasticity modulus of the material on filling degree with spirally reinforced elements (solid lines – $\delta/R_1 = 0.059$, dotted – 0, 1). *1* – organic-carbon plastic; *2* – carbon-glass-reinforced plastic; *3* – glass-glass-fiber-reinforced plastic

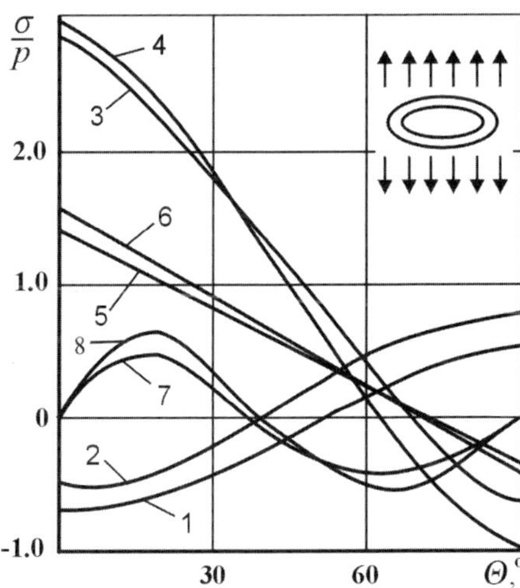

Fig. 3.16. Variations of stress tensor components in response to angle θ magnitude. *1* – $\sigma_{r1} = \sigma_{r21}$; *2* – $\sigma_{r3} = \sigma_{r23}$; *3* – $\sigma_{\theta 21}$; *4* – $\sigma_{\theta 23}$; *5* – $\sigma_{\theta 3}$; *6* – $\sigma_{\theta 1}$; *7* – $\sigma_{r\theta 1}$; *8* – $\sigma_{r\theta 3}$

In Fig. 3.17, dependencies are presented of nondimensional tangentially reduced elastic moduli E_x^*, E_y^* and G_{xy}^* upon filling degree with spirally reinforced elements for two types of materials – carbon-glass-reinforced plastic and organic-carbon plastic. Note that the structural element cross-section is an ellipse whose semiaxes ratio is $a/b = 1.5$, and $\delta/a = 0.2$. As appears from the diagram, with growing filling degree of the material its elastic characteristics become much improved in the transverse direction, even at an insignificant content of the main reinforcement.

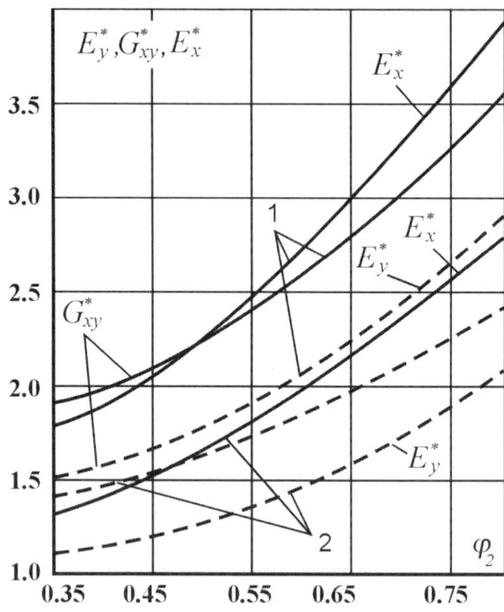

Fig. 3.17. Transversal elastic characteristics of the material versus filling degree. *1* – carbon-organic plastic; *2* – carbon-glass-reinforced plastic

The above – presented numerical analysis indicates that the method proposed for calculations of composite characteristics is highly useful in studying its stress – strain state and elastic properties, with allowance for the anisotropy of intermediate layers. Furthermore, a conclusion has been derived that the spirally reinforced element configuration and properties exert a perceptible influence on the stress distribution inside the structure and on the values of reduced elastic constants.

3.4 Stress-state Analysis of Composites Filled with Thin Isotropic Layers of Auxiliary Reinforcement

The analytical model for fillers displaying weak anisotropic properties can be simplified by considering the element core material of the layer and that of the matrix to be isotropic. When the layer is rather thin, the model can be still more simplified through the introduction of a supposition that the element core to matrix contact occurs along the layer medial line γ (Fig. 3.2). Let us also assume that the layer element excised by two neighboring planes OXY is a curvilinear rod and displacements at transition through contour γ are assumed to be uninterrupted

$$u^{(1)} + iv^{(1)} = u^{(3)} + iv^{(3)} = u + iv. \tag{3.34}$$

The index in (3.34) expresses the function as an attribute of corresponding zone S_k.

Bonding conditions of the main element are described in [54, 220]:

$$\delta_1 U + F_2 - \frac{1}{\rho}\int_0^S F_1\,dS = -\frac{CR}{\rho}\,;$$

$$\left(\delta_1 + \delta_2\frac{R^2}{\rho^2}\right)U - \delta_2\frac{R^2}{\rho}\dot V + F_2 = 0\,, \tag{3.35}$$

where $U(S)$, $V(S)$ are the chief unknown functions related to displacements u and v on contour γ in the manner

$$u + iv = \int (U + iV)t\,dS\,; \tag{3.36}$$

ρ – contour γ curvature radius; R – some characteristic dimension of contour γ; $F_1(S)$, $F_2(S)$ are expressed on γ through complex potentials $\varphi_1(t)$ and $\varphi_3(t)$ [54]:

$$\dot t(F_1+iF_2)=\frac{1}{2G_3 R}[(1+\kappa_3)\varphi_3(t)-(1+\kappa_1)\varphi_1(t)]-\frac{1-\lambda}{R}(u^{(1)}+iv^{(1)}); \tag{3.37}$$

$\delta_1(S), \delta_2(S)$ – reduced rigidity of the rod excised by two parallel sections calculated as

$$\delta_1 = \left(1 - \frac{E_2}{E_3}\right)\frac{\delta(S)}{R}(1 - v_3)\,;$$

$$\delta_2 = \frac{1 + v_3}{12}\left(1 - \frac{E_2}{E_3}\right)\frac{\delta^3(S)}{R^3}\,,$$

where E_2, E_3 elastic moduli of the layer and matrix; $\lambda = G_1/G_3$.

Complex potentials $\varphi_3(t)$ and $\psi_3(t)$ will be chosen from (3.3), so functions $\varphi_3(t)$ and $\psi_3(t)$ are analytic in zone S_3 (Fig. 3.1) of the hexagonal lattice. Angle α will further be assumed to be equal to zero.

Complex potentials in zone S_1 are chosen as follows

$$\varphi_1(t) = 2RG_1 \sum_{k=1}^{N_1} a_k\xi^{2k-1}\,;$$

$$\psi_1(t) = 2RG_1 \sum_{k=1}^{N_1} b_k\xi^{2k-1};\quad \left(\xi = \frac{t}{R}\right). \tag{3.38}$$

Displacements on contour γ are calculated using (3.2) with account of (3.38), (3.3) and conditions of (3.34):

$$\left\{\begin{matrix} u^{(3)} \\ v^{(3)} \end{matrix}\right\} = \sum_{k=1}^{N_1}\left[a_k\frac{\mathrm{Re}}{\mathrm{Im}}\{\kappa_1 t_1^{2k-1} - (2k-1)t_1\bar t^{2k-2}\} - b_k\frac{\mathrm{Re}}{\mathrm{Im}}\{\bar t_1^{2k-2}\}\right], \tag{3.39}$$

where $u^{(3)}$, $v^{(3)}$ are expressed through yet unknown parameters α_k, β_k,

$$t_1 = x_1 + iy_1 = \frac{x + iy}{R}\,.$$

To find additional boundary conditions on contour γ, let us calculate values F_1 and F_2 from (3.35) by the formula (3.37):

$$
\begin{Bmatrix} F_1 \\ F_2 \end{Bmatrix} = (\alpha_0 + \Gamma)(1 + \kappa_3)\frac{\mathrm{Re}}{\mathrm{Im}}\{\dot{t}t_1\} + (1 + \kappa_3)\sum_{k=0}^{N_3} R^{2k+2}\alpha_{2l+2}
$$
$$
\times \frac{\mathrm{Re}}{\mathrm{Im}}\left\{\dot{t}\left[-\frac{t_1^{-2k-1}}{2k+1} + \sum_{j=0}^{\infty} r_{j,k}\frac{t_1^{2j+1}}{2j+1}\right]\right\}
$$
$$
-\sum_{k=1}^{N_3} a_k \frac{\mathrm{Re}}{\mathrm{Im}}\left\{\dot{t}[(\lambda + \kappa_1)t_1^{2k-1} - (1 - \lambda)(2k - 1)t_1\bar{t}_1^{-2k-2}]\right\}
$$
$$
-(1 - \lambda)\sum_{k=1}^{N_1} b_k \frac{\mathrm{Re}}{\mathrm{Im}}\left\{\dot{t}\bar{t}^{2k-1}\right\}. \tag{3.40}
$$

It remains to find U and V interrelated with vector of displacements $u + iv$ through formula (3.36), and we obtain

$$
U + iV = (\dot{u} + i\dot{v})\dot{\bar{t}} \quad \text{or}
$$
$$
\begin{Bmatrix} U \\ V \end{Bmatrix} = \sum_{k=1}^{N_1} a_k \frac{\mathrm{Re}}{\mathrm{Im}}\left\{(2k - 1)[\kappa_1 t_1^{2k-2} - \bar{t}^{2k-2} - (2k - 2)\dot{\bar{t}}^2\bar{t}\,t_1^{2k-3}]\right\}
$$
$$
-\sum_{k=1}^{N_1} b_k(2k - 1)\frac{\mathrm{Re}}{\mathrm{Im}}\left\{\dot{\bar{t}}^2\bar{t}^{2k-2}\right\}. \tag{3.41}
$$

From the latter relation one finds $\dot{V} = \dfrac{d}{dS}V$:

$$
R\dot{V} = \sum_{k=1}^{N_1} a_k(2k - 1)(2k - 2)
$$
$$
\times \mathrm{Im}\left\{\kappa_1 \dot{t}t_1^{2k-3} - \dot{\bar{t}}\bar{t}^{2k-3} - (2k - 3)\dot{\bar{t}}^3\bar{t}^{2k-4} - 2\dot{t}\bar{t}_1\bar{t}^{2k-3}R\right\}
$$
$$
-\sum_{k=1}^{N_1} b_k(2k - 1) \times \mathrm{Im}\left\{2R\dot{\bar{t}}\ddot{\bar{t}}_1 t_1^{2k-2} + (2k - 2)\dot{\bar{t}}^3\bar{t}^{2k-4}\right\}. \tag{3.42}
$$

By substituting (3.40), (3.41) and (3.42) into boundary conditions (3.35) and using the collocation method, we obtain a system of equations

$$
\sum_{k=1}^{N_1}\left(A_{k(i)}^{(j)}a_k + B_{k(i)}^{(j)}b_k\right) + \sum_{k=0}^{N_3}\left(L_{ka(i)}^{(j)}\alpha_{2k+2} + L_{kb(i)}^{(j)}\beta_{2k+2}\right) + C_{(i)}^{(j)}C = G_{(i)}^{(j)}.
$$
$$
\tag{3.43}
$$

The coefficients of system (3.43) are of the form

$$A^{(j)}_{k(1,2)} = \frac{\mathrm{Re}}{\mathrm{Im}} \left\{ \kappa_3 t_j^{2k-1} + (2k-1) t_j \bar{t}^{2k-2} \right\},$$

$$B^{(j)}_{k(1,2)} = \frac{\mathrm{Re}}{\mathrm{Im}} \left\{ \bar{t}^{2k-1} \right\};$$

$$L^{(j)}_{ka(1,2)} = \frac{\mathrm{Re}}{\mathrm{Im}} \left\{ -\kappa_3 \frac{t_j^{-2k-1}}{2k+1} - t_j \bar{t}_j^{-2k-2} \right.$$
$$\left. + \sum_{n=0}^{\infty} \left[r_{nk} \left(\kappa_3 \frac{t_j^{2n+1}}{2n+1} - t_j \bar{t}_j^{2n} \right) \right] - S_{nk}(2k+2) \frac{\bar{t}_j^{2n+1}}{2n+1} \right\},$$
$$k \geq 1;$$

$$L^{(j)}_{0a(1,2)} = \frac{\mathrm{Re}}{\mathrm{Im}} \{ \bar{t}_j \} \frac{\pi}{4\sqrt{3}} L^2;$$

$$L^{(j)}_{0b(1,2)} = \frac{\mathrm{Re}}{\mathrm{Im}} \{ t_j \} \frac{\pi}{4\sqrt{3}} L^2;$$

$$L^{(j)}_{kb(1,2)} = \frac{\mathrm{Re}}{\mathrm{Im}} \left\{ \frac{\bar{t}_j^{-2k-1}}{2k+1} - \sum_{n=0}^{\infty} r_{nk} \frac{\bar{t}_j^{2n+1}}{2n+1} \right\}, \ k \geq 1;$$

$$A^{(j)}_{k3} = \delta(t_j) U_{ak}(t_j) - \frac{\delta_2(t_j)}{\rho(t_j)} V_{ak}(t_j) + F_{ak}(t_j);$$

$$B^{(j)}_{k3} = \delta(t_j) U_{bk}(t_j) - \frac{\delta_2(t_j)}{\rho(t_j)} V_{bk}(t_j) + F_{bk}(t_j);$$

$$L^{(j)}_{ka(3)} = F_{\alpha k}(t_j), \ k \geq 1;$$

$$L^{(j)}_{kb(3)} = F_{\beta 1}(t_j), \ k \geq 1;$$

$$L^{(j)}_{0a(3,4)} = L^{(j)}_{0b(3,4)} = 0;$$

$$A^{(j)}_{k4} = \delta_1(t_j) U_{ak}(t_j) + F_{ak}(t) + \frac{D_{ak}(t_j)}{\rho(t_j)};$$

$$B^{(j)}_{k4} = \delta_1(t_j) U_{bk}(t_j) + F_{bk}(t) + \frac{D_{bk}(t_j)}{\rho(t_j)};$$

$$L^{(j)}_{ka(4)} = F_{\alpha,k}(t_j) + \frac{D_{\alpha,k}(t_j)}{\rho(t_j)}, \ k \geq 1;$$

$$L^{(j)}_{bk(4)} = F_{\beta 1}(t_j) + \frac{D_{\beta 1}(t_j)}{\rho(t_j)}, \ k \geq 1;$$

$$\delta(t) = \delta_1(t) + \frac{\delta_2(t)}{\rho^2(t)};$$

L – distance between centers of the elements.

$$V_{ak}(t) = (2k-1)(2k-2)$$
$$\times \mathrm{Im}\left\{\kappa_1 t^{2k-3}\dot{t} - 2\bar{t}^{2k-3}\dot{\bar{t}} - (2k-3)t\bar{t}^{2k-4}\dot{\bar{t}}^3 - 2t\bar{t}^{2k-3}\dot{\bar{t}}\ddot{\bar{t}}\right\};$$

$$V_{bk}(t) = -\mathrm{Im}\left\{(2k-2)\bar{t}^{2k-3}\dot{\bar{t}}^3 + 2\bar{t}^{2k-2}\dot{\bar{t}}\ddot{\bar{t}}\right\};$$

$$U_{ak}(t) = (2k-1)\mathrm{Re}\left\{\kappa_1 t^{2k-2} - \bar{t}^{2k-2} - (2k-2)t\bar{t}^{2k-3}\dot{\bar{t}}^2\right\};$$

$$U_{bk}(t) = -(2k-1)\mathrm{Re}\left\{\bar{t}^{2k-2}\dot{\bar{t}}^2\right\};$$

$$F_{\alpha,k}(t) = (1+\kappa_3)\mathrm{Im}\left\{\dot{\bar{t}}\left(-\frac{t^{-2k-1}}{2k+1} + \sum_{n=0}^{\infty} r_{nk}\frac{t^{2n+1}}{2n+1}\right)\right\};$$

$$F_{\beta 1}(t) = \frac{\pi l^2}{4\sqrt{3}}(1+\kappa_3)\mathrm{Im}\left\{\dot{\bar{t}}t\right\};$$

$$F_{ak}(t) = -\mathrm{Im}\left\{\dot{\bar{t}}[1+\kappa_1(1-\lambda)\kappa_1]t^{2k-1} - (2k-1)t\bar{t}^{2k-2}\right\};$$

$$F_{bk}(t) = \mathrm{Im}\left\{\dot{\bar{t}}(1-\lambda)t^{2k-1}\right\};$$

$$F_G(t) = \frac{P_x + P_y}{4}(1+\kappa_3)\mathrm{Im}\left\{\dot{\bar{t}}t\right\};$$

$$D_{\alpha k}(t) = (1+\kappa_3)\int_0^S \mathrm{Re}\left\{\dot{\bar{z}}\left(-\frac{t^{-2k-1}}{2k+1} + \sum_{n=0}^{\infty} r_{nk}\frac{z^{2n+1}}{2n+1}\right)\right\}dS;$$

$$D_{\beta 1}(t) = \frac{\pi L^2}{4\sqrt{3}}(1+\kappa_3)\int_0^S \mathrm{Re}\left\{\dot{\bar{t}}t\right\}dS;$$

$$D_{ak}(t) = -\int_0^S \mathrm{Re}\left\{\dot{\bar{t}}[1+\kappa_1+(1-\lambda)\kappa_1]z^{2k}\right\}dS;$$

$$D_{\beta k}(t) = (1-\lambda)\int_0^S \mathrm{Re}\left\{\dot{\bar{t}}\bar{t}^{2k-1}\right\}dS;$$

$$D_G(t) = \frac{P_x + P_y}{4}(1+\kappa_3)\int_0^S \mathrm{Re}\left\{\dot{\bar{t}}t\right\}dS;$$

S – contour arc length corresponding to t;

$$r_{nk} = \frac{(2n+2k+1)!}{(2n)!(2k+1)!}g_{n+k+1}\left(\frac{1}{2}\right)^{2n+2k+2};$$

$$S_{nk} = \frac{(2n + 2k + 2)!}{(2n)!(2k + 2)!} \rho_{n+k+1} \left(\frac{1}{2}\right)^{2n+2k+2} ;$$

$$\lambda = L/2 .$$

Considering that parameters $\alpha_0, \beta_0, \alpha_{2k+2}, \beta_{2k+2}$ are interrelated, then the number of unknowns in (3.39) and (3.43) will be:

$$N = 2N_1 + 2N_3 + 2 .$$

The unknown parameters

$$\alpha_k, \quad \beta_k \ (k = 1, \ldots, N_1), \quad \alpha_{2k+2}, \quad \beta_{2k+2} \ (k = 0, \ldots, N_3)$$

are found from (3.39) and (3.43) by the collocation method.

3.4.1 Numerical Computation of Low-filled Material

When the distance between spirally reinforced elements is large enough and a continuous layer of auxiliary reinforcement is laid on their side surface, and the element contours are the circles of (3.43), closed-type solutions are to be used. Study of the problems can result in a series of qualitative solutions disclosed in [131].

By varying values of infinite loads P_1 and P_2, one can obtain tensile, compressive and shear stress distribution. Stress values are computed under varying ratio of the element core shear modulus to that of the matrix within $\lambda = 3 - 21$ range and different rigidity of the auxiliary reinforcement layer $E_2/E_3 = 0 - 15$. During computations, layer thickness was varied within rather wide limits $\delta/R = 0 - 0.2$. Also, stresses on the layer and element core contours under varying θ angle within $\theta = 0 - 90°$ limits were calculated at tensile and shearing loads in a plane perpendicular to the main reinforcement.

In Fig. 3.18, data on stress $\sigma_{\theta 1}$ variations (in the element core) are presented under tension as dependent on the interlayer properties $\theta = \dfrac{\pi}{2}$ and $\lambda = 10$.

Curves 1, 2 and 3 in Fig. 3.18 correspond to various layer thicknesses. Analysis of the results obtained suggests that variation of the layer properties results in perceptible changes in stress concentration in the element core. As the layer thickness diminishes, the variation becomes weak. Analogous data are given in Fig. 3.19, where the dependence of the same stresses is shown as a function of the layer properties at $\theta = 0°$. Stress concentration changes in the range studied from 1.5 to 2 times, which is the evidence of the considerable effect of the interlayer properties.

Dependences of maximum stresses $\sigma_{\theta 1}$ (curves 1 and 2) and $\sigma_{\theta 3}$ (curves 3 and 4) without the layer (curves 1, 3 and 5) and with it at $\delta/R = 0.18$ and $E_2/E_3 = 15$ (curves 2, 4 and 6) are presented in Fig. 3.20. It is evident that improved characteristics of the element core augment the stresses under consideration. However, under $\lambda \geq 15$, the variations are insignificant. Normal

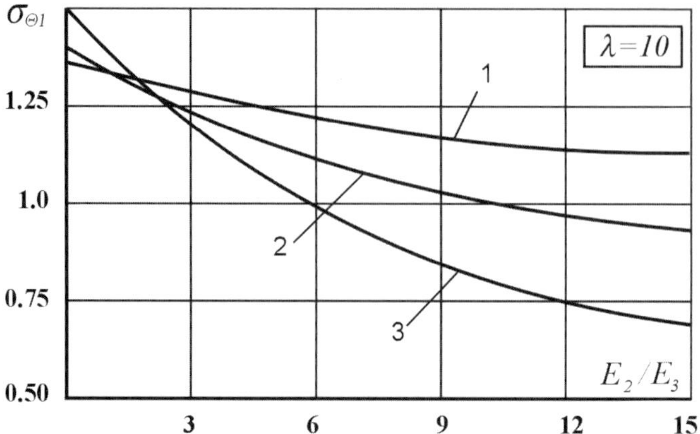

Fig. 3.18. Dependence of $\sigma_{\Theta 1}$ on interlayer properties under uniaxial tension $(\theta = 90°)$

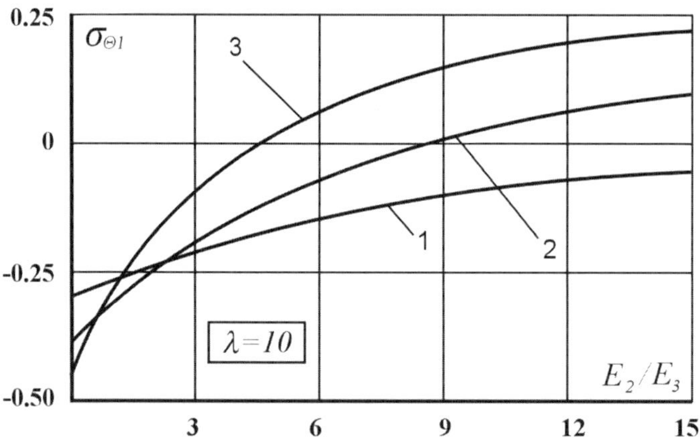

Fig. 3.19. Stress $\sigma_{\Theta 1}$ variation as dependent on layer properties at uniaxial tension $(\theta = 0°)$. Notations correspond to Fig. 3.18

stresses σ_r behave similarly (curves 5 and 6). It should be underlined that the introduction of the interlayer leads in all cases to reduced stress concentration.

In Fig. 3.21, data on tangential stress variations in the element core $\sigma_{\theta 1}$ and matrix $\sigma_{\theta 3}$ are illustrated under shearing as dependent upon interlayer properties and at $\lambda = 10$. The dependence is extremely strong for stresses $\sigma_{\theta 1}$ (curves 1, 2 and 3). For example, when the layer is $\delta/R = 0.06$, stresses reduce by 57% and when its thickness is $\delta/R = 0.2$ they reduce by 250% within variation $E_2/E_3 = 0 - 15$ of the interlayer properties. At the same time, stress $\sigma_{\theta 3}$ (curves 4, 5 and 6) vary only slightly, by just 5–10% within the studied range.

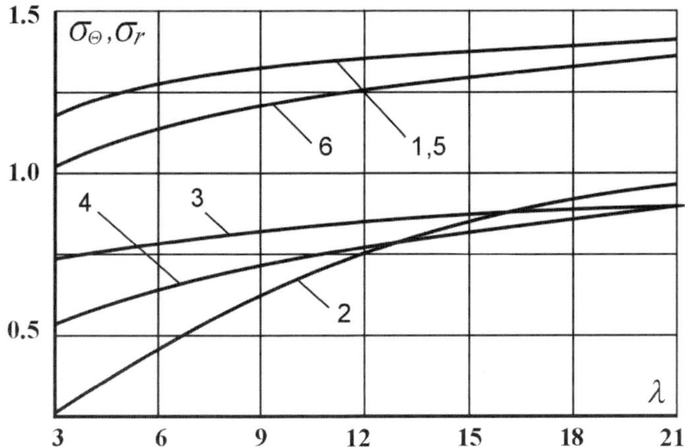

Fig. 3.20. Maximum stress versus element core properties under uniaxial tension. $1,\,2 - \sigma_{\Theta 1}$ $(\Theta = 90°)$; $3,\,4 - \sigma_{\Theta 3}$ $(\Theta = 0°)$; $5,\,6 - \sigma_r$; $1,\,3,\,5 - E_2/E_3 = 0,\,\delta/R = 0,$ $2,\,4,\,6 - E_2/E_3 = 15,\,\delta/R = 0.18$

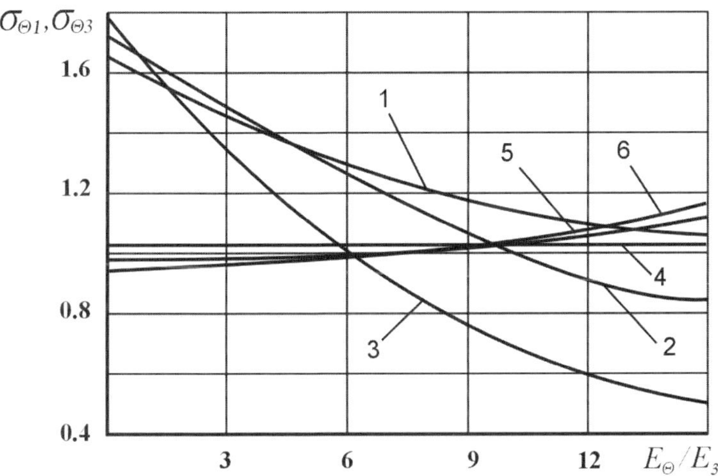

Fig. 3.21. Stress variations in the element core and matrix under shear depending on interlayer properties. $1,\,2,\,3 - \sigma_{\Theta 1}$ $(\Theta = 90°)$; $4,\,5,\,6 - \sigma_{\Theta 3}$ $(\Theta = 0°)$; $5,\,6 - \sigma_r$; $1,\,4 - \delta/R = 0.06$; $2,\,5 - \delta/R = 0.12$; $3,\,6 - \delta/R = 0.2$

As an example, some diagrams of normal and tangential stress distribution are presented in Fig. 3.22 over the interlayer contour at $\lambda = 10$, $E_2/E_3 = 15$, $\delta/R = 0.12$. The diagrams prove that the most critical areas of destruction turn out to be the $\theta = 0°$ and $\theta = \pi/2$ points, where both normal and tangential stresses reach their maximum. Matrix failure is most probable at point $\theta = 0°$ that of the element material at point $\theta = \pi/2$.

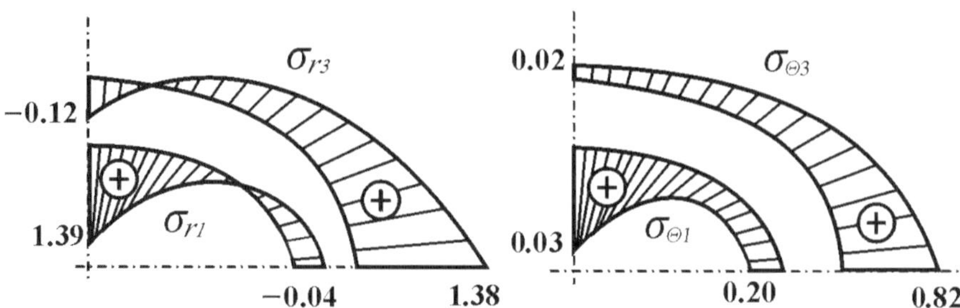

Fig. 3.22. Stress distribution diagrams over interlayer contour

Analysis of stress concentration dependencies on the interlayer thickness and properties has revealed that under uniaxial transversal loading and shear of the model proposed, the geometrical dimensions of the interlayer and its physico-mechanical properties exert an essential effect on the composite stress-strain state. The larger portion of the effect is directed towards the distribution of tangential stresses.

The results obtained are consistent with experimental evidence set forth in [288] where the effect of a stratum at the interface with a rigid inclusion on the model strength has been estimated. As the interlayer, soft (rubber) and stiff (epoxide resin) materials were used. By the method of photoelasticity it was found that the introduction of this interlayer perceptibly influences stress redistribution and changes the type and behavior of composite destruction. Thus, when the rubber is used, destruction runs at the interface with the core, whereas with epoxide compound the cracks propagate in the interlayer. With the rising shear modulus of the interlayer radial stresses σ_r attain their maximums at $\theta = 0°$ and continue growing to 1.3–1.5. In contrast, tangential stresses reach their maximum at $\theta = \pi/2$ and diminish with increasing shear modulus of the layer down to 0.3–0.7 while the layer thickness is in this case $0.0665R$.

As shown in 2, the form of the spirally reinforced filler cross-section can, in a general case, be other than a circle. Consider the stress state of an infinite space filled with cylindrical unbroken spirally reinforced filler whose cross-section shape is random under the action of forces perpendicular to the filler axis. We will assume that a sufficiently thin continuous layer is laid on the side surface of the spirally reinforced element.

For numerical computation the element core shape was chosen in accordance with recommendations described in part 2. In Fig. 3.23, computation results are shown of maximum stresses $\sigma_{\theta 1}^{\max}$ appearing in the element core under the uniaxial tension of the composite taking account of the interlayer properties. It turns out that for the case of the noncircular (elliptical) contour the layer affects the stress state of the material more notably compared to the circular contour (see Fig. 3.18). The general tendency is seen in stress con-

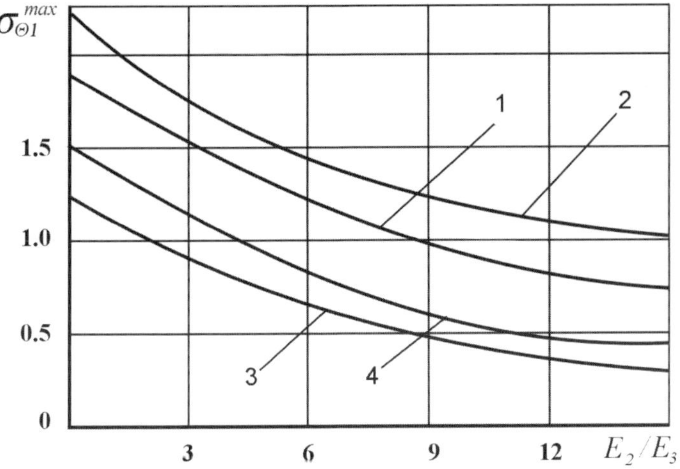

Fig. 3.23. Maximum stress dependence in the spirally reinforced element core upon interlayer properties. $1 - \lambda = 10$, $a/b = 0.67$; $2 - \lambda = 20$, $a/b = 0.67$; $3 - \lambda = 10$, $a/b = 1.50$; $4 - \lambda = 20$, $a/b = 1.50$

centration reduction with growing elastic characteristics of the layer. With the increasing rigidity of the layer the highest stresses in the element core diminish by 50% on average.

The orientation of the ellipsoidal element core relative current loading is very important. Depending on this, e.g. stresses $\sigma_{\theta 1}$ may vary within considerably wide limits (Figs. 3.24 and 3.25). Consequently, stress concentration

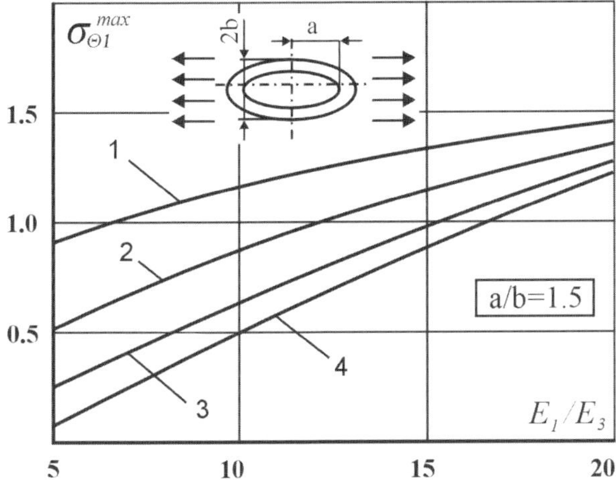

Fig. 3.24. Maximum stress dependence in the element core on its properties. $1 - E_2/E_3 = 0$; $2 - E_2/E_3 = 5$; $3 - E_2/E_3 = 10$; $4 - E_2/E_3 = 15$

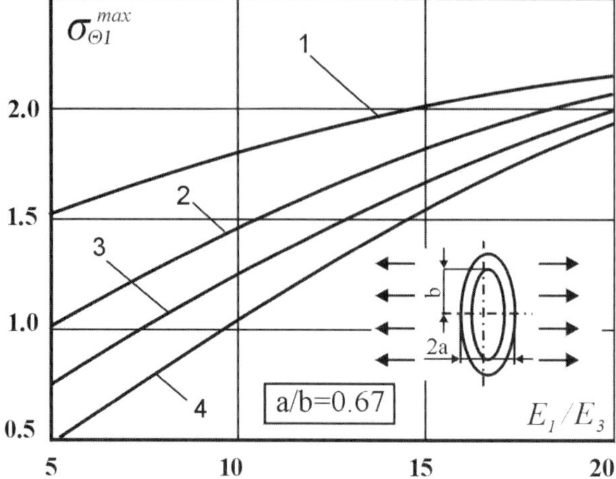

Fig. 3.25. Maximum stress $\sigma_{\theta 1}$ dependence on element core properties at varying orientation. Notations are similar to Fig. 3.24

changes significantly at similar element core dimensions but in a different orientation relative to outer loading.

An analogous conclusion can be derived from studying stress $\sigma_{\theta 1}$ variations over the element core contour (Fig. 3.26). As earlier, stress concentration depends strongly upon properties of the interlayer and reduces with incrementing stress, and so on the orientation of the element. In the latter case, stress reduction corresponds to such an element's location when its major semiaxis is oriented along the load action line.

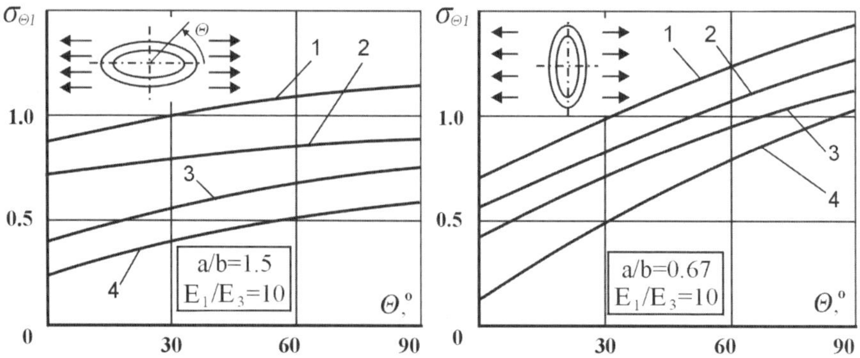

Fig. 3.26. Stress $\sigma_{\theta 1}$ variation in the element core as dependent on angle θ. Notations are similar to Fig. 3.24

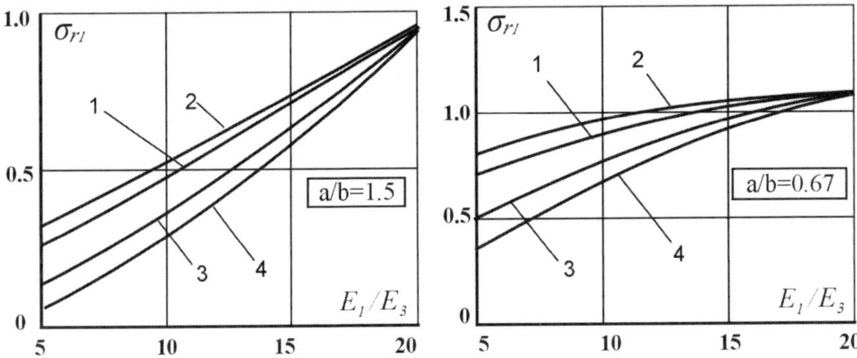

Fig. 3.27. Dependence of highest radial stresses in the element core on its properties under various E_2/E_3 and a/b. Notations are similar to Fig. 3.24

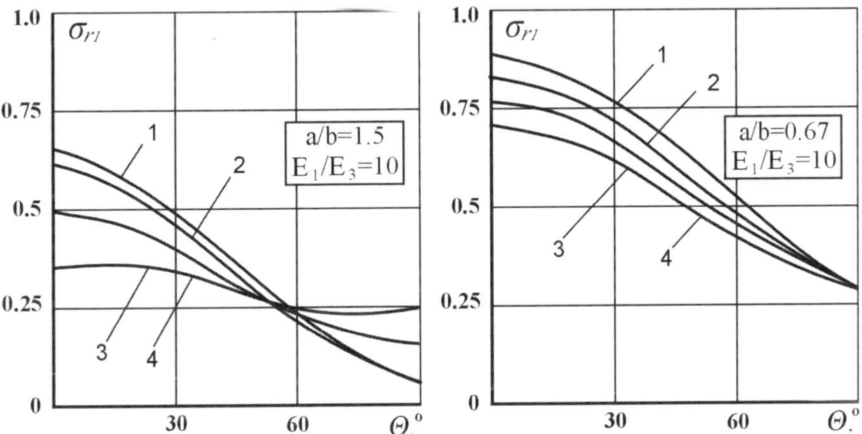

Fig. 3.28. Radial stress variation on the element core contour. Notations correspond to Fig. 3.2

One can make similar conclusions when examining radial stresses σ_{r1} in the element core (Figs. 3.27 and 3.28), though absolute values turn out to be a little lower. This is also true for different values of a/b (see Fig. 3.29).

It has been proved earlier that the composite with spirally reinforced filler most probably undergoes destruction between the elements, but not inside them. In this connection, stress variations have been studied on the interlayer contour neighboring the matrix. Data on variations of tangential $\sigma_{\theta3}$ and radial σ_{r3} stresses as dependent on the element core orientation and properties are shown in Figs. 3.30 and 3.31. It is evident that the dependence of stresses in the matrix on properties of the layer and element core is less than in the element core alone.

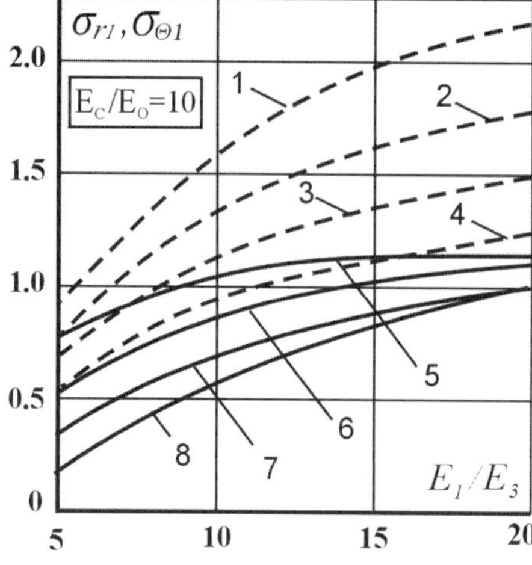

Fig. 3.29. Stress σ_{r1} (*solid lines*) and σ_{θ_1} (*dotted*) dependence on the element core properties

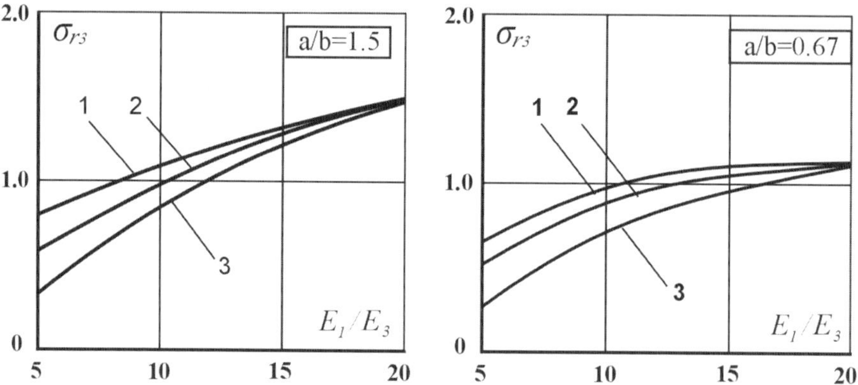

Fig. 3.30. Variations of radial stresses in the matrix depending on the element core properties. $1 - E_2/E_3 = 15$; $2 - E_2/E_3 = 10$; $3 - E_2/E_3 = 0$; $4 - E_2/E_3 = 5$

The distribution of tangential and radial stresses over the element contour in the matrix is depicted in Fig. 3.32. It appears that the most critical points in the matrix are $\theta = 0°$, in which radial stresses are the highest, and $\theta = \pi/2$ where tangential stresses reach the maximum.

It follows that orientation of the ellipsoidal spirally-reinforced element, as well as its elastic characteristics and interlayer properties strongly influence stress redistribution in the material structure.

Analogous investigations were carried out for the case where the intermediate layer was absent and round fibers were used [221]. With this aim stress at the composite interface was estimated with allowance for its filling degree. The result obtained agreed well with the above-presented evidence.

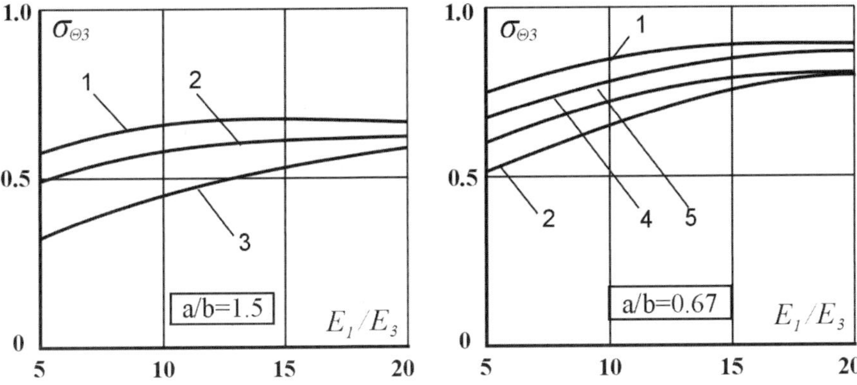

Fig. 3.31. Dependence of tangential stresses in the matrix on the element core properties. $1 - E_2/E_3 = 15$, $2 - E_2/E_3 = 0$; $3 - E_2/E_3 = 7$; $4 - E_2/E_3 = 5$; $5 - E_2/E_3 = 10$

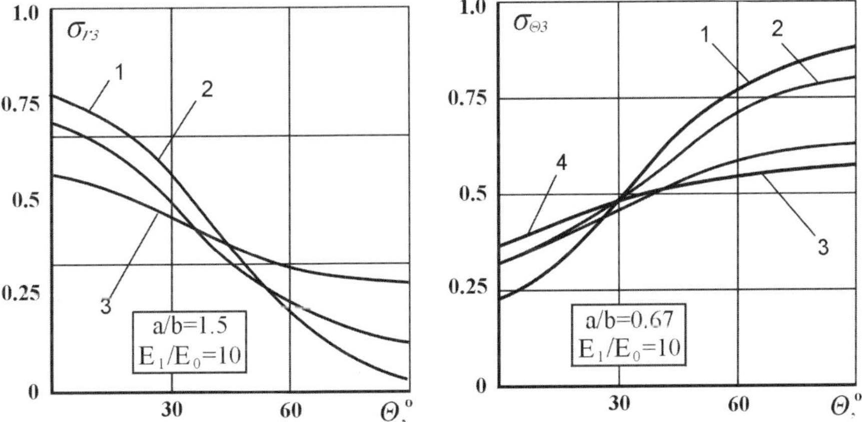

Fig. 3.32. Variations of radial and tangential stresses in the matrix over the element contour. $1 - E_2/E_3 = 15$; $2 - E_2/E_3 = 10$; $3 - E_2/E_3 = 0$; $4 - E_2/E_3 = 5$

3.4.2 Composite Design with Allowance for Filling Degree

To study the effect of the structural parameters of the above-considered model of the composite based on spirally reinforced filler a specific design routine has been elaborated. The calculation accuracy was estimated based on computation results of examples considered in [210]. The results obtained by this method coincided with those from [210] to about 0.5% accuracy for the case without the interlayer. Part of the results is expounded in [222].

As an example, the design of a carbon plastic with spirally reinforced filler based on an epoxide binder has been presented, where the elements were of a

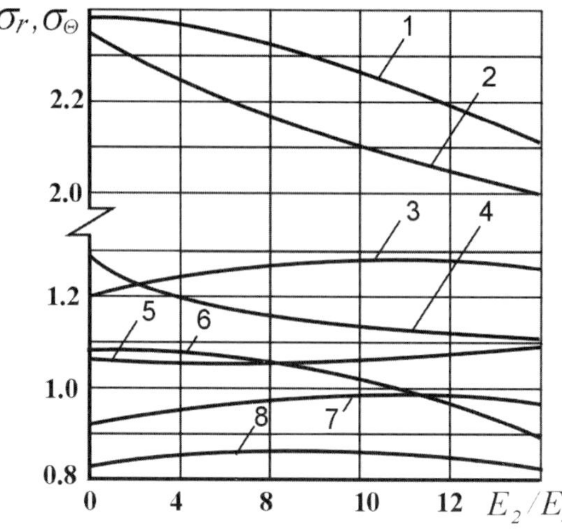

Fig. 3.33. Normal radial and tangential stresses versus the relationship of the elastic moduli of the layer to the matrix

circular form, for which the radial elasticity modulus of the core E_1 relation to that of the matrix was 9, Poisson's ratio of the element core was $\gamma_1 = 0.22$, and that of the matrix $\gamma_3 = 0.35$. For the case where the main reinforcing material, namely carbon fibers, was wound by one glass fiber, the relation of the layer thickness to the element radius was $\delta/R = 0.06$.

Let us take the case of a biaxial tension of the material. In Fig. 3.33 the dependencies of normal radial σ_r and tangential σ_θ stresses on the elastic moduli of the layer E_2 and matrix E_3 are given. Curves 1, 2, 7 and 8 show stress σ_θ distribution in the element core (curves 1, 2) and matrix (curves 7, 8) for two magnitudes of filling with reinforcing elements $\varphi_2' = 0.58$ (curves 2, 8) and $\varphi_2'' = 0.82$ (curves 1, 7). Similarly, distribution of σ_r in the element core is described by curves 4 and 6, while in the matrix by 3 and 5. Note that curves 5 and 6 correspond to $\varphi_2' = 0.58$ and curves 3 and 4 to $\varphi_2'' = 0.82$ values. It turns out that lessening of stresses σ_θ and σ_r in the element core constitutes for values $\varphi_2' = 0.58$, correspondingly, 11.5% and 1%, and for $\varphi_2'' = 0.82$–9.2% and 11.8%. Stresses in the matrix appeared not to change at all (σ_θ) or negligibly (σ_r).

The dependence of the highest stresses in the matrix σ_r and element core σ_θ (curves 1, 2) on the filling degree is shown in Fig. 3.34 for two values $E_2/E_3 = 1$ (i.e. without the interlayer) presented by curves 1 and 3, and $E_2/E_3 = 23$ – curves 2 and 4. It is apparent that the highest stresses in the matrix of the composite based on spatially reinforced filler differ slightly from those of a common composite. However, introduction of a spatial reinforcement and changed filling degree significantly reduce the highest stresses in the element core.

In Fig. 3.35 one can trace σ_r (curves 2, 4, 6 and 8) and σ_θ variations (curves 1, 3, 5 and 7) for the case of a uniaxial composite tension at $\varphi_2 = 0.58$ depending on angle θ. Curves 1, 2, 5 and 6 correspond to stress distribution

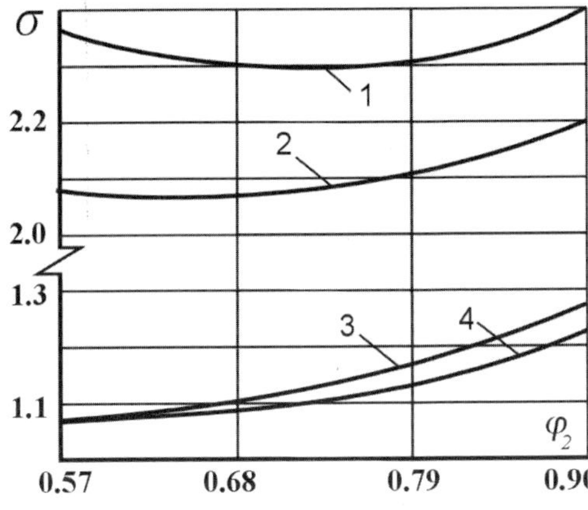

Fig. 3.34. Variations of highest stresses in the material as a function of filling degree

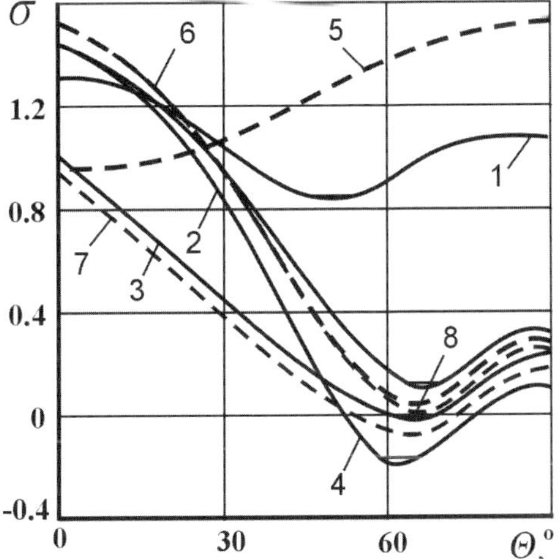

Fig. 3.35. Dependence of radial and tangential stresses in material ingredients upon angle Θ under uniaxial tension

in the core, and curves 3, 4, 7 and 8 – in the matrix. Dotted lines conform to $E_2/E_3 = 23$ and solid ones to $E_2/E_3 = 5$. As it follows from the information presented above that only stresses σ_Θ in the core differ essentially (curves 1, 5) for the two values of the elasticity modulus of the layer, whereas the rest stresses remain almost intact.

With the changing shape of the spirally reinforced element, stress distribution in the material ingredients also changes. Thus, in Fig. 3.36 stress variations are presented in the core and matrix under the biaxial tension of the composite in response to variations in filling degree φ_2. Notice that the layer thickness is the same as in the previous cases $E_2/E_3 = 10$; $E_1/E_3 = 20$ and the semiaxes ratio is $a/b = 1.5$. Proceeding from the results obtained it follows that with increasing filling degree tangential stresses diminish while radial ones increase.

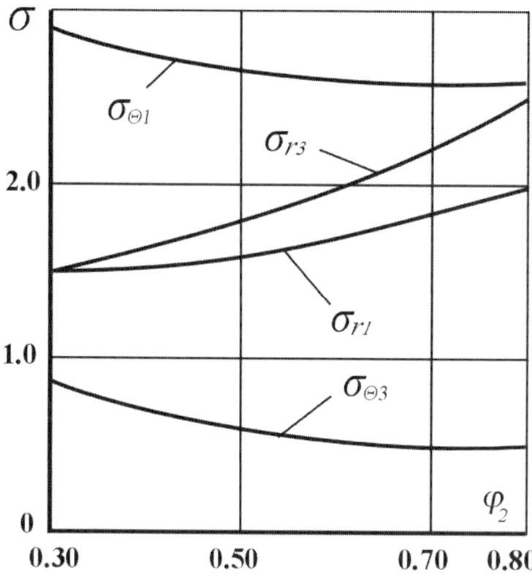

Fig. 3.36. Variations in radial and tangential stresses in material ingredients dependent on filling degree

Stress variations over the element contour under uniaxial tension for a material with analogous parameters and $\varphi_2 = 0.73$ is illustrated in Fig. 3.37. Most critical here are points $\theta = 0;\ \pi/2$, where radial and tangential stresses reach their limiting values in the matrix, thus leading to its failure. As the layer properties increase, stresses in the matrix rise too, while in the core they diminish. Data shown in Fig. 3.38 correspond to a material having $E_1/E_3 = 20$ and $\varphi_2 = 0.73$.

The dependence of elastic constants E_x^* (solid lines) and E_y^* (dotted) upon element core properties $a/b = 15;\ E_2/E_3 = 10$ at various filling degrees is presented in Fig. 3.39. It can be seen that with improving element core properties the elastic characteristics of the composite increase in both directions.

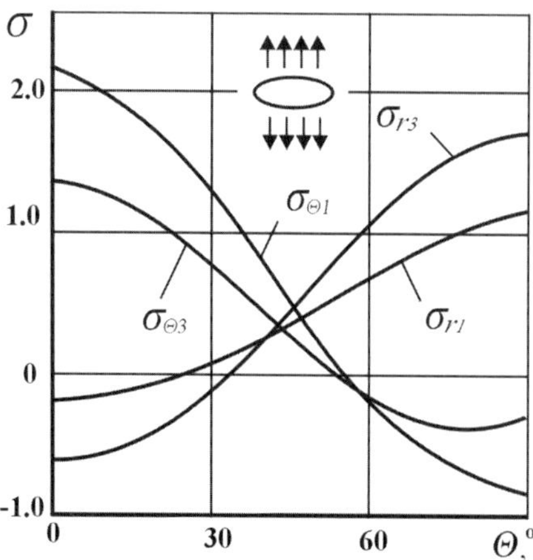

Fig. 3.37. Stress dependence of material ingredients on angle Θ under uniaxial tension

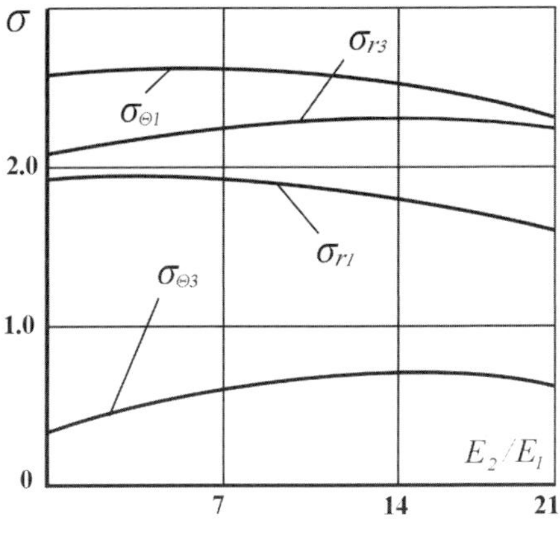

Fig. 3.38. Stress dependence of material ingredients on interlayer properties

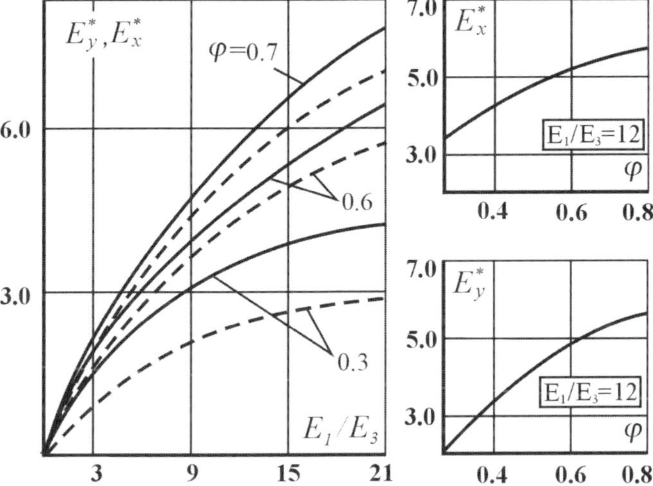

Fig. 3.39. Variations of reduced elastic moduli of the material as a function of element core properties and filling degree

The developed simplified calculation procedure of the stress-strain state of materials enables efficient and rapid estimation of the stress state and computation of reduced elastic characteristics. The proposed method, however, fits only within some variation range of the auxiliary reinforcing layer thickness. At $\delta/R = 0.09$ the solution error of test problems is 8%, which reduces to 5% with increasing number of collocation points N_3. In the course of numerical experiments an application boundary of the simplified method has been found, that is $\delta/R \leq 0.075$. For a greater layer thickness the error is more perceptible.

Thus, it has been established that stress distribution in the material ingredients is strongly dependent on the elastic characteristics and shape of the intermediate layer. As the layer rigidity increases, only tangential stresses in the element core augment, while in the matrix the variation is negligible. Hence, it is feasible to attain optimum stress distribution in ingredients of a high-filled composite based on a spirally reinforced filler through selection of physico-chemical parameters of the interlayer and the shape of the element.

3.5 Investigation of the Stress-Strain State of a Composite with Two Intermediate Layers

It has been proved in 2 that to lessen thermal stresses in a material it is expedient to make its structure such as to lay two layers of auxiliary reinforcement on the main reinforcement fibers. Choice of the properties of each layer and direction of their packing will result in an equilibrium structure of the composite. A natural requirement imposed on the interlayer material is the thickness of the auxiliary reinforcement. The thinner each layer is, the better the composite characteristics in this direction. By using glass and organic fibers of a small metric criterion for winding one can produce interlayers whose thickness ratio to the element radius is 0.06. Therefore, the layers in the calculation model can be taken as sufficiently thin and isotropic. Proceeding from conditions of the technological process of manufacturing, such a spirally reinforced filler requires high enough forces when winding reinforcement in the main direction. This results in a high filling degree in the element core and makes the core round, which practically does not change during forming.

The calculation model assumes the interlayer to be hollow, isotropic cylinders, while the matrix and element core are isotropic. Loading applied to the material model is analogous to that described earlier. To study plane strain of the composite let us present complex potentials in zone S_3 (Fig. 3.40) in the form of (3.3). To solve the problem it is sufficient to consider the stress state of a representative element of the structure.

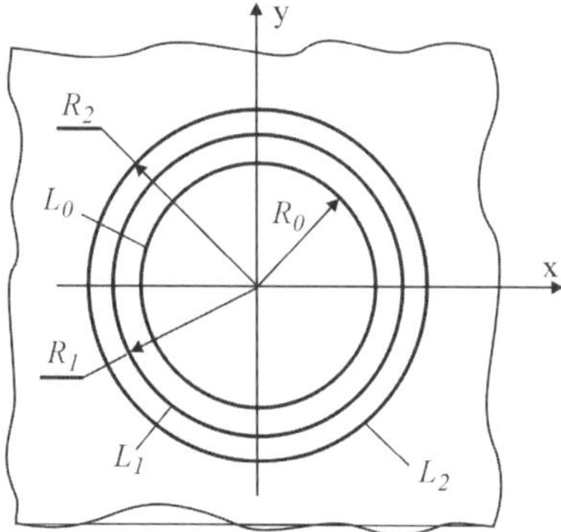

Fig. 3.40. A representative element of material model with two intermediate layers

Complex potentials in zones S_j will be presented approximately taking account of the problem symmetry as

$$\varphi_j(\xi) = 2G_j R \sum_{k=0}^{N_j} \left(a_{j,k}\xi^{2k+1} + b_{j,k}\xi^{-2k-1}\right) ;$$

$$\Psi_j(\xi) = 2G_j R \sum_{k=0}^{N_j} \left(c_{j,k}\xi^{2k+1} + d_{j,k}\xi^{-2k-1}\right) ;$$

$$(j = 0, 1, 2) . \tag{3.44}$$

For $j = 0$ coefficient $b_{0,k} = d_{0,k} = 0$.

On contours L_j boundary conditions are recorded

$$u_r^{(j)} = u_r^{(j+1)}; \quad u_\theta^{(j)} = u_\theta^{(j+1)} ;$$
$$\sigma_r^{(j)} = \sigma_r^{(j+1)}; \quad \tau_{r\theta}^{(j)} = \tau_{r\theta}^{(j+1)} . \tag{3.45}$$

Included in the boundary conditions functions will be calculated as follows [203]:

$$2G_j(u_r^{(j)} + iu_\theta^{(j)}) = \sigma^{-1}[\kappa_j\varphi_j(\xi) - z\overline{\varphi'(\xi)} - \overline{\Psi_j(\xi)}] ; \tag{3.46}$$

$$\sigma_r^{(j)} = 2\operatorname{Re}\{\varphi_j'(\xi)\} - \operatorname{Re}\{\sigma^2[\overline{z}\varphi_j''(\xi) + \Psi_j'(\xi)]\} ;$$
$$\tau_{r\theta}^{(j)} = I_m\{\sigma^2[\overline{z}\varphi_j''(\xi) + \Psi_j'(\xi)]\} . \tag{3.47}$$

Here $\sigma = \exp i\theta$, θ is polar angle; $\xi = \dfrac{z}{R}$.

On contours L_j: $\xi_j = \dfrac{z_j}{R} = r_j\sigma$; $r_j = \dfrac{R_j}{R}$; R is some characteristic dimension. Furthermore, it is taken that $R = R_2$.

Using (3.46) and (3.47) as well as (3.44) and (3.36) it is possible to find displacement and stress components on contours L_j. Components of deformation and strength are given in the addendum.

By equating coefficients of similar arguments or corresponding stresses and displacements one gets a system of linear algebraic equations about the following parameters

$$a_{j,k}, c_{j,k} \ (j = 0; \ k = 0, \dots, N_j);$$
$$a_{j,k}, c_{j,k}, b_{j,k}, d_{j,k} \ (j = 1, 2; \ k = 0, \dots, N_j);$$

$\alpha_{2k+2}, \beta_{2k+2}$ $(k = 0, \dots, N_3)$ with total number N:

$$N = 2[N_0 + 2(N_1 + N_2) + N_3 + 6] . \tag{3.48}$$

The set of equations for definition of coefficients (3.48) have been instanced in the application.

3.5.1 Numerical Computation of Low-filled Material Model with Spirally Reinforced Elements

Proceeding from the above, certain qualitative estimates can be obtained for the low-filled material model. This approach has been treated in [223], where, according to the criterion of deformation energy, maximal equivalent stresses were determined on contours L_0, L_1, L_2 in each component to simplify the analysis

$$\sigma_{ij} = [\sigma_{rij}^2 + \sigma_{\theta ij}^2 + 3\sigma_{r\theta ij}^2]^{1/2} . \tag{3.49}$$

Furthermore function H_{ij} is found

$$H = \frac{\sigma_{ij}}{\sigma_j^*} , \tag{3.50}$$

where σ_j^* is the material strength in the jth zone referred to the matrix strength.

The value of function H_{ij} is calculated for the material having an element core shear modulus ratio to that of the matrix $\mu_0/\mu_3 = 30$. Notice that the thickness of each interlayer is identical $\delta/R = 0.2$, whereas their properties change within 0.2–14 limits. The transversal loading of the model under shear loading is also considered.

The calculations have shown that the most risky points of stress concentrations are at $\theta = 0°$, $\pi/4$, $\pi/2 \ldots$ In Figs. 3.41 and 3.42 data on function H_{ij} variations are illustrated as dependent on the properties of the interlayers. From Fig. 3.41 it follows that function H_{10} in the element core greatly increases when the properties of the first layer remain invariable ($\mu_1/\mu_2 = 2$), while in the matrix stresses become reduced. An analogous phenomenon is observed as the first layer properties change (Fig. 3.42) and those of the second remain intact ($\mu_2/\mu_3 = 2$). In this case H_{32} rises and H_{10} stays practically invariable. Hence, there is an opportunity to control variation of the

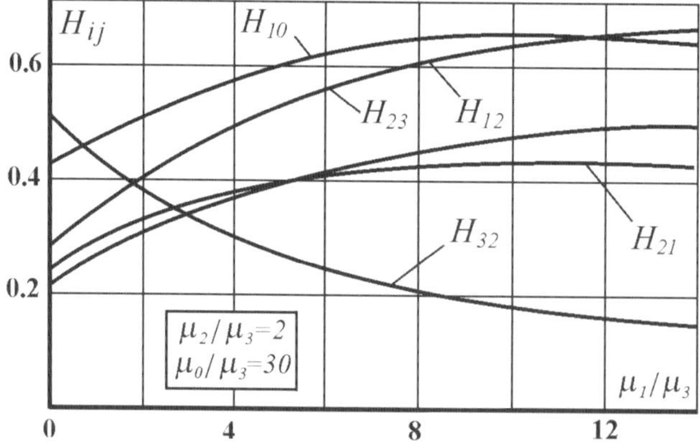

Fig. 3.41. Function H_{ij} dependence on the second layer properties

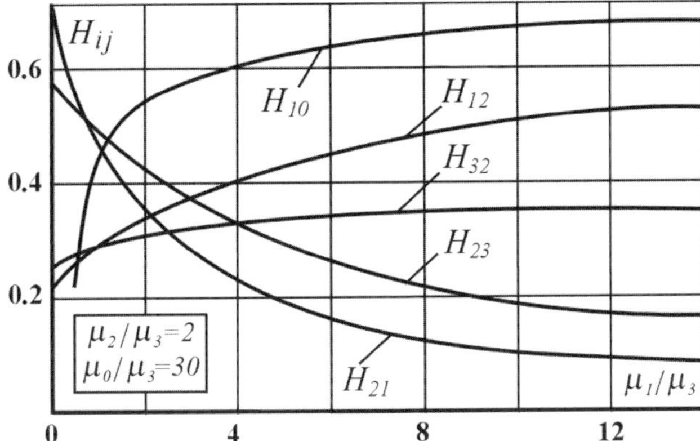

Fig. 3.42. Variations in H_{ij} function depending on the first – layer properties

stress state of the composite under study by choosing the corresponding correlations of the properties of the interlayers.

It is of interest to analyze functions H_{21}, H_{12} from the viewpoint of probable exfoliation of the material along the bonding line between two layers. If one changes the properties of the second layer, both functions will behave similarly, but variations in the second layer reduce H_{21} and raise H_{12}. Consequently, the enhanced characteristics of the first interlayer may bring about, under respective conditions, delamination in the composite.

Dependencies describing function H_{ij} variations in both layers are presented in Figs. 3.43–3.46. Thus, a conclusion can be made that within the considered range of characteristics of the interlayers the most loaded turns out to be the first layer portion neighboring the element core, as well as the

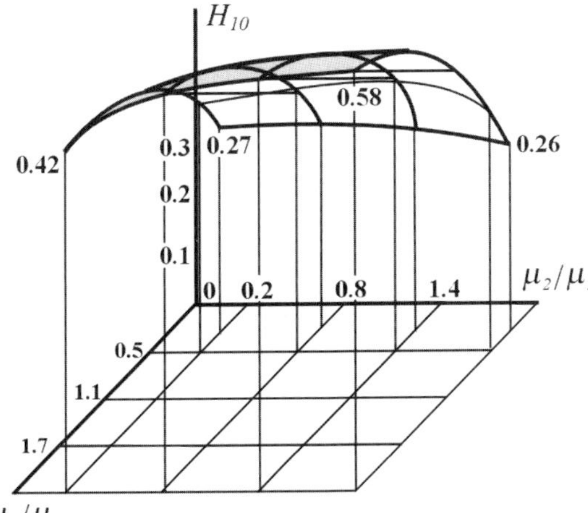

Fig. 3.43. Function H_{10} dependence on the interlayer properties

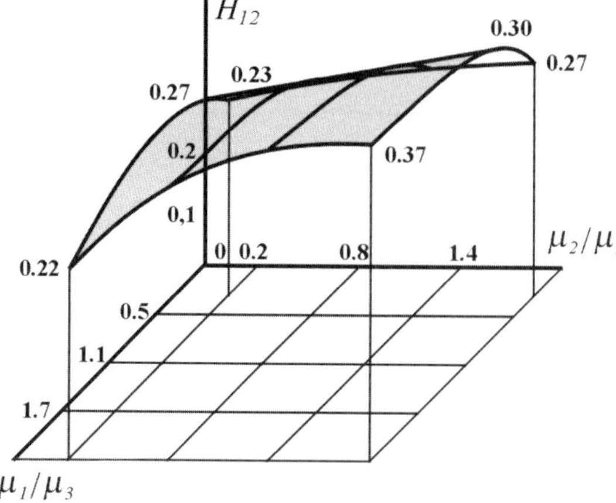

Fig. 3.44. Function H_{12} dependence on the properties of the interlayers

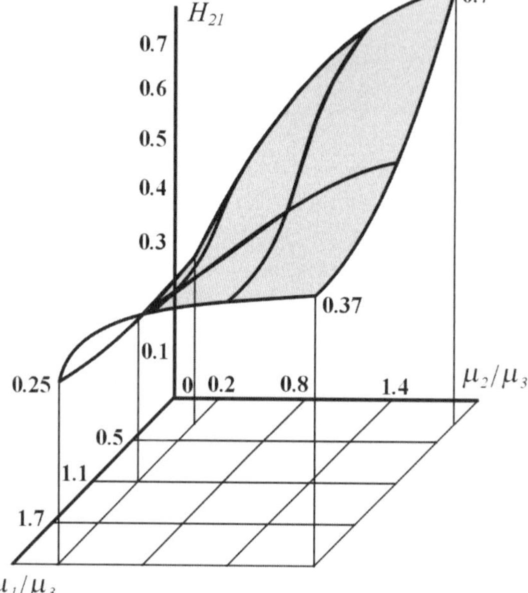

Fig. 3.45. Function H_{21} dependence on characteristics of auxiliary reinforcement layers

second layer contact line bordering the first one. This is also proof that under certain conditions failure of the system is probable through violation of adhesive bonds between the first layer and the element core.

In cases where the mechanical characteristics of the layers are low, then the matrix experiences the main load (Fig. 3.47) and the system tends towards the stress state of a plate with a round hole.

The above-described data are consistent with the results cited in [124] for a model with few intermediate layers and low filling ($\varphi = 0.3$). Notice

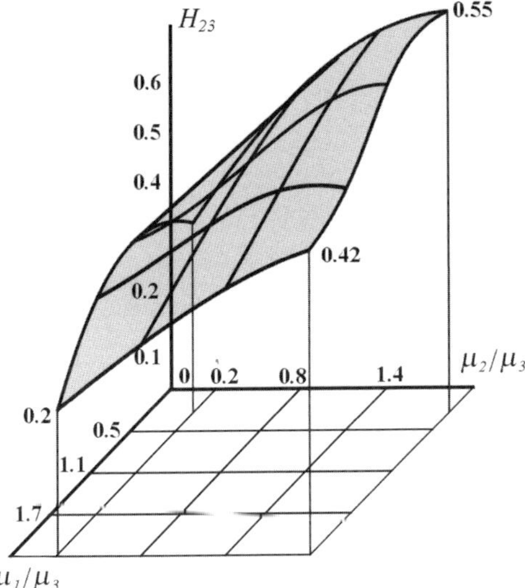

Fig. 3.46. Variations of function H_{23} depending on characteristics of interlayers

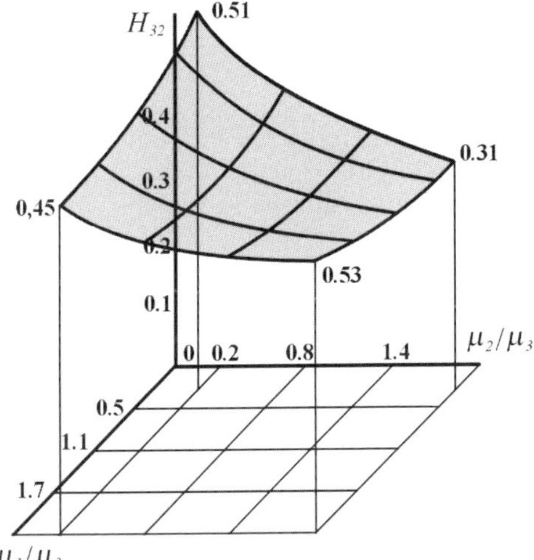

Fig. 3.47. Function H_{32} dependence on characteristics of interlayers

that the total volume of the layers in the model was only 5%. Computation results showed that variations in characteristics of the layers greatly affected the material stress state and its efficient elastic constants.

3.5.2 Material Model Design with Allowance for Filling

The diagram illustrating stresses is shown in Fig. 3.48.

The computation of stresses for two materials is illustrated as an example. The materials used are with similar and different properties of interlayers. In

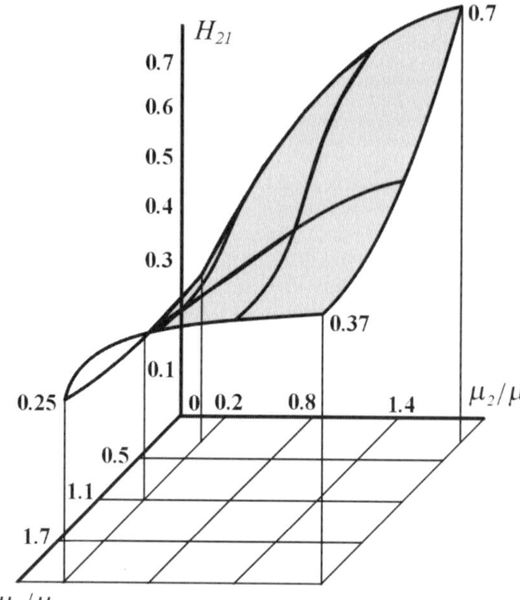

Fig. 3.48. Diagram of structural stresses acting in the material

the first case $E_0/E_3 = 6.29$; $E_1/E_3 = E_2/E_3 = 19.57$. In the second case $E_0/E_3 = 6.29$; $E_1/E_3 = 31.43$; $E_2/E_3 = 19.57$. The stress state of the material was analyzed for two loading schemes, namely, the uniaxial tension at which the filling degree with spirally reinforced elements was $\varphi = 0.71$ and the shear of the material with filling degree $\varphi = 0.50$. In Figs. 3.49–3.51, stress variations on element contours L_0, L_1, L_2 are described as dependent on angle θ under uniaxial tension. The dotted lines correspond to the material with similar interlayers. Proceeding from the above, the most evident variations are observed in stresses $\sigma_{\theta 12}, \sigma_{\theta 10}, \sigma_{\theta 23}$. Moreover, introduction of an additional layer augments stresses $\sigma_{\theta 12}$ and $\sigma_{\theta 10}$ in the layer and element core. In contrast, stress $\sigma_{\theta 23}$ in the matrix somewhat reduces. The variation character of the rest stresses changes only insignificantly on the introduction of the additional layer.

Analogous data on stress values arising at material shearing are given in Figs. 3.52–3.54. It is evident that stresses $\sigma_{\theta 12}, \sigma_{\theta 21}, \sigma_{\theta 10}, \sigma_{r 10}, \sigma_{r 23}, \sigma_{\theta 23}$ undergo the strongest variations. As soon as an additional interlayer is introduced, tangential stresses in the first layer go down, while in the second layer and in the core they rise. In contrast, in the matrix they reduce by 30–50%. It is to be noted that tangential stresses display the highest values at the first to second layer interface, between the second layer and the matrix.

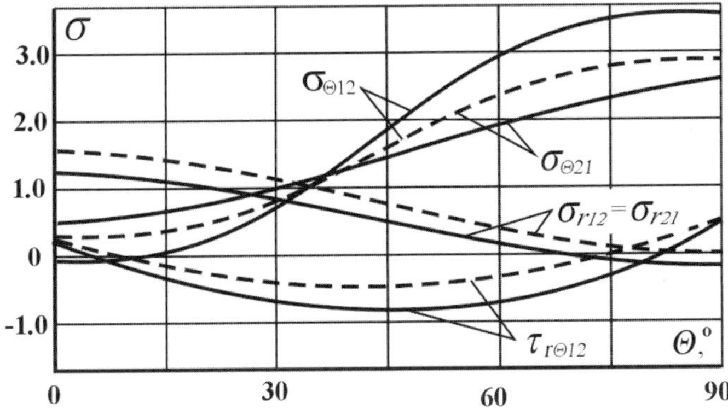

Fig. 3.49. Dependence of stresses on contour L_1 upon angle Θ under uniaxial loading

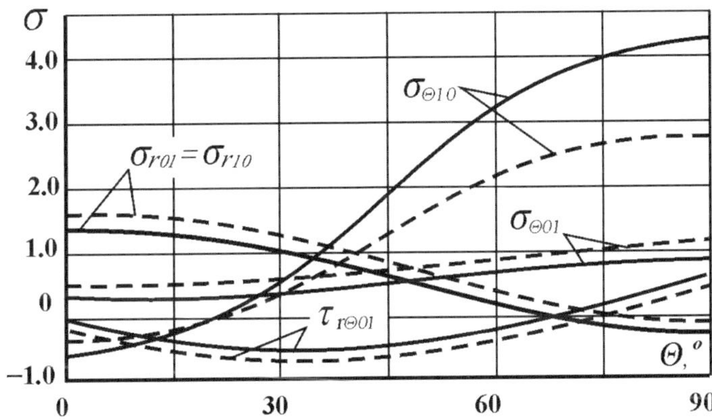

Fig. 3.50. Stress variation on contour L_0 under uniaxial tension of the composite

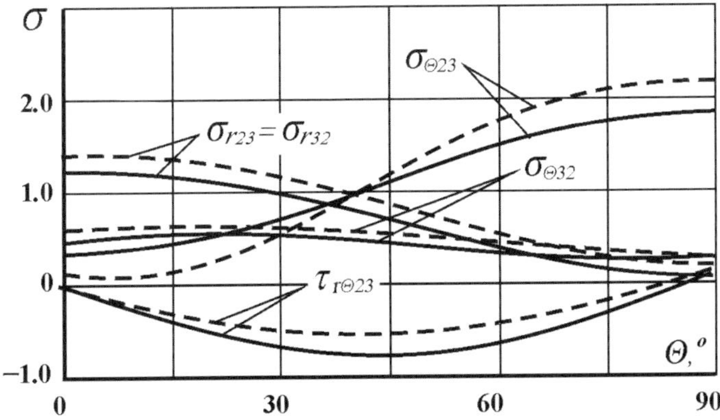

Fig. 3.51. Stress dependence on contour L_2 upon angle θ at uniaxial tension

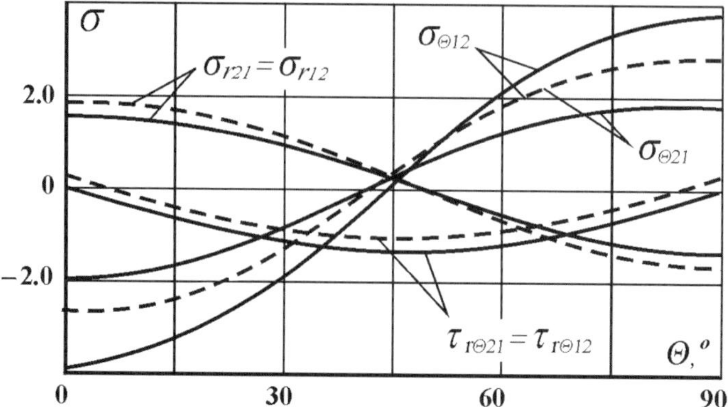

Fig. 3.52. Stress variation on contour L_1 dependent on angle θ under shear

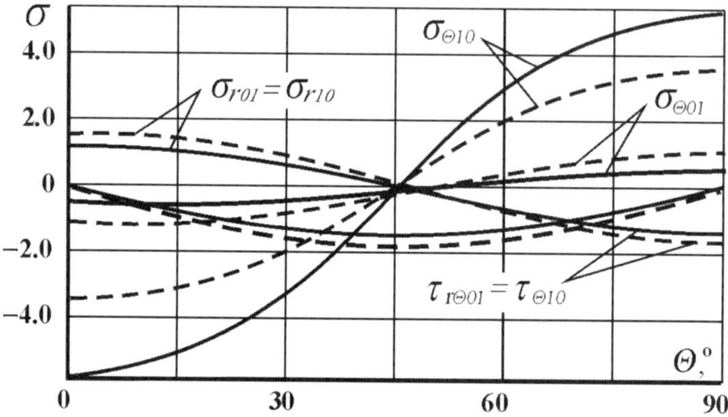

Fig. 3.53. Stress variations on contour L_0 depending on angle θ at shearing

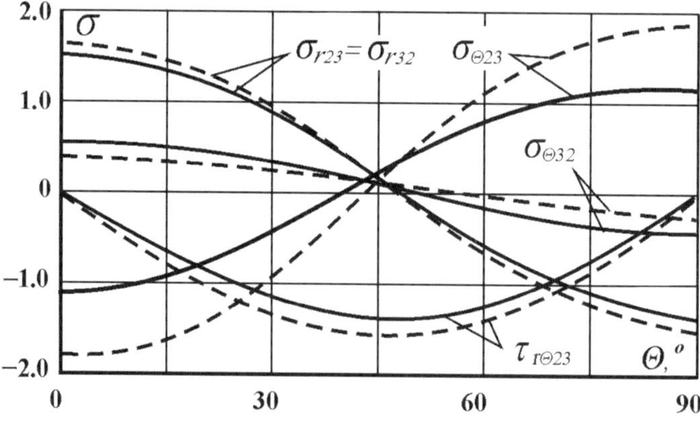

Fig. 3.54. Stress dependence on contour L_2 upon angle θ at shear

Consequently, introduction of an additional intermediate layer and monitoring of the physico-mechanical properties of each layer can serve for a targeted variation of the stress state in composites based on spirally reinforced fillers.

3.6 Conclusions

1. It has been established during investigations that introduction of inhomogeneous intermediate layers into the composite brings about considerable variations in the material properties and changes microstresses in its structure. To enhance the physico–mechanical characteristics of the materials in a transverse direction specific recommendations have been devised

2 Applicability limits of softened conditions have been defined at the interface of the element core with the matrix and the interlayer. This simplified the problem solution and determination of the stress-strain state and elastic characteristics of the material.

3. It has been found that by way of monitoring geometrical parameters and physico–mechanical properties of each of the two intermediate layers introduced in the material structure it is possible to regulate the stress–strain state of the composite based on spirally reinforced filler.

4. As a result of numerical studies of the composites the stress–strain state of the material based on spirally reinforced fillers has been evaluated and its transversal elastic characteristics have been estimated.

4 Stress-Strain State of the Composite Based on Spirally Reinforced Filler Under Loading in the Direction of Reinforcement

4.1 Composite Tension in the Reinforcing Direction

The material under study is based on spirally reinforced filler whose cross-section presents a system of double-periodic elements. Let us isolate a representative element of the material S that consists of a matrix, core and an intermediate layer (Fig. 4.1). Zone S is arranged so that the material cross-section is superimposed by a double-periodic system of the zones coinciding at the overlap with S. In the case of hexagonal packing, the area of zone S for any kind of the element cross-section is $\Omega = 2\sqrt{3}$, i.e. it is the area of a rhombus with its vertices in the centers of the four neighboring elements. Areas of the core, matrix and the layer found within S zone are then, correspondingly $\Omega_1, \Omega_3, \Omega_2$. Let the medium strain in the axis z-direction be $\langle \varepsilon_z \rangle$, then [233].

$$\langle \sigma_z \rangle = \frac{E_z^{(1)}\Omega_1 + E_z^{(2)}\Omega_2 + E_3\Omega_3}{\Omega} \langle \varepsilon_z \rangle + \langle \sigma_z^{(1)} \rangle + \langle \sigma_z^{(2)} \rangle + \langle \sigma_z^{(3)} \rangle , \qquad (4.1)$$

where $E_z^{(1)}, E_z^{(2)}, E_3$ are elastic moduli of each constituent; $\langle \sigma_1^{(1)} \rangle, \langle \sigma_z^{(2)} \rangle, \langle \sigma_z^{(3)} \rangle$ – mean stresses in zones S_1, S_2, S_3:

$$\langle \sigma_Z^{(i)} \rangle = \frac{1}{\Omega_i} \int \int_{S_I} \sigma_Z^{(i)} dS_i , \quad (i = 1, 2, 3) . \qquad (4.2)$$

Stresses $\sigma_z^{(i)}$, presented in (4.2), are calculated in terms of stresses in the section plane [249]:

$$\begin{aligned}
\sigma_R^{(1)} &= v'(\sigma_x^{(1)} + \sigma_y^{(1)}) ; \\
\sigma_z^{(2)} &= v_{zr}\sigma_r^{(2)} + v_{z\Theta}\sigma_\Theta^{(2)} ; \\
\sigma_z^{(3)} &= v_3(\sigma_x^{(3)} + \sigma_y^{(3)}) .
\end{aligned} \qquad (4.3)$$

Analytic functions $\varphi_1(z), \psi_1(z)$ in zone S_1 will be presented as (3.19):

$$\varphi_1(z) = \langle \varepsilon_z \rangle \sum_{K=1}^{N_1} \Phi_{1K} z^{2K-1} ; \quad \psi_1 = \langle \varepsilon_z \rangle \sum_{K=1}^{N_1} \Phi_{2K} z^{2K-1} . \qquad (4.4)$$

Stresses and displacements are calculated using (4.4) and dependencies (3.20). Using (3.15) one can then define stresses σ_n and τ. The stresses, displacements $u^{(1)}$ and $v^{(1)}$ are determined by formulas similarly to Chap. 3.

Stresses $\sigma_r^{(2)}, \sigma_\theta^{(2)}, \tau_{r\theta}^{(2)}$ and displacements $u_r^{(2)}, u_\theta^{(2)}$, in the layer are also determined as in, e.g. Chap. 3, the displacements will be of the form

$$u_r^{(2)} = \sum_{K=0}^{N_2} \sum_{j=1}^{4} C_{jK} 2k \left[\beta_{11}^{(2)} + \gamma - (\gamma - \beta_{12}^{(2)}) \right] \lambda_{jK} \langle \varepsilon_z \rangle ;$$

$$u_\Theta^{(2)} = \sum_{K=0}^{N_2} \sum_{j=1}^{4} C_{jK} \left[\beta_{22}^{(2)} \lambda_{jK}^2 - (\beta_{11}^{(2)} + 4\gamma k^2) \right] \langle \varepsilon_z \rangle . \tag{4.5}$$

Note that $\beta_{jk}^{(2)}$ are constants in contrast to those presented in Chap. 3. Stresses $\sigma_n^{(2)}, \tau^{(2)}$ and displacements $u^{(2)}$ and $v^{(2)}$ are found by formulas (3.16) and (3.29).

Complex Kolosov–Muskhelishvily's potentials within the region of matrix S_3 are determined as

$$\frac{\Phi(z)}{2G_3} = \langle \varepsilon_z \rangle \left\{ \alpha_0 z + \sum_{K=0}^{\infty} \alpha_{2K+2} \left[z^{-2K-2} + \sum_{j=0}^{\infty} r_{j,K} z^{2j} \right] \right\} R^{2K+2} ;$$

$$\frac{\Psi(z)}{2G_3} = \langle \varepsilon_z \rangle \left\{ \beta_0 z + \sum_{K=0}^{\infty} \beta_{2K+2} \left[z^{-2K-2} + \sum_{j=0}^{\infty} r_{jK} z^{2j} \right] \right. $$

$$\left. - \sum_{K=0}^{\infty} (2K + 2)\alpha_{2K+2} \sum_{j=0}^{\infty} z^{2j} S_{jK} \right\} , \tag{4.6}$$

where $r_{jK}; S_{jK}; \alpha_0; \beta_0$ are found from (3.4) and (3.5).

Using (4.6) one can find displacements

$$\frac{u^{(3)}}{\langle \varepsilon_z \rangle} = \sum_{K=0}^{N_3} \alpha_{2K+2} A_{2K+2}^{(4)}(x, y) + \sum_{K=0}^{N_3} \beta_{2K+2} B_{2K+2}^{(4)}(x, y) + C^{(4)}(x, y) ;$$

$$\frac{v^{(3)}}{\langle \varepsilon_z \rangle} = \sum_{K=0}^{N_3} \alpha_{2K+2} A_{2K+2}^{(5)}(x, y) + \sum_{K+0}^{N_3} \beta_{2K+2} B_{2K+2}^{(5)}(x, y) + C^{(5)}(x, y) .$$

$$\tag{4.7}$$

Coefficients of (4.7) are given in the addendum.

Stresses $\sigma_n^{(3)}$ and $\tau^{(3)}$ are determined by (4.6) and (3.15), as follows

$$\frac{\sigma_n^{(3)}}{E_3} = \sum_{K=0}^{N_3} \left\{ \alpha_{2K+2} \left[A_{2K+2}^{(1)} \cos^2 \alpha_1 + A_{2K+2}^{(2)} \sin^2 \alpha_1 + A_{2K+2}^{(3)} \sin 2\alpha \right] \right.$$

$$\left. + \beta_{2K+2} \left[B_{2K+2}^{(2)} \sin^2 \alpha_1 + B_{2K+2}^{(1)} \cos^2 \alpha_1 + B_{2K+2}^{(3)} \sin 2\alpha_1 \right] \right\} \frac{\langle \varepsilon_z \rangle}{1 + v_3} ;$$

$$\frac{\tau^{(3)}}{E_3} = \sum_{K+0}^{N_3} \left\{ \alpha_{2K+2} \left[\frac{A_{2K+2}^{(2)} - A_{2K+2}^{(1)}}{2} \sin 2\alpha_1 + A_{2K+2}^{(3)} \cos 2\alpha_1 \right] \right.$$

$$\left. + \beta_{2K+2} \left[\frac{B_{2K+2}^{(2)} - B_{2K+2}^{(1)}}{2} \sin 2\alpha_1 + B_{2K+2}^{(3)} \cos 2\alpha_1 \right] \right\} \frac{\langle \varepsilon_z \rangle}{1 + v_3}, \quad (4.8)$$

where α_1 – angle between the normal to the outer contour of the layer and the x-axis.

4.2 Boundary Conditions and Determination of Elastic Constants

Stresses $\sigma_n^{(j)}, \tau^{(j)}$ and displacements $u^{(j)}, v^{(j)}$ $(j = 1, 2, 3)$ are related on contours L_K $(k = 1, 2)$ through dependencies (3.14):

$$u^{(i-1)} = u^{(i)}; \quad \sigma_n^{(i-1)} = \sigma_n^{(i)}; \quad v^{(i-1)} = v^{(i)}; \quad \tau^{(i-1)} = \tau^{(i)}; \quad (i = 2, 3). (4.9)$$

Boundary conditions (4.9) serve to find unknown parameters

$$\Phi_{1K}, \Phi_{2K} \ (k = 1, \ldots, N_1), \quad C_{10}, C_{20}, C_{jK} \ (j = 1, \ldots, 4; \ k = 1, \ldots, N_2),$$
$$\alpha_{2K+2}, \beta_{2K+2} \ (k = 0, \ldots, N_3). \quad (4.10)$$

Coefficients (4.10) are calculated by a system of linear algebraic equations, for which the conditions of (4.9) are given through allocations. Based on these parameters the material characteristics are defined.

By denoting

$$\langle \sigma_z^{(i)} \rangle = \langle \varepsilon_z \rangle \langle E_z^{(i)} \rangle, \quad (i = 1, 2, 3) \quad (4.11)$$

let us calculate mean displacements in the sections of the element core, layer and matrix using (4.2). Thus, the reduced elastic modulus in reinforcement direction (by 4.1) will be

$$\langle E_z \rangle = \frac{E_z^{(1)} \Omega_1 + E_z^{(2)} \Omega_2 + E_3 \Omega_3}{\Omega} + \frac{\langle \sigma_z^{(1)} \rangle}{\langle \varepsilon_z \rangle} + \frac{\langle \sigma_z^{(2)} \rangle}{\langle \varepsilon_z \rangle} + \frac{\langle \sigma_z^{(3)} \rangle}{\langle \varepsilon_z \rangle}. \quad (4.12)$$

As previously, we will denote the material filling degree by the main reinforcement as φ_0 and by the auxiliary as φ_a, then instead of (4.12) we have

$$\langle E_z \rangle = E_z^{(1)} \varphi_0 + E_z^{(2)} \varphi_a + (1 - \varphi_0 - \varphi_a) E_3 + \frac{\langle \sigma_z^{(1)} \rangle}{\langle \varepsilon_z \rangle} + \frac{\langle \sigma_z^{(2)} \rangle}{\langle \varepsilon_z \rangle} + \frac{\langle \sigma_z^{(3)} \rangle}{\langle \varepsilon_z \rangle}, \quad (4.13)$$

where the first three members correspond to the additivity rule.

Thus, displacements in the matrix are, on the one hand, defined by [233]

$$u^{(3)} + iv^{(3)} = x\langle \varepsilon_x \rangle + y\langle \varepsilon_y \rangle = \frac{z + \bar{z}}{2} \langle \varepsilon_x \rangle + \frac{z - \bar{z}}{2} \langle \varepsilon_y \rangle, \quad (4.14)$$

and, on the other hand,

$$u^{(3)} + iv^{(3)} = -\nu_3 z\langle\varepsilon_z\rangle + \frac{1}{G_3}[\chi_3\varphi(z) - \varphi'(z)z - \overline{\psi}(z)]\,, \tag{4.15}$$

are defined by (4.7), whereas $\varphi(z) = \int \Phi(z)dz$, $\psi(z) = \int \Psi(z)dz$.

Let us find the increment in displacements in the adjacent cells of the reinforced medium using (4.14). Considering that $z_j - z_{j-1} = \omega_j$, where ω_j is the material lattice spacing, we obtain

$$\frac{\omega_j + \overline{\omega}_j}{2}\langle\varepsilon_x\rangle + \frac{\omega_j - \overline{\omega}_j}{2}\langle\varepsilon_y\rangle\,, \tag{4.16}$$

and from (4.15) with account of potentials (4.6) it turns out that

$$\nu_3\omega_j\langle\varepsilon_z\rangle + [\chi_3(\omega_j\alpha_0 - \alpha_2 x_j) - \omega_j\alpha_0 - \beta_0\overline{\omega}_j + \beta_2\overline{x}_j + \overline{\gamma}\alpha_2]\langle\varepsilon_z\rangle\,. \tag{4.17}$$

The parameters of the latter relation, e.g. of the hexagonal packing, are of the kind [233]

$$\omega_1 = 2;\quad \omega_2 = 1 + \sqrt{3};\quad x_1 = \frac{\pi}{\sqrt{3}};\quad x_2 = \frac{\pi(1 - i\sqrt{3})}{2\sqrt{3}};$$

$$\gamma_1 = \gamma_2 = 0;\quad \alpha_0 = \frac{\pi}{4\sqrt{3}}\beta_2;\quad \beta_0 = \frac{\pi}{2\sqrt{3}}\alpha_2\,.$$

By equating displacement increments of (4.16) and (4.17), a relation is found of $\langle\varepsilon_x\rangle, \langle\varepsilon_y\rangle, \langle\varepsilon_z\rangle$ (at $j = 1$, $j = 2$):

$$2\langle\varepsilon_x\rangle = \langle\varepsilon_z\rangle\left[-2\nu_3 - \alpha_2\left(\frac{\pi}{\sqrt{3}}\chi_3 + \frac{\pi}{\sqrt{3}}\right) + \beta_2\left(\frac{\pi\chi_3}{2\sqrt{3}} - \frac{\pi}{2\sqrt{3}} + \frac{\pi}{\sqrt{3}}\right)\right];$$

$$\langle\varepsilon_x\rangle + i\sqrt{3}\langle\varepsilon_y\rangle = \langle\varepsilon_z\rangle\left\{-\nu_3(1 + i\sqrt{3}) - \alpha_2\left[\frac{\pi(1 - i\sqrt{3})}{2\sqrt{3}}\chi_3 + (1 - i\sqrt{3})\frac{\pi}{2\sqrt{3}}\right]\right.$$

$$\left. + \beta_2\left[\frac{\pi\chi_3}{4\sqrt{3}}(1 + i\sqrt{3}) - \frac{\pi}{4\sqrt{3}}(1 + i\sqrt{3}) + \frac{\pi(1 + i\sqrt{3})}{2\sqrt{3}}\right]\right\}\,. \tag{4.18}$$

The imaginary parts of the latter equality yield the following relation

$$\sqrt{3}\langle\varepsilon_y\rangle = \langle\varepsilon_z\rangle\left[-\nu_3\sqrt{3} + \alpha_2\left(\frac{\pi\chi_3}{2} + \frac{\pi}{2}\right) + \beta_2\left(\frac{\pi\chi_3}{2} - \frac{\pi}{4} + \frac{\pi}{2}\right)\right]\,. \tag{4.19}$$

Hook's law for an orthotropic medium is the presumption that $\langle\sigma_x\rangle = \langle\sigma_y\rangle = 0$ is

$$\langle\varepsilon_x\rangle = -\frac{\langle\nu_{zx}\rangle}{\langle E_z\rangle}\langle\sigma_z\rangle;\quad \langle\varepsilon_y\rangle = -\frac{\langle\nu_{zy}\rangle}{\langle E_z\rangle}\langle\sigma_z\rangle\,. \tag{4.20}$$

Let us present conditions (4.18) and (4.19) in a subset form

$$\langle\varepsilon_x\rangle = k_x\langle\varepsilon_z\rangle;\quad \langle\varepsilon_y\rangle = k_y\langle\varepsilon_z\rangle\,. \tag{4.21}$$

Substituting them into (4.20) we obtain

$$k_x\langle\varepsilon_x\rangle = -\frac{\langle\nu_{zy}\rangle\langle\sigma_z\rangle}{\langle E_z\rangle};\quad k_y\langle\varepsilon_z\rangle = -\frac{\langle\nu_{zy}\rangle\langle\sigma_z\rangle}{\langle E_z\rangle}\,.$$

Proceeding from (4.1), i.e. $\langle\sigma_z\rangle = \langle\varepsilon_z\rangle\langle E_z\rangle$ it follows that

$$\langle v_{zx}\rangle = v_3 + \alpha_2 \frac{\pi}{2\sqrt{3}}(\chi_3 + 1) - \beta_2 \frac{\pi}{4\sqrt{3}}(\chi_3 + 1)\,;$$

$$\langle v_{zy}\rangle = v_3 - \alpha_2 \frac{\pi}{2\sqrt{3}}(\chi_3 + 1) - \beta_2 \frac{\pi}{4\sqrt{3}}(\chi_3 + 1)\,. \qquad (4.22)$$

Using the characteristics thus obtained, one has

$$\langle v_{yz}\rangle = \frac{\langle E_y\rangle\langle v_{zy}\rangle}{\langle E_z\rangle}\,; \qquad \langle v_{xz}\rangle = \frac{\langle E_x\rangle\langle v_{zx}\rangle}{\langle E_z\rangle}\,. \qquad (4.23)$$

4.3 Longitudinal Shear of the Composite Structure Based on Spirally Reinforced Filler

As earlier, we represent the material structure as a system of transverse isotropic elements of a random cross-section on whose surface anisotropic intermediate layers are disposed, inside an isotropic binder-matrix. For simplification, on hexagonal arrangement of the spirally reinforced elements is presumed.

In the case of pure shear of the medium under consideration, the following relations are obtained for displacements

$$\gamma_{zx} = \frac{\partial W}{\partial x}\,; \qquad \gamma_{zy} = \frac{\partial W}{\partial y}\,, \qquad (4.24)$$

and for stresses

$$\frac{\partial \tau_{zx}}{\partial x} + \frac{\partial \tau_{zy}}{\partial y} = 0\,; \qquad \frac{\partial \tau_{zx}}{\partial z} + \frac{\partial \tau_{zy}}{\partial z} = 0\,. \qquad (4.25)$$

From the above relations and with an allowance for Hook's law it follows that $\Delta W = 0$.

The solution of the latter equation for the zone of element S_1 core will be

$$W_1(x, y) = \varphi_1(z) + \overline{\varphi}_1(z)\,, \qquad (4.26)$$

where analytic function $\varphi_1(z)$ within the zone S_1 can be presented as an expansion in Taylor's series (bearing in mind problem symmetry about the x and the y axes)

$$\varphi_1(z) = \sum_{k=0}^{\infty}(a_k + ib_k)\cdot z^{2k+1}\,. \qquad (4.27)$$

It is evident that

$$\tau_{xz}^{(1)} - i\cdot\tau_{yz}^{(1)} = 2G_1^{(1)}\varphi_1'(z)\,. \qquad (4.28)$$

Wherefrom,

$$\frac{\tau_{xz}^{(1)}}{G_3} = 2\delta_1 \sum_{k=0}^{\infty} \left[a_k \operatorname{Re}\{z^{2k}\} - b_k I_m\{z^{2k}\} \right] (2k+1) ;$$

$$\frac{\tau_{yz}^{(1)}}{G_3} = -2\delta_1 \sum_{k=0}^{\infty} \left[a_k I_m\{z^{2k}\} + b_k \operatorname{Re}\{z^{2k}\} \right] (2k+1) . \tag{4.29}$$

Here $\delta_1 = G_1/G_3$, and G_3 is the shear modulus of the matrix.

From (4.26) and (4.27) we obtain:

$$W_1(x,y) = 2 \sum_{k=0}^{\infty} \left[a_k \operatorname{Re}\{z^{2k+1}\} - b_k I_m\{z^{2k+1}\} \right] . \tag{4.30}$$

For the case of an orthogonal intermediate layer with shear modulus G_{rz} and $G_{\Theta z}$, Hook's law is of the kind

$$\gamma_{\Theta z}^{(2)} = -\frac{\tau_{\Theta z}^{(2)}}{G_{\Theta z}} = \frac{1}{2}\frac{\partial W_2}{\partial \Theta} ; \quad \gamma_{rz}^{(2)} = -\frac{\tau_{rz}^{(2)}}{G_{rz}} = \frac{1}{2}\frac{\partial W_2}{\partial r} . \tag{4.31}$$

For the case of a pure orthogonal shear the equations are

$$\frac{\partial \tau_{rz}}{\partial z} = \frac{\partial \tau_{\Theta z}}{\partial z} = 0 ;$$

$$\frac{\partial \tau_{rz}}{\partial r} + \frac{1}{r}\frac{\partial \tau_{\Theta z}}{\partial \Theta} + \frac{\tau_{rz}}{r} = 0 . \tag{4.32}$$

Substituting (4.31) into the latter relation we have

$$G_{rz} \left(\frac{\partial^2 W_2}{\partial r^2} + \frac{1}{r}\frac{\partial W_2}{\partial r} \right) + \frac{G_{\Theta r}}{r^2}\frac{\partial^2 W_2}{\partial \Theta^2} = 0 . \tag{4.33}$$

Obviously, under pure shear the displacement component should meet condition $W_2(r,\Theta) = -W_2(r,\Theta+\pi)$, where r and Θ are polar coordinates in the zone of layer S_2.

Solution of the differential expression in partial derivatives (4.33) is sought as

$$W_2(r,\Theta) = \sum_{k=0}^{\infty} W_k(r) \cos(2k+1)\Theta . \tag{4.34}$$

By substituting (4.34) into (4.33) and equating the cosine factors of similar arguments, one gets differential equations about $W_k(r)$

$$W_k''(r) + \frac{W_k'(r)}{r} - \varepsilon_k^2 \frac{W_k(r)}{r^2} = 0 , \tag{4.35}$$

where $\varepsilon_k^2 = \dfrac{G_{\Theta z}}{G_{rz}}(2k+1)^2$.

Euler equations (4.35) are solved by a substitution of $r = \ell^\rho$ and are reduced to

$$\frac{\partial^2 W_k}{\partial \rho^2} - \varepsilon_k^2 W_k = 0 .$$

Solutions of the latter are functions of the kind

$$W_k(r) = C_{1k}r^{\varepsilon k} + C_{2k}r^{-\varepsilon k} \,. \tag{4.36}$$

Hence, we have

$$W_2(r,\Theta) = \sum_{k=0}^{\infty}(C_{1k}r^{\varepsilon k} + C_{2k}r^{-\varepsilon k})\cos(2k+1)\Theta \,. \tag{4.37}$$

Stresses $\tau_{\Theta z}^{(2)}$ and $\tau_{rz}^{(2)}$ are found from (4.31)

$$\frac{\tau_{rz}^{(2)}}{G_3} = \delta_2 \sum_{k=0}^{\infty} \varepsilon_k(C_{1k}r^{\varepsilon k-1} - C_{2k}r^{-\varepsilon k-1})\cos(2k+1)\Theta \,;$$

$$\frac{\tau_{\Theta z}^{(2)}}{G_3} = -\delta_3 \sum_{k=0}^{\infty}(2k+1)(C_{1k}r^{\varepsilon k-1} + C_{2k}r^{-\varepsilon k-1})\sin(2k+1)\Theta \,, \tag{4.38}$$

where $\delta_2 = \dfrac{G_{rz}}{G_3}$; $\delta_3 = \dfrac{G_{\Theta z}}{G_3}$.

For the isotropic binder-matrix the relations (4.24) and (4.25) are also true with displacement W_3:

$$W_3 = 2\,\mathrm{Re}\{\varphi_3(z)\} \,, \tag{4.39}$$

where $\varphi_3(z)$ is of the form [233]:

$$\varphi_3(z) = (\alpha_0 + i\beta_0)z + \sum_{k=0}^{\infty}(\alpha_{2k+2} + i\beta_{2k+2})$$

$$\times \left[-\frac{z^{-2k-1}}{2k+1} + \sum_{j=0}^{\infty} r_{j,k}\frac{z^{2j+1}}{2j+1} \right] + \Re - i\Re' \,. \tag{4.40}$$

Here $\Re$ and $\Re'$ characterize the external loading.

Upon corresponding substitution of indices in (4.28), we obtain stress values in zone S_3:

$$\frac{\tau_{xz}^{(3)}}{G_3} = 2(\alpha_0 + \Re) + 2\sum_{k=0}^{\infty}[\alpha_{2k+2}\,\mathrm{Re}\{f(z)\} - \beta_{2k+2}\,\mathrm{Im}\{f(z)\}] \,;$$

$$\frac{\tau_{yz}^{(3)}}{G_3} = -2(\beta_0 - \Re') - 2\sum_{k=0}^{\infty}[\alpha_{2k+2}\,\mathrm{Im}\{f(z)\} + \beta_{2k+2}\,\mathrm{Re}\{f(z)\}] \,, \tag{4.41}$$

where $f(z) = \left(z^{-2k-2} + \sum_{j=0}^{\infty} r_{j,k}z^{2j}\right)\cdot R^{2k+2}$, and parameters $r_{j,k}$ (see Chap. 3).

The value of

$$T_3 = \tau_{xz}^{(3)}\cos\alpha_1 + \tau_{yz}^{(3)}\sin\alpha_1 \tag{4.42}$$

is expressed by [222]:

$$\frac{T_3}{G_3} = 2\,\mathrm{Re}\left\{\varphi_3'(z)e^{i\alpha_1}\right\} , \tag{4.43}$$

where α_1 is as earlier the angle between the x-axis and the normal to zone S contour.

Using (4.41) and (4.42) we obtain

$$\frac{T_3}{G_3} = 2\alpha_0 \cos\alpha_1 + 2\Re \cos\alpha_1 - 2\beta_0 \sin\alpha_1 + \Re \sin\alpha_1$$

$$+2\sum_{k=0}^{\infty}\Big[\alpha_{2k+2}(\mathrm{Re}\{f(z)\}\cos\alpha_1 - \mathrm{Im}\{f(z)\}\sin\alpha_1)$$

$$-\beta_{2k+2}(\mathrm{Im}\{f(z)\}\cos\alpha_1 + \mathrm{Re}\{f(z)\}\sin\alpha_1)\Big] . \tag{4.44}$$

Displacements (4.39) can be written proceeding from (4.40) as follows

$$W_3(x,y) = 2(\alpha_0 + \Re)x - (\beta_0 - \Re')y$$

$$+2\sum_{k=0}^{\infty}\left[\alpha_{2k+2}\,\mathrm{Re}\{F(z)\} - \beta_{2k+2}\,\mathrm{Im}\{f(z)\}\right] , \tag{4.45}$$

where $F(z) = \left(-\dfrac{z^{-2k-1}}{2k+1} + \displaystyle\sum_{j=0}^{\infty} r_{j,k}\dfrac{z^{2j+1}}{2j+1}\right)R^{2k+2}.$

4.4 Boundary Conditions and Determination of Elastic Constants at a Longitudinal Shear

Boundary conditions for the contours correspond to the equalities of tangential stresses

$$T_{j-1} = T_j; \quad (j = 2,3) \tag{4.46}$$

and displacements

$$W_{j-1} = W_j; \quad (j = 2,3) , \tag{4.47}$$

where, analogous to (4.43):

$$\frac{T_1}{G_3} = 2\delta_1 \sum_{k=0}^{\infty}\Big[a_k(\mathrm{Re}\{z^{2k}\}\cos\alpha_1 - \mathrm{Im}\{z^{2k}\}\sin\alpha_1)$$

$$-b_k(\mathrm{Im}\{z^{2k}\}\cos\alpha_1 + \mathrm{Re}\{z^{2k}\}\sin\alpha_1)\Big](2k+1);$$

$$\frac{T_2}{G_3} = \tau_{rz}\cos\beta_1 + \tau_{\Theta z}\sin\beta_1 . \tag{4.48}$$

β_1 – an angle between polar radius r and the normal to the contour of zone S_3.

Based on (4.38) we have

$$\frac{T_2}{G_3} = \sum_{k=0}^{\infty} \Big\{ C_{1k} r^{\varepsilon_k - 1} \big[\delta_2 \varepsilon_k \cos(2k+1)\Theta \cos\beta_1$$
$$- (2k+1)\delta_3 \sin(2k+1)\Theta \sin\beta_1 \big]$$
$$- C_{2k} r^{-\varepsilon k - 1} \big[\delta_2 \varepsilon_k \cos(2k+1)\Theta \cos\beta_1$$
$$+ (2k+1)\delta_3 \sin(2k+1)\Theta \sin\beta_1 \big] \Big\} . \tag{4.49}$$

Stresses T_3 are determined by (4.44). The displacements included in condition (4.47) are defined by (4.30), (4.37) and (4.45). The collocation method reduces conditions (4.46) and (4.47) to a system of linear algebraic equations with respect to factors

$$a_k, b_k \ (k = 1, \ldots, N_1), \quad C_{1k}, C_{2k} \ (k = 0, \ldots, N_2),$$
$$\alpha_{2k}, \beta_{2k} \ (k = 0, \ldots, N_3) .$$

The symmetry of the problem under study means that the mean longitudinal shear moduli $\langle G_{23} \rangle$ and $\langle G_{13} \rangle$ are equal. Thus far, it is sufficient to find the relation between mean shear angles $\langle \gamma_{xz} \rangle$ and mean tangential stresses $\langle \tau_{xz} \rangle$. The stresses are calculated as

$$\langle \tau_{xz} \rangle = \frac{1}{\Omega} \iint_s \tau_{xz} ds . \tag{4.50}$$

Equation (4.50) can be presented as:

$$\langle \tau_{xz} \rangle = \frac{1}{\Omega} \left\{ \iint_{S_1} \tau_{xz}^{(1)} dx dy + \iint_{S_2} \tau_{xz}^{(2)} dx dy + \iint_{S_3} \tau_{xz}^{(3)} dx dy \right\} , \tag{4.51}$$

where $\tau_{xz}^{(1)}$ and $\tau_{xz}^{(3)}$ are determined by (4.29) and (4.41) and parameter $\tau_{xz}^{(2)}$ is found from

$$\tau_{xz}^{(2)} = \tau_{r\Theta}^{(2)} \cos\Theta + \tau_{\Theta r}^{(2)} \sin\Theta .$$

Thus, $\langle \tau_{xz} \rangle$ will be sought through the factors determined by the collocation procedure

$$a_k, b_k \ (k = 1, \ldots, N_1), \quad C_{1k}, C_{2k} \ (k = 0, \ldots, N_2),$$
$$\alpha_{2k}, \beta_{2k} \ (k = 0, \ldots, N_3) .$$

Let us express the mean shear angle $\langle \gamma_{xz} \rangle$ through factors α_k and β_k. For this purpose, an increment in displacements W in congruent points z and $z + \omega j$ should be calculated [233]:

$$W(z + \omega j) - W(z) = \omega_j(\alpha_0 + i\beta_0) + \overline{\omega}_j(\alpha_0 + i\beta_0)$$
$$- x_j(\alpha_2 + i\beta_2) - \overline{x}_j(\alpha_2 + i\beta_2) , \tag{4.52}$$

where $x_1 = \dfrac{\pi}{\sqrt{3}}; \ x_2 = \dfrac{\pi(1 - i\sqrt{3})}{2\sqrt{3}} .$

At $j = 1$ from above formula we obtain:

$$\langle \gamma_{xz} \rangle = 2\alpha_0 - \frac{\pi}{\sqrt{3}}\alpha_2 \,. \qquad (4.53)$$

Based on the known strain to stress dependence

$$\langle \gamma_{xz} \rangle = \frac{\langle \tau_{xz} \rangle}{\langle G_{xz} \rangle}$$

and using relations (4.51) and (4.53), we obtain the reduced shear modulus $\langle G_{xz} \rangle$.

4.5 Determination of Elastic Constants of the Composite Based on Spirally Reinforced Filler Under Loading in the Direction of the Main Reinforcement

In accordance with the above-described design procedure of the elastic constants for the composite with spirally reinforced filler, the elastic modulus in the direction of the main reinforcement, longitudinal shear modulus and Poisson's ratio have been calculated.

In Figs. 4.1–4.3, variations of the elastic characteristics of the material are shown as a function of its filling with a reinforcement arranged lengthwise only φ_0. The choice of φ_0 enables the comparison of the material properties achieved to those of unidirectional, single-component and hybrid composites based on the same reinforcing fillers. Introduction of a spiral reinforcement

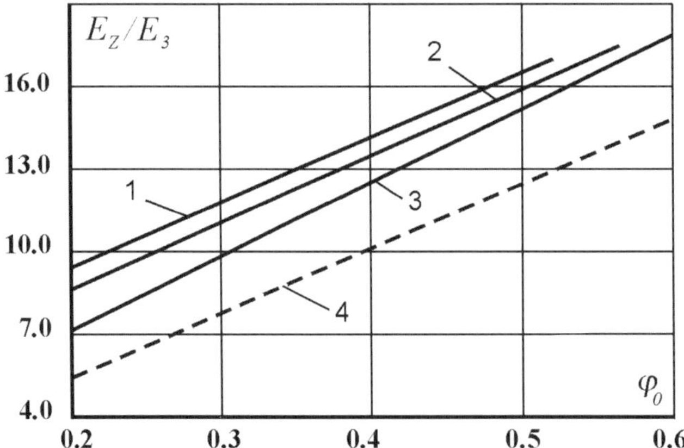

Fig. 4.1. Longitudinal elastic modulus of glass-organic-reinforced plastics versus filling degree with round elements. 1 – 15–17%. $\delta/R = 0.06$; $2 - \delta/R = 0.1$; $3 - \delta/R = 0.3$; 4 – unidirectional plastic [222]

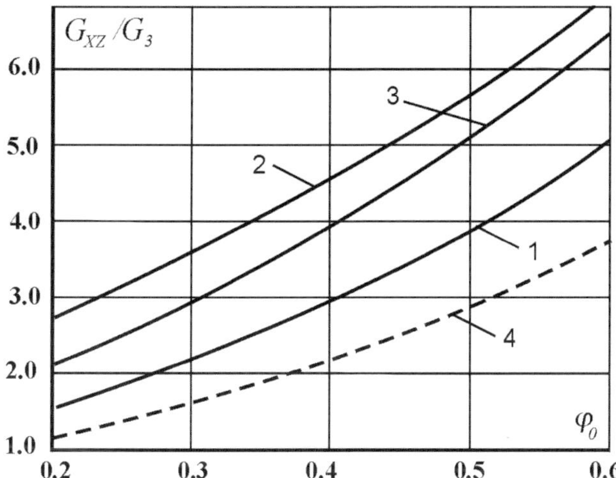

Fig. 4.2. The longitudinal shear modulus as a function of filling degree (notations as in Fig. 4.1)

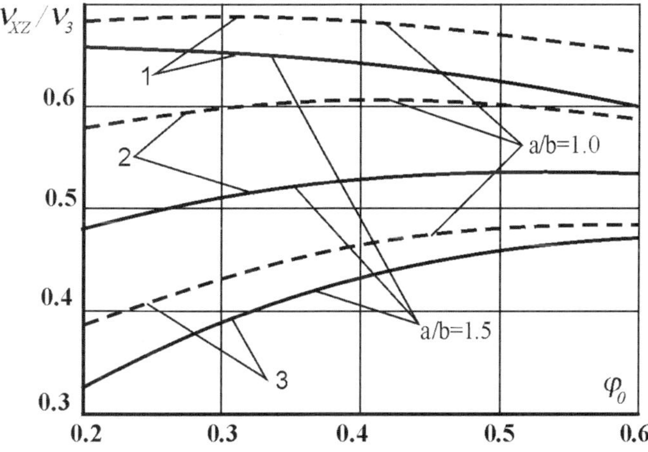

Fig. 4.3. Poisson's ratio versus filling degree (notations correspond to Fig. 4.1)

makes it possible to raise the longitudinal elastic modulus as compared to the unidirectional one by 15–17%.

As seen in Fig. 4.1, with increasing interlayer thickness the elastic modulus rises somewhat (e.g. curves 1 and 3 with $\delta/R = 0.06$ and 0.2, correspondingly). The longitudinal shear modulus shows a stronger dependence on the intermediate layer parameters. As seen from Fig. 4.2, increasing the thickness of the intermediate layer from $\delta/R = 0.06$ (curve 1) to $\delta/R = 0.2$ (curve 3) increases the longitudinal shear modulus by 30–60%, and the parameter increases further compared to unidirectional material (curve 4).

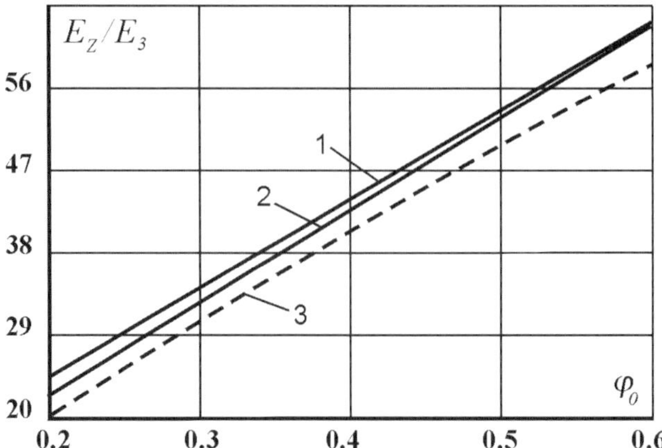

Fig. 4.4. Variation of longitudinal elastic modulus in coal-glass-reinforced plastics depending on filling degree with an auxiliary reinforcement and round form of the element. *1* – $\varphi_3 = 0.7$; *2* – $\varphi_3 = 0.4$; *3* – unidirectional coal-plastic

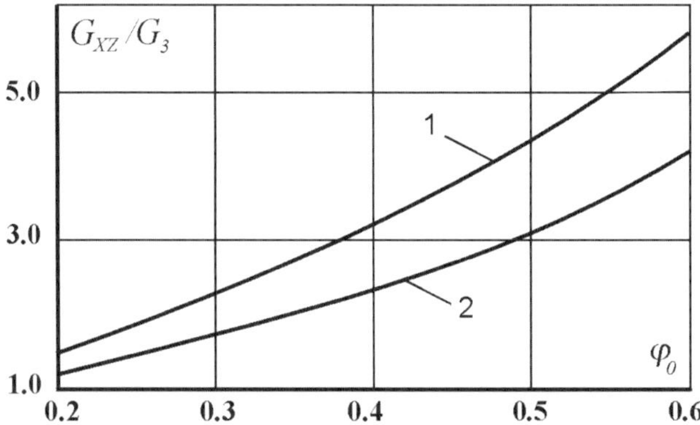

Fig. 4.5. Dependence of longitudinal shear modulus on filling degree for various filling degrees of the layer. *1* – $\varphi_3 = 0.4$; *2* – $\varphi_3 = 0.7$

The behavior of Poisson's ratio is illustrated in Fig. 4.3. In contrast to shear modulus and longitudinal elastic modulus, Poisson's ratio is also dependent upon the shape of the spirally reinforced element.

Variation of the interlayer filling degree φ_3 does not exert any perceptible effect on the longitudinal elastic modulus (Fig. 4.4).

At the same time, both longitudinal elastic modulus and Poisson's ratio (Fig. 4.6) are to some extent dependent upon the filling degree of the layer, i.e. upon its elastic characteristics.

It follows from Fig. 4.6 that increased filling of the layer to 0.7 raises the longitudinal shear modulus of the coal-plastic by 10–40%.

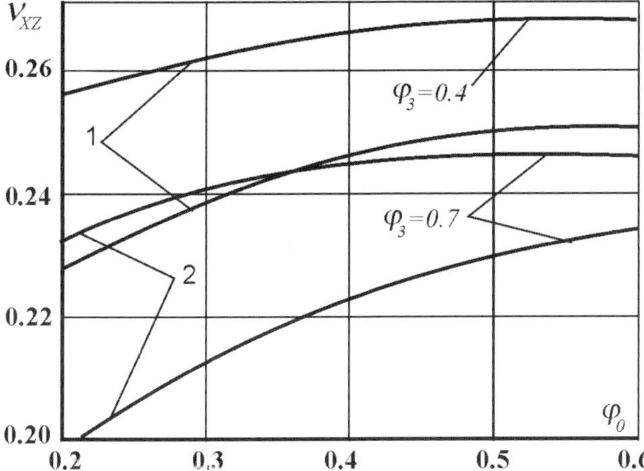

Fig. 4.6. Variation of Poisson's ratio of carbon-glass-reinforced plastics depending on filling degree for various winding reinforcements. *1* – glass fibers; *2* – organic fibers

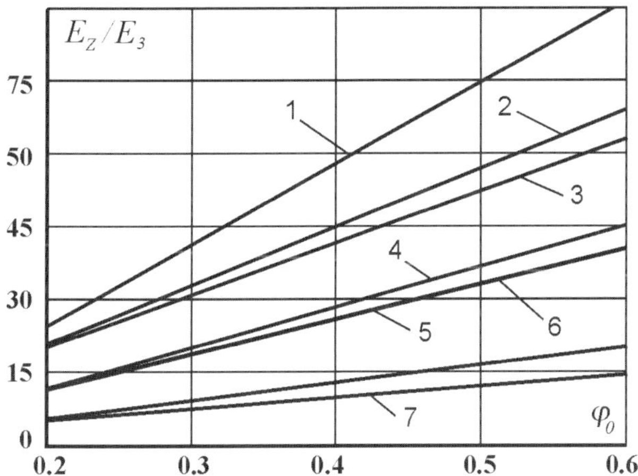

Fig. 4.7. Longitudinal elastic modulus of composites with various combinations of filler ingredients. *1* – boron-glass-reinforced plastic; *2* – carbon-glass-reinforced plastic; *3* – carbon-organo plastic; *4* – organic-organoplastic; *5* – organic- glass-reinforced plastic; *6* – glass-glass-fiber reinforced plastic; *7* – glass-organic-reinforced plastic

Analogous changes also occur with Poisson's ratio. In this case the type of the winding reinforcement plays a certain role (e.g. compare curves 1 and 2 in Fig. 4.6).

Figures 4.7–4.9 illustrate the effect of the kind of ingredients on the composite characteristics in the direction of the main reinforcement. Their vari-

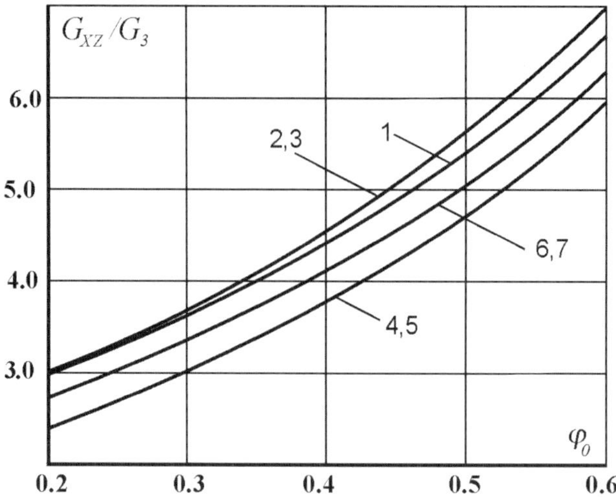

Fig. 4.8. Longitudinal shear modulus of various composites (notations are similar to Fig. 4.7)

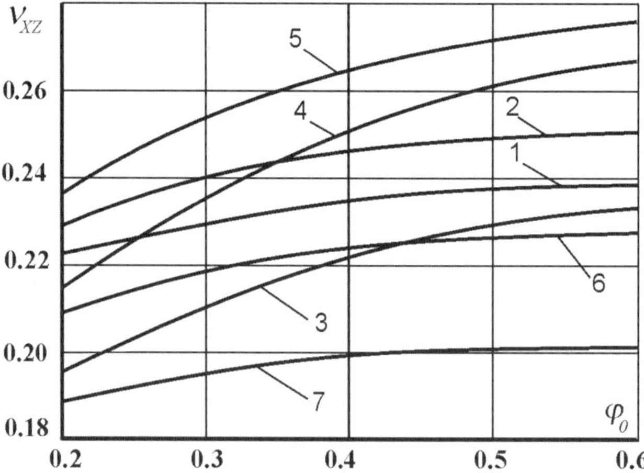

Fig. 4.9. Variation of Poisson's ratio depending on filling degree for materials with different combinations of components (notations are similar to Fig. 4.7)

ational behavior in response to filling growth is analogous to the same characteristics of unidirectional materials. The comparison of obtained data with those for unidirectional materials has proved that [20, 45, 233] in all cases the elastic modulus in the direction of the main reinforcement is not below that of the unidirectional material. On the contrary, the longitudinal shear modulus of unidirectional materials is below that of the composites based on spirally reinforced fillers.

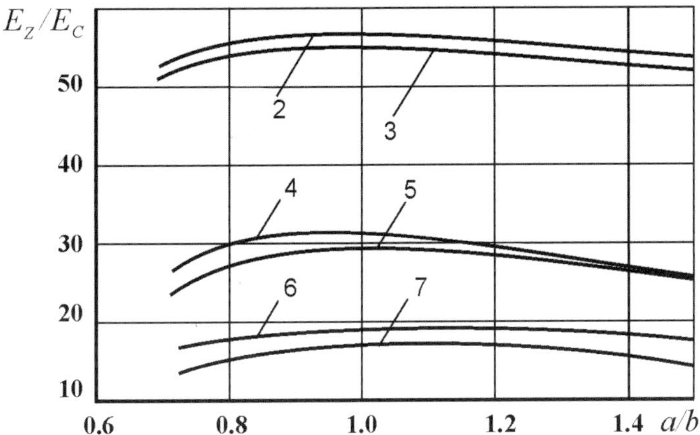

Fig. 4.10. Longitudinal elastic modulus versus semiaxes relation of the elliptical element (filling degree of the element core is 0.8, and of the layer 0.7). (notations are similar to Fig. 4.7)

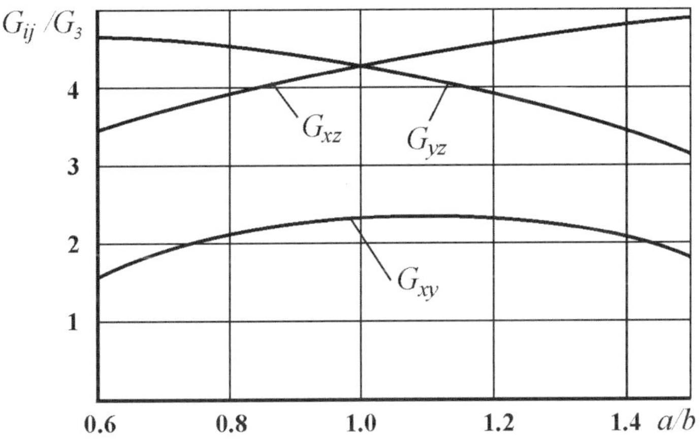

Fig. 4.11. Variations of longitudinal and transverse shear moduli of coal plastics as a function of the characteristic dimensions of the element ($\delta/R = 0.1$; $\varphi = 0.55$)

It follows from Chap. 2, that the shape of the element can be approximated by an ellipse whose semiaxes ratio is $a/b = 0.66$–1.5.

In Fig. 4.10, a dependence is shown of the longitudinal elastic modulus versus relation a/b for various combinations of ingredients and a constant filling degree of the interlayer. It turns out from above that the longitudinal elastic modulus is in fact independent of relation a/b. In contrast, the longitudinal shear modulus is conditioned by both correlation of the characteristic dimensions of the element and the orientation of relative loads. Fig. 4.11 presents data on variations of shear modulus within the XOZ and

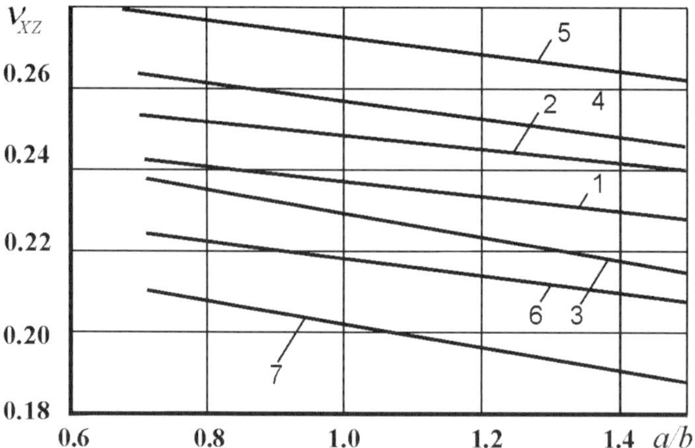

Fig. 4.12. Poisson's ratio versus the semiaxes relation of the elliptical element (notations are similar to Fig. 4.7)

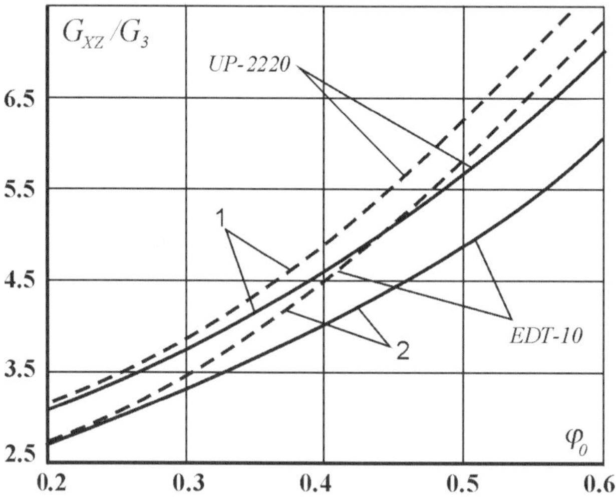

Fig. 4.13. Longitudinal shear modulus versus filling degree and binder type. *1 –* coal-glass plastic; *2 –* organic-glass-reinforced plastic

XOY planes dependent on relation a/b. As the relation increases, G_{XZ} increases and G_{YZ} diminishes. At $a/b = 1$ the moduli become equal. The effect of a/b on the transverse shear modulus G_{XY} calculated by the earlier described formulas is also shown in the figure. In this case, the shear modulus remains practically invariable. Moreover, it has been determined that with increasing a/b relation the value of Poisson's ratio diminishes (Fig. 4.12).

Insofar as the use of matrices with higher-grade characteristics in common composites raises material properties under shear, it is of interest to clarify

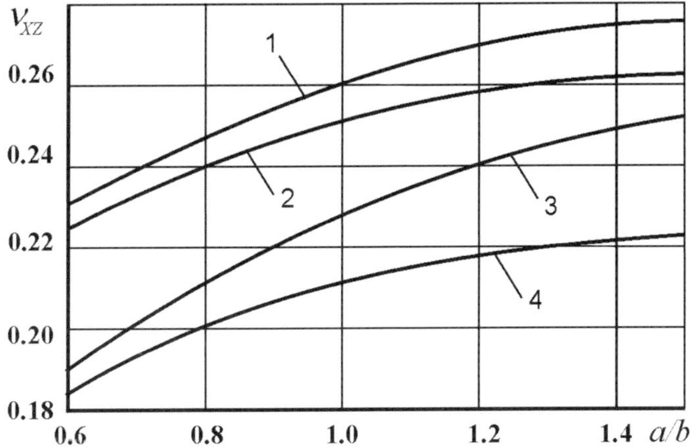

Fig. 4.14. Poisson's ration variation depending on filling degree and binder type. *1, 3* – organic-glass-reinforced plastic; *2, 4* – coal-glass-reinforced plastic; *1, 2* – EDT-10 binder; *3, 4* – UP-2220

the effect of the type of binder employed on composite properties with spirally reinforced filler. For this purpose UP-2220 and EDT-10 (epoxy filler) types of binders, distinguished by their perfect physico-mechanical parameters, were used. The effect of the binder type on the shear modulus of the coal-glass plastic and organic-glass-reinforced plastic is illustrated in Fig. 4.13.

As expected, the use of improved binder leads to some increase in shear modulus, which is, however, less expressed than in common composite materials. Corresponding changes in Poisson's ratio are shown in Fig. 4.14.

4.6 Optimization of Elastic Properties and Geometrical Parameters of the Composite Based on Spirally Reinforced Filler

The principal part of the above-described microstructural investigations is devoted to examination of stress fields in the composite with given geometrical and elastic properties of material components. However, the great variety of available original materials, their structural, geometrical and physical parameters evokes a question on optimum designing of composites with spiral reinforced filler. Today, a number of investigations are dealing with the problem of optimum designing of materials, and their structures [130, 225, 226].

It is common knowledge that failure of a composite material occurs chiefly due to an exhausted bearing ability of the material or matrix nucleus outside a spirally reinforced element. Therefore, the goal function is to make provision for equivalent stresses both in the element core and in the matrix.

Therefore, the goal function should take account of equivalent stresses just as in the element core, as well as in the matrix. Hence, the goal function $G(x)$ can be presented as

$$G(\overline{x}) = [g_1(\overline{x}) - kg_2(\overline{x})]^2 \,, \tag{4.54}$$

where $g_j(\overline{x})$ is the maximum values of the relations.

$$g_j(\overline{x}) = \frac{\sigma_{xj}^2}{[R_j^+]^2} + \frac{\sigma_{yj}^2}{[R_j^-]^2} + \frac{\tau_{xyj}^2}{[T_j]^2} - 2v_j\frac{\sigma_{xj}\sigma_{yj}}{[R_j^+]^2} \,. \tag{4.55}$$

Thus, $[R_j^+], [R_j^-], [T_j]$ are related to strength stresses at infinity for, correspondingly, tension, compression and shear; v_j are Poisson's ratios.

Values with index $j = 1$ belong to the element core material, and $j = 2$ to the matrix.

Dependencies (4.55) correspond to the strength criterion for materials that respond differently under tension and compression [20]. The weighting factor k in (4.54) depends on the contribution of the binder-matrix failure to the total breakage of the composite, where $\overline{x}$ is the vector of the control parameters. Since composite failure proceeds mainly on attaining matrix strength, it is further taken as $k \geq 1$.

The representative element in the proposed optimum composite model (Fig. 4.15) consists of an element core, layer and matrix that occupy, correspondingly, zones S_1, S_2, S_3. The layer of zone S_2 is bounded by contours L_1 and L_2, the distance between which $\delta = $ const is calculated as a segment of polar radius $r(\delta = AB)$. Let γ be a locus bisecting segment AB and written in a parametric form using function $\omega(\sigma)$ as follows

$$x + iy = \omega(\sigma) = R \sum_{j=N_1}^{N_2} \alpha_j\sigma^{-j} \,. \tag{4.56}$$

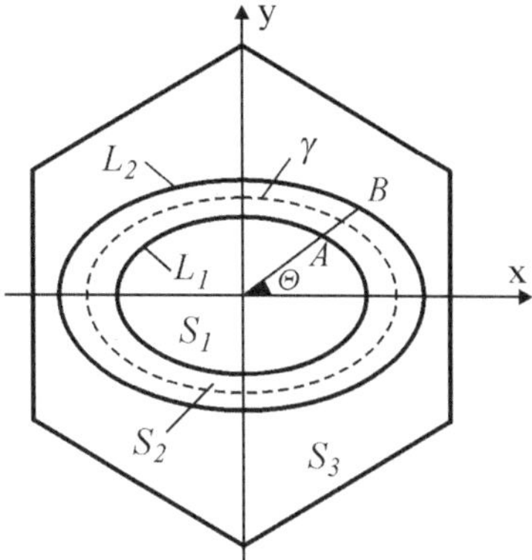

Fig. 4.15. An element of the composite model

When α_j and δ are given, then contours L_1 and L_2 are fully defined. R in the latter relation expresses some parameters that depend upon the filling degree of the composite.

Taking account of the above, for control parameters that are elements of vector $\bar{x}$ (4.54) one may take the following

$$x_1 = \alpha_{-N_1};\ x_2 = \alpha_{1-N_1};\ \ldots x_{N_1+1} = \alpha_0;\ x_{N_1+2} = \alpha_1;$$

$$x_{N_1+N_2+1} = \alpha_{N_2};\ x_{N_1+N_2+2} = \varphi_1;\ x_{N_1+N_2+3} = \varphi_3\,, \tag{4.57}$$

where φ_1 – filling degree of the element core material; φ_3 – filling degree of the layer material.

By giving φ_3 or φ_1 it is possible to determine all elastic characteristics of certain materials of the main and auxiliary reinforcement based on dependencies [20, 132, 135].

For the element core

$$G_{xy} = G_c \frac{\kappa_c + \varphi_1 + (1-\varphi_1)G_c/G_b}{(1-\varphi_1)\kappa_c + (1+\varphi_1\kappa_c)G_c/G_b}\,;$$

$$E_z = \varphi_1 E_g + (1-\varphi_1)E_c + \frac{8}{t}(v_b - v_c)^2 \varphi_1(1-\varphi_1)G_c\,;$$

$$\frac{1}{E_x} = \frac{v_{xz}^2}{E_z} + \frac{1}{8G_c t}\left[2(1-\varphi_1)(\kappa_c-1)+(\kappa_g-1)(\kappa_c+2\varphi_1-1)\frac{G_c}{G_b}+\frac{2tG_c}{G_{xy}}\right]\,;$$

$$v_{xz} = v_c - \frac{1}{t}\left[\varphi_1(\kappa_c + 1)(v_b - v_c)\right]\,;$$

$$\frac{v_{xy}}{E_x} = -\frac{v_{xz}^2}{E_z} + \frac{1}{8G_c t}\left[\frac{2tG_c}{G_{xy}} - (\kappa_g - 1)(\kappa_c + 2\varphi_1 - 1)\right.$$

$$\left.\times \frac{G_c}{G_b} - 2(1-\varphi_1)(\kappa_c - 1)\right]. \tag{4.58}$$

For the layer

$$E_\theta = \varphi_3 E_b^1 + (1-\varphi_3)E_c + \frac{8}{t'}(v_b - v_c)^2 \varphi_3(1-\varphi_3)G_c\,;$$

$$G_{z\theta} = G_c \frac{\kappa_c + \varphi_3 + (1-\varphi_3)G_c/G_b'}{(1-\varphi_3)\kappa_c + (1+\varphi_3\kappa_c)G_c/G_b'}\,;$$

$$\frac{1}{E_r} = \frac{v_{r\theta}^2}{E_\theta} + \frac{1}{8G_c t'}\left[2(1-\varphi_3)(\kappa_c-1)+(\kappa_b'-1)(\kappa_c+2\varphi_3-1)\frac{G_c}{G_b'}+\frac{2t'G_c}{G_{z\theta}}\right]\,;$$

$$G_{\theta r} = G_c \frac{1 + \varphi_3 + (1+\varphi_3)G_c/G_b'}{(1-\varphi_3) + (1+\varphi_3)G_c/G_b'}\,;$$

$$v_{z\theta} = v_c - \frac{1}{t'}\left[\varphi_3(\kappa_c + 1)(v_b' - v_c)\right]\,, \tag{4.59}$$

where E_b, G_b, v_b are elastic constants of the main reinforcement fibers used to fill the element core material; E_b', G_b', v_b' – the same for the layer material; E_c, G_c, v_c are elastic constants of the binder;

$$\kappa = 3 - 4v\,;$$

$$t = 2 + \varphi_1(\kappa_c - 1) + (1 - \varphi_1)(\kappa_b - 1)G_c/G_b\,;$$

$$t' = 2 + \varphi_3(\kappa_c - 1) + (1 - \varphi_3)(\kappa_b - 1)G_c/G'_b\,.$$

Thus, by giving the original materials for constituents of the spirally reinforced material, it is easy to calculate all elastic constants for any filling value of either the layer φ_3 or the element core φ_1 using dependencies (4.58) and (4.59). Note that the total filling degree of a material by the spirally reinforced materials remains invariable.

Evidently, certain limitations should be imposed on parameters $x_{N_1+N_2+2}$ and $x_{N_1+N_2+3}$

$$\varphi_1^0 \le x_{N_1+N_2+2} \le \varphi_1^m\,;$$

$$\varphi_3^0 \le x_{N_1+N_2+3} \le \varphi_3^m\,, \tag{4.60}$$

where $\varphi_1^0, \varphi_3^0, \varphi_1^m, \varphi_3^m$ are, respectively, the least and most filling values of the element core and layer materials.

It is possible to avoid consideration of the limitations (4.60) by the introduction of new parameters x'_k ($k = N_1 + N_2 = 2;\ N_1 + N_2 + 3$) related to the previous ones as follows [227]

$$x'_{N_1+N_2+2} = \frac{\varphi_1^0 + \varphi_1^m}{2} + \frac{\varphi_1^m - \varphi_1^0}{2}\sin x_{N_1+N_2+2}\,;$$

$$x'_{N_1+N_2+3} = \frac{\varphi_3^0 + \varphi_3^m}{2} + \frac{\varphi_3^m - \varphi_3^0}{2}\sin x_{N_1+N_2+3}\,. \tag{4.61}$$

Variation of control parameters (4.57) or (4.61), i.e. characteristics of the material structural components, may change not only the transversal properties of the composite microstructure but also the longitudinal elastic modulus and shear modulus.

In this connection, the following limitations are imposed on parameters (4.57) or (4.61)

$$\frac{E^*}{E_0} \ge 1;\quad \frac{G^*}{G_0} \ge 1\,, \tag{4.62}$$

where E^* and G^* are reduced elastic moduli of the material in the direction of the main reinforcement packing; E_0, G_0 are the same elastic characteristics for a unidirectional material with an optimum filling degree for a given type of fibrous filler. Thus, limitations (4.62) are reduced to the provision for the elastic properties of the material to be optimized in the direction of the main reinforcement on a par with the analogous characteristics of a unidirectional composite.

One can employ Fisher's strength criterion for a layer with orthotropic parameters to account for the difference between elastic and strength characteristics of a material in the direction of the symmetry axes of the mechanical properties. The method is considered to be a special case of the generalized criterion of Goldenblat – Kopnov [130]:

$$\frac{\sigma_r^2}{[R_r^+]^2} + \frac{\sigma_\theta^2}{[R_\theta^+]^2} + \frac{\tau_{r\theta}^2}{[T]^2} - k_1 \frac{\sigma_r \sigma_\theta}{[R_r^+][R_\theta^+]} \leq 1 \,, \tag{4.63}$$

where $[R_r^+], [R_\theta^+]$ are related to the load at strength infinity at the layer material tension in directions r and θ

$$k_1 = \frac{E_r(1 + v_{\theta r}) + E_\theta(1 + v_{r\theta})}{2[E_r E_\theta (1 + v_{\theta r})(1 + v_{r\theta})]^{1/2}} \,.$$

Relation (4.63) is one more limitation imposed on control parameters (4.57) and (4.61).

Thus, the search for the optimum parameters of composite elements is reduced to a general problem of mathematical programming, i.e. there is a need to find the values $(N_1 + N_2 + 3)$ of parameters x_k, making allowance for the least goal function $G(\overline{x})$ when limitations (4.62) and (4.63) are met. To determine values of the goal function (4.54), direct problem solution methods are used (see Chap. 3). Hence, the formulated problem is related to the problems of nonlinear programming.

Since the relation between control parameters (4.57), (4.61) and function (4.54) is rather complex, investigation of the goal function with respect to convexity, single-extremity and so on seems improbable. Therefore, to solve the problem a variant of the random search algorithm with tuition is used [228]. The algorithm includes two stages on each of which parameters x_k^* of vector $\overline{x}$ are determined at the ith step by the formula

$$x_k^* = x_k + \beta_k^{(j)} \xi_k^{(i)} \,, \tag{4.64}$$

where $\beta_k^{(j)}$ – scaling multipliers related to the stage number $j = 1, 2$; $\xi_k^{(i)}$ – random numbers distributed evenly within the interval $[-0, 5 \ldots + 0, 5]$.

Scaling multipliers allow for the extent of the effect of corresponding parameters x_k on variations of the goal function (4.54). It is apparent that parameters $\beta_{k1}^{(j)}$ and $\beta_{k2}^{(j)}$ at large k_1 and k_2 ($k_1 \leq N_1$; $k_2 \leq N_1 + N_2 + 1$) will be small.

During the first stage, $\beta_k^{(1)}$ and $x_k = x_k^{(0)}$ are considered as given and constant, i.e. the search is in the vicinity of reference point $\overline{x}^{(0)}$ and j_1 tests are made. Furthermore, the least of the j_1 values of the goal function and the corresponding x_k parameters are remembered, being taken as references at the second stage.

During the second stage ($j = 2$) coefficients $\beta_k^{(j)}$ are halved, i.e., the volume of the hyper parallelepiped where the search is performed, decreases $2^{N_1+N_2+3}$ times. In the $\overline{x}^{(0)}$ vicinity j_2 tests are fulfilled and value W is calculated, allowing the probability of successful advancement to an optimum point

$$W = \frac{j^*}{j_2} \,,$$

where j^* is the number of points in which the goal function value is less than at the initial stage. As was proved in [228], at $W = 0.2$, values of $\beta_k^{(2)}$ were chosen correctly. In the case where $W < 0.2$ or $W > 0.2$, to exercise an optimum control over dimensions of the search area, the following is proposed

$$\beta_k^{(2)} = \begin{cases} \dfrac{5}{6}\beta_k^{(2)}, & \to W < 0.2; \\[2mm] \beta_k^{(2)}, & \to W = 0.2; \\[2mm] \dfrac{6}{5}\beta_k^{(2)}, & \to W > 0.2. \end{cases}$$

Upon correction of parameters $\beta_k^{(2)}$, the second stage is repeated. The algorithm makes provision for suspension of computation when the repetition of the second stage is j_3 times, there will be $W = 0$.

For a practical solution of the problem of optimizing the goal function (4.54) it is assumed that parameters α_j in (4.56) are real numbers and indices j in (4.56) take only odd values. This agrees with the material model that is symmetrical about the X and Y axes (see Chap. 3). It is assumed that $N_1 = 1\,(\alpha_{-1} = 1)$, $N_2 \leq 5$, for which the number of varying parameters of vector $\underline{x}$ does not exceed 5.

Since the degree of the material filling is considered to be given and invariable, it is necessary to determine in a respective manner parameter R in Eq. (4.56). Presuming for simplification the hexagonal arrangement of the spirally reinforced elements, let us determine the area of the whole representative element (Fig. 4.15),

$$\Omega = 2\sqrt{3}\,, \tag{4.65}$$

then areas Ω_1 (roughly) and Ω_2 of the regions are found in turn between contours γ and L_2 and inside contour γ [130]:

$$\Omega_1 = R\frac{\delta}{2} \int_0^{2\pi} |\omega'(\theta)|\,d\theta\,; \tag{4.66}$$

$$\Omega_2 = \frac{1}{2} \int_0^{2\pi} \big[\, \mathrm{Re}\{\omega(\theta)\}I_m\{\omega'(\theta)\} - \mathrm{Re}\{\omega'(\theta)\}I_m\{\omega(\theta)\}\,\big]\,d\theta$$

$$= \pi R^2 \left(1 - \sum_{j=1}^{N_2} j\alpha_j \right)\,. \tag{4.67}$$

Making allowance for factors α_j $(j \geq 1)$, further simplification of the integral (4.66) is introduced

$$\Omega_1 \approx R\pi\delta\,. \tag{4.68}$$

Bearing in mind that the filling degree of the spirally reinforced element is φ_2, then from equality

$$\varphi_2 \Omega = \Omega_1 + \Omega_2 \tag{4.69}$$

we define R through α_j as a root of equation

$$\pi R^2 \left(1 - \sum_{j=1}^{N_2} j\alpha_j\right) + R\delta\pi = 2\sqrt{3}\varphi_2\,, \tag{4.70}$$

This means that

$$R = \frac{-\delta\pi + \sqrt{\delta^2\pi^2 + 8\sqrt{3}\left(1 - \sum_{j=1}^{N_2} j\alpha_j\right)\varphi_2}}{2\pi\left(1 - \sum_{j=1}^{N_2} j\alpha_j\right)}\,. \tag{4.71}$$

If one presumes that $\delta = \beta R \ (\beta \le 0,1)$, then from (4.71) we have:

$$R = \left[\frac{2\sqrt{3}\varphi_2}{\pi\left(1 - \sum_{j=1}^{N_2} j\alpha_j - \beta\right)}\right]^{1/2}\,. \tag{4.72}$$

In accordance with the above-described procedure, three types of transverse loading at infinity were calculated, namely 1 – one-sided tension, 2 – biaxial tension, 3 – transversal shear. All the necessary strength characteristics and elastic moduli were chosen from the known data. Filling degree was $\varphi_2 = 0.8$, layer width ratio to the ellipse larger semi-axis was 0.1, elastic characteristics of the equivalent unidirectional material E_0 and G_0 were determined, with allowance for the optimum filling degree, using dependencies cited in [233] and taking account of the type of the reinforcement. The largest and the least values of the characteristics of Eq. (4.60) were calculated on the basis of the processing capabilities of producing the element core and layer materials.

The conducted calculations for various types of material components have confirmed that the γ contours given by the function (4.56) are close to elliptical ($N_2 = 3$) and slightly dependent upon the types of main and auxiliary reinforcement used. They are principally conditioned by the character of the applied loading. As for instance, for coal-glass-reinforced plastic based on EDT-10 binder, the following type of γ contour has been derived:

– under a uniaxial tension (Fig. 4.16a)

$$\omega(\sigma) = R\left(\sigma^2 + \frac{0.24}{\sigma} + \frac{0.07}{\sigma^3}\right)$$

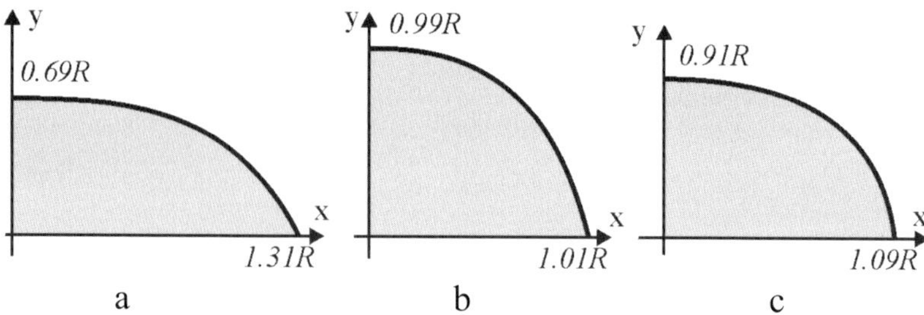

Fig. 4.16. Optimum form of spirally reinforced elements for various conditions of transversal loading of the composite

– under biaxial tension (Fig. 4.16b)

$$\omega(\sigma) = R\left(\sigma + \frac{0.01}{\sigma^3}\right);$$

– under shear (Fig. 4.16c)

$$\omega(\sigma) = R\left(\sigma - \frac{0.08}{\sigma} + \frac{0.01}{\sigma^3}\right).$$

Values of the optimum design variables of the named materials are presented for fewer than three types of loading in Table 4.1.

Table 4.1. Optimum Rated Parameters of the Composite Structure

Parameters	Notation	Type of loading		
		Uniaxial tension	Biaxial tension	Shear
Filling degree of element core by the main reinforcement	φ_1	0.71	0.67	0.65
Filling degree by auxiliary reinforcement	φ_2	0.64	0.58	0.73
Form parameters of the element	α_1	0.24	0	−0.08
Form parameters of the element	α_2	0.07	0.01	0.01

4.7 Conclusions

1. The developed design procedure has made a provision for defining the elastic characteristics of the composite based on spirally reinforced filler under loading along the chief reinforcement length and longitudinal shear, with allowance for the anisotropy of the components used.
2. The completeded research has shown an essential dependence of the longitudinal elastic modulus and shear modulus of the material on the characteristics of the intermediate layer. This ensured improvement of specified characteristics by 30–60% by means of appropriate selection of geometrical parameters and the properties of the layer, in contrast to the optimum reinforced unidirectional composite.
3. Solution of the problem of optimizing the structure of the composite based on hybrid spatially-reinforced fillers enables selection of an optimum form of the element and the optimum filling degree of each component. This serves as a basis for choosing the technological parameters of manufacture and for achieving a material with the most favorable combination of elastic and strength characteristics.
   Analysis of the results of the undertaken investigations has shown that properties of the intermediate layer and its dimensions influence the longitudinal shear modulus of the material and Poisson's ratio most profoundly. Thus, shear modulus increases by 30–60% with growing filling degree.

5 Description of Production Technology and Physico-Mechanical Properties of Composites Based on Spirally Reinforced Fillers

5.1 Stages of Material Production Technology

5.1.1 Technological Peculiarities and Basic Process Parameters of Spiral Winding

As it follows from the process flow diagram illustrated in Fig. 5.1, production process of materials with spirally reinforced fillers differs from other processes of manufacturing composites, e.g. unidirectional ones. The distinguishing feature consists in an additional operation of spiral winding of the main reinforcement. When large diameter fibers are used in spirally reinforced elements, e.g. boron, they are impregnated upon winding, further processing stages being analogous to those of unidirectional filler. In the case of a fine-fibrous filler on the base of carbon, organic or glass fibers, operation regimes of manufacturing spirally reinforced elements and choice of winding operation order within the technological sequence influence other process operations, including impregnation, reinforcement tension, product forming and solidification.

Spiral winding with a dry fine-fibrous reinforcement produces spirally reinforced elements of high packing density which strongly hampers penetration of the binder into the interfiber clearances. Even vacuumization of the spirally reinforced filler followed by impregnation under pressure won't yield any single-piece structure. Therefore, it's expedient to wind the fine-fiber filler upon impregnation. The amount of the binder contained in the winding prepreg is to be estimated with allowance for the winding fiber saturation.

The additional operation of spiral winding of the main reinforcement ensures a set of required structural and geometrical characteristics of the spirally reinforced filler. The chief technological parameters of the operation are the main reinforcement drawing rate, rotation velocity of the winder, tension force of the main reinforcement at winding and tension force of the winding fiber [229]. The requirements imposed on thus obtained spirally reinforced filler and on winding process parameters may vary in response to the kind of article produced and reinforcement type. Most common among them is the stability of winding pitch which guarantees engagement of coils of neigh-

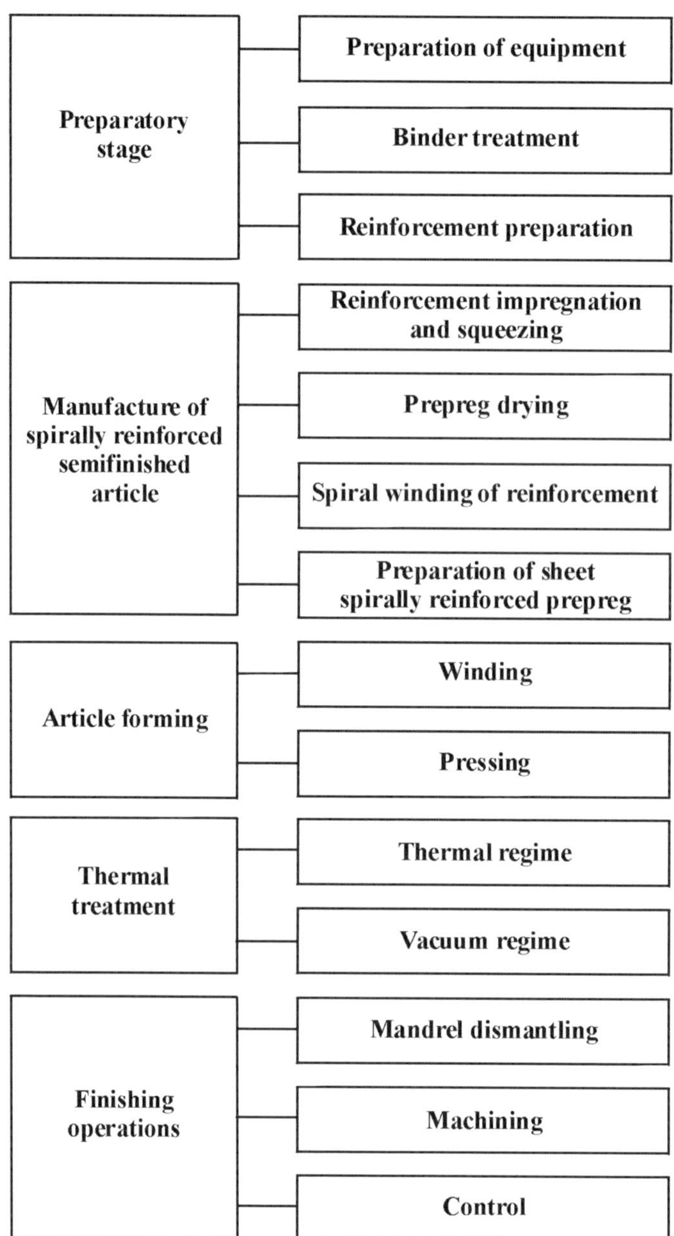

Fig. 5.1. Process flow diagram

boring elements. The winding pitch t_0 relation to the spirally reinforcement process is expressed by the formula

$$t_0 = \frac{V_d}{n_0}.$$ (5.1)

Hence, theoretical stability of winding pitch depends upon winder rotation velocity n_0 and drawing rate V_d. The winder represents a rapidly rotating system which along with driving strain and friction forces is effected by the tension force of the winding fiber. The moment of resistance of the winding fiber is, however, much less than the inertia moment of the winder rotor, so the winder effects but insignificantly pitch instability. The effect of drawing rate is related to chosen technological regimes of producing spirally reinforced materials. In case only one layer is wound on a cylindrical mandrel or a preliminary fabricated spirally reinforced filer is used, the drawing rate remains constant. In contrast, when the process of manufacturing thin-walled or shaped wound products is continuous, drawing rate is conditioned by geometrical parameters and shape of the product. To eliminate instability in winding pitch a special automatic system for correction of winder rotation is used able to adjust it to the drawing rate of the main reinforcement.

Moreover, pitch stability is affected by the magnitude and tension force constancy in the winding fiber. When the tension force is low, friction forces keeping winding fibers in an equilibrium state on the main reinforcement surface are also low and winding might go off course as the semifinished spirally reinforced filler passes braking and guiding units. In the presence of binder spots on the being wound reinforcement may stick the fibers, so this is to be eliminated too. The prepreg is usually wound after a preliminary molding-on by passing through a heated spinneret where the binder contained in the prepreg is softened and the winding fiber is bonded with the main reinforcement. As soon as the fiber tension force is violated, the fiber runout angle from the pickup changes and the fiber winding pitch becomes variable. The defect is eliminated by application of special winders and devices preserving tension force of the winding fibers constant.

A peculiarity of the process of manufacturing materials with spirally reinforced fillers consists in probable formation of porous inclusions at the interfaces of spirally reinforced elements. The inclusions lower significantly mechanical characteristics of the composite by violating its continuity and impair cohesion between elements. Mechanism of their formation is attributed to capture of air inclusions during material forming from spirally reinforced fillers and to isolation of volatile products at material heating and binder polymerization. It's hard to remove them from the material since coils of the spiral winding and dense packing of the elements whose dimensions exceed much the inclusions hamper their motion along boundaries of spirally reinforced elements both in longitudinal and transversal directions. As a result, a situation occurs when time needed for full removal of gaseous inclusions surpasses that of binder thickening brought about by polymerization. A part of the volatile matter still remains in the material that is why it's recommended to use vacuumization of the products at the initial stage.

5.1.2 Computation of Technological Process Parameters

Insofar a typical process underlies the technique for manufacturing articles of composites based on spirally reinforced fillers, all recommendations necessary for selecting corresponding types of reinforced plastics can be used for calculations of technological parameters under study. Exclusions are characteristics governing the spiral-reinforcement process and thermal – treatment regimes of the material with spirally reinforced fillers under low temperatures. During processing named parameters turn to be interrelated and are conditioned by the process stability and quality of the resultant product. In contrast to impregnation and squeezing of the main reinforcement for which the amount of binder is to be foreknown, winding process of the prepreg with dry fibers of auxiliary reinforcement is considered to be more technologically practicable.

The filling mass factors of a braid saturated with the binder is determined as follows

$$\varphi'_F = \frac{m_B + m_L}{m_G}, \tag{5.2}$$

where m_B, m_G, m_L are masses of binder, saturated braid and volatile fractions of solvent, respectively.

By representing the denominator in (5.2) as a sum of component masses and dividing the numerator and dominator by the mass of produced hardened composite, we get

$$\varphi'_F = \frac{\varphi_B + m_L/m}{\varphi^0_{MR} + \varphi_B + m_L/m}, \tag{5.3}$$

where φ^0_{MR} and φ_B are mass factors of composite filling with the main reinforcement and binder.

m_L/m ratio can be presented as

$$\frac{m_L}{m} = \frac{\varphi_B \cdot \varphi'_L}{1 - \varphi_L}. \tag{5.4}$$

$\varphi'_L = \dfrac{m_L}{m_B + m_L}$ – relative solvent content in the ready binder. Then, taking account of (5.4) relation (5.3) can be presented as

$$\varphi'_F = \frac{\varphi_B}{\varphi_B + \varphi^0_{MR} \cdot (1 - \varphi'_L)}. \tag{5.5}$$

In the main reinforcement prepreg an amount of binder necessary for saturation of winding layers calculated by (5.5) is introduced.

Force parameters of the spiral winding process are calculated proceeding from conditions of preserving linearity of the main reinforcement whose violation [230] results in impaired strength and elasticity characteristics of composites and can be even lower than analogous characteristics of materials with linearly disposed fibers.

Let's study the behavior of the main reinforcement under the action of forces arising during winding. The main reinforcement acquires after passing through the forming or squeezing spinneret a circular cross-section of diameter D and experiences the action of axial tensile force P_0. The winding fiber is wound round the main reinforcement at pitch t_0, angle α and tension force P thus creating evenly distributed along the circular helix loading

$$q = \frac{2 \cdot P}{D} \sin^2 \alpha \,. \tag{5.6}$$

By taking the braid as a beam with free ends of reinforced plastic, of length $t_0/2$ in plane YZ within the range of friction forces, load q can be presented as a sinusoidal one

$$q^* = q \cdot \sin \frac{2 \cdot \pi \cdot z}{t_0} \,. \tag{5.7}$$

If to simultaneously take account of the longitudinal forces and transversal load it's possible to derive a system of equations expressing longitudinal-transverse bending of the anisotropic beam

$$\begin{cases} \dfrac{\partial^2 W}{\partial z^2} + \dfrac{1}{\beta^2} \cdot \dfrac{\partial^2 W}{\partial y^2} = 0 \,; \\[2ex] (D \cdot G_{ZY} + P_0) \cdot \dfrac{\partial^2 V}{\partial z^2} + G_{ZY} \cdot \displaystyle\int_{-D/2}^{D/2} \dfrac{\partial^2 W}{\partial z \cdot \partial y} \cdot dy = -q^* \,. \end{cases} \tag{5.8}$$

The solution of the system for the case of simply supported edges of the beam is of the kind

$$W = \frac{q}{\lambda^2 \cdot (P_0 + \varphi_m \cdot \lambda^2 \cdot E_Z \cdot J)} \cdot \frac{sh(\kappa \cdot 2 \cdot y/D)}{\rho \cdot ch\kappa} \cdot \cos(\lambda z) \,; \tag{5.9}$$

$$V = \frac{-q}{\lambda^2 \cdot (P_0 + \varphi_m \cdot \lambda^2 \cdot E_Z \cdot J)} \cdot \sin \lambda z \,; \tag{5.10}$$

where $\lambda = \dfrac{2 \cdot \pi}{t_0}$; $\beta = \sqrt{\dfrac{E_Z}{G_{ZY}}}$; $\kappa = \dfrac{\pi \cdot D}{t_0} \cdot \beta$; $\varphi_m = \dfrac{3 \cdot (\kappa - th\kappa)}{\kappa^3}$.

With the supposition that $\varepsilon_y = 0$, we have

$$\tau_{ZY} = G_{ZY} \left(\frac{\partial W}{\partial y} + \frac{\partial V}{\partial z} \right) \,. \tag{5.11}$$

Wherefrom, based on strain relations, we get

$$\tau_{ZY} = \frac{q G_{ZY}}{\lambda (P_0 + \varphi_m \lambda^2 E_Z J)} \left(1 - \frac{ch\kappa \cdot \dfrac{2y}{D}}{ch\kappa} \right) \cos \lambda z \,. \tag{5.12}$$

Maximum tangential stresses are reached at $y = 0$. Since for unidirectional composites $1/ch\kappa \to 0$, then the condition of preserving linearity of the braid being wound can be like

$$\tau_{\max} = \frac{qG_{ZY}}{\lambda(P_0 + \varphi_m\lambda^2 E_Z J)} \leq [\tau_0];\tag{5.13}$$

$$\frac{P \cdot t \cdot G_{ZY}\sin^2\alpha}{\pi \cdot D \cdot \left[P_0 + \dfrac{3 \cdot D \cdot G_{ZY} \cdot (\pi \cdot D \cdot \sqrt{E_Z/G_{ZY}} - t_0)}{16 \cdot \sqrt{E_Z/G_{ZY}}}\right] \cdot [\tau_0]} \leq 1,\tag{5.14}$$

where $[\tau_0]$ are limiting stresses in the braid induced by friction or cohesion forces between fibers in, correspondingly, dry or impregnated braid.

Consequently, the spirally reinforced filler can be distorted during production under a certain correlation of geometrical and force parameters due to relative shear of the fibers (Fig. 5.2).

The maximum tangential stress values are unlikely to depend on the tension force of the main reinforcement since $P_0 = (0 - 0.3)P_F$, which is less by two orders of magnitude l the value of the second summand in the denominator. Besides, $t_0 \leq \pi \cdot D \cdot \sqrt{E_Z/G_{ZY}}$. Therefore, an approximate relation can be used for practical solutions

$$\frac{16 \cdot P \cdot t_0}{3 \cdot \pi^2 \cdot D^3 \cdot [\tau_0]} \cdot \sin^2\alpha \leq 1.\tag{5.15}$$

Wherefrom, the tension force of the winding fiber is determined by

$$P \leq \frac{3 \cdot \pi^2 \cdot D^3 \cdot [\tau_0]}{16 \cdot t_0 \cdot \sin^2\alpha}.\tag{5.16}$$

When a dry braid is wound and friction value is conditioned by the tension force, then

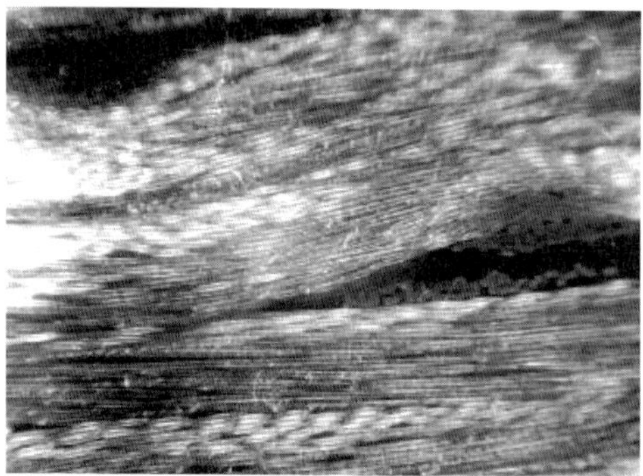

Fig. 5.2. Distorted fibers of the main reinforcement in the material with the spirally reinforced filler

$$[\tau_0] = \frac{2Pf_{tr}\sin\alpha}{Dt_0}, \tag{5.17}$$

Equation (5.15) leads to

$$\frac{8}{3f_{tr}}\left(\frac{t_0}{\pi D}\right)^2 \sin\alpha \leq 1, \tag{5.18}$$

i.e. distortion conditions of the woundaround braid aren't dependent in this case on the tension force of the winding fiber but are determined fully by geometrical parameters of the spirally reinforced element.

Under the torsion torque generated by the winding fiber

$$M_{tr} = \frac{PD}{2}\sin\alpha, \tag{5.19}$$

deviations from linearity of peripheral fibers due to their twisting are observed. Presuming that a bunch of fibers fed to the winding zone upon leaving the squeezing or forming device is devoid of sufficient twisting stiffness, then the torque is considered to be balanced owing to only tangential constituents of the tension force of the main reinforcement evenly distributed across its section.

Assuming that the deviation angle of the fibers from linearity η is a function of radius r, then the total balancing moment in the braid equals to

$$M_y = \frac{8P_0}{D^2}\int_0^{D/2} r^2\,\mathrm{tg}\,\eta\,dr. \tag{5.20}$$

Admitting that angle η dependence can be written as a power function $\eta = \eta_0(2r/D)^m$ where η_0 is deviation angle of peripheral fibers, and taking the angle as small as $(\mathrm{tg}\,\eta = \eta)$, we can determine by (5.20)

$$M_y = \frac{P_0\eta_0 D}{m+3}. \tag{5.21}$$

By equating (5.19) and (5.21) we get a formula interrelating the larger angle deviations of the main reinforcement from linearity, including force parameters of the reinforcement process

$$\eta_0 = \frac{(m+3)P\sin\alpha}{2P_0}. \tag{5.22}$$

Exponent m can be determined from (5.22) under known η_0. The experimental dependence of η_0 on relation P/P_0 is illustrated in Fig. 5.3. Under $P = 0.1\,\mathrm{N}$, $P_0 = 25\,\mathrm{N}$ and $\sin\alpha = 0.995$, we have $\eta_0 = 0.008$, $m = 1$.

By setting limiting values of $[\eta_0]$ and P, the needed tension force of the main reinforcement during winding can be found

$$P_0 \geq \frac{2P\sin\alpha}{[\eta_0]}. \tag{5.23}$$

Under the action of strained winding fibers the wound braid of the main reinforcement deviates some distance and the magnitude of the distance is

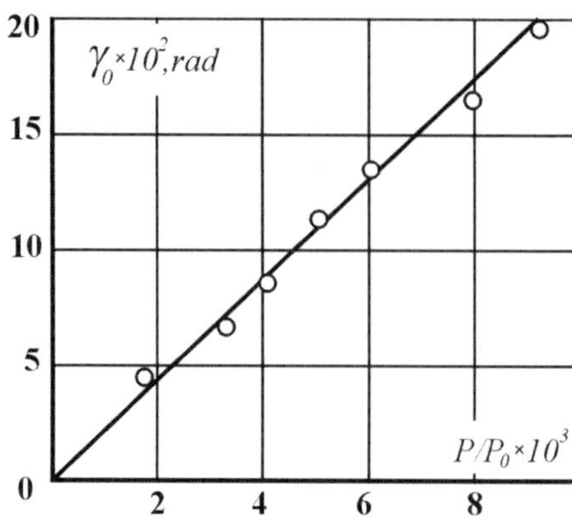

Fig. 5.3. Dependence of twist angle of peripheral fibers of the main reinforcement upon correlation of force parameters during spiral winding

conditioned by that of the counteracting tension force of the main reinforcement. The direction of the tension force of the winding fiber varies in response to the frequency equal to angular velocity of the winder rotation ω

$$P^* = P \sin \alpha \sin \omega t.$$
(5.24)

So, the being-wound braid oscillates with amplitude y whose maximum value is calculated by the dynamic equation

$$\varphi_i'' + p_i^2 \varphi_i = \frac{2P}{\gamma F L} \sin \frac{i\pi C}{L} \sin \omega t$$
(5.25)

where L – distance between supports; C – distance from support to force application point; $p_i = \dfrac{i\pi}{L} \sqrt{P_0/\gamma F}$ – frequency of natural vibration of the braid stretched by force P_0; γ – braid material density; F – cross-section area; φ_i – known time function related to oscillation amplitude through dependence

$$y = \sum_{i=1}^{\infty} \varphi_i \sin \frac{i\pi z}{L}.$$
(5.26)

Solution of (5.25) with allowance for forced oscillations only of the flexible braid under the harmonic transverse load is of the kind [231]:

$$y = \frac{2P^*}{\gamma F L} \sum_{i=1}^{\infty} \frac{\sin \dfrac{i\pi C}{L} \sin \dfrac{i\pi z}{L}}{p_i^2 - \omega^2} \sin \omega t.$$
(5.27)

Keeping to only the first term of the series, the maximum oscillation amplitude is found at $z = L/2$:

$$y_{\text{max}} = \frac{2P^* L \sin \dfrac{\pi C}{L}}{\pi^2 P_0 (1 - \omega^2/p_i^2)}.$$
(5.28)

A set of conclusions can be derived from above equation important at developing and designing winding equipment and choosing technological parameters of the process.

The dependencies of the maximum oscillation amplitude of the wrapped braid are presented in Figs. 5.4 and 5.5 as against rotation velocity of the winder n and distance between bearing points L. The values are given for an impregnated braid of 0.2 mm diameter under $C = 10$ mm and $P^*/P = 0.03$.

The analysis of presented diagrams proves that there's a region of L and n values where the frequency of forced oscillations approaches natural oscillations of the being wound braid and the oscillation amplitude is observed to be in resonant growing. There's a direct dependence between ratios C/L and P^*/P_0, and any changes don't effect the dependence behavior but result in scale variation of the diagrams along y_{max} axis. Under high enough oscillation amplitudes the wrapped braid beat against the guides and supports or ro-

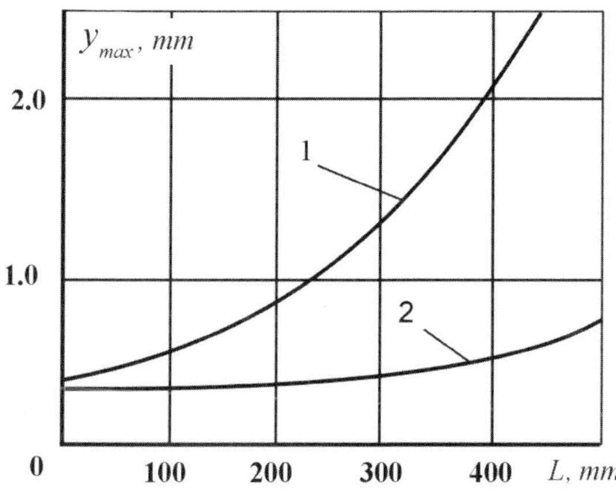

Fig. 5.4. Dependence of the maximum oscillation amplitude of the wound braid upon distance between its supports

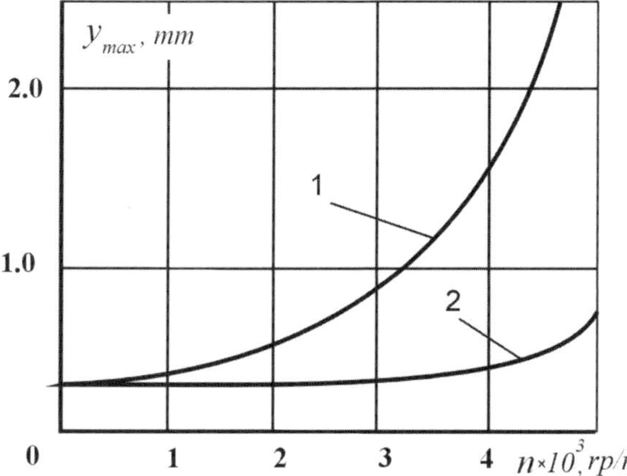

Fig. 5.5. Variations in maximum oscillation amplitude of the braid versus winder rotation velocity

tating elements of the winder leads to its damaging. So far it's recommended to take maximum acceptable oscillation amplitude $y_{\max} \leq 0.5\,\mathrm{mm}$. This amplitude ensures sufficient stability at perfect quality of produced spirally reinforced semifinished material. This can be attained either by contracting distance between supports of the wound braid or by incrementing tension of the main reinforcement which results in either on increase in natural oscillations of the braid or bringing the bearing point most closely to the winding zone, i.e. in reducing parameter C.

By keeping conditioning $C \ll L$, (5.28) transforms to the form

$$y_{\max} = \frac{2P^*C}{P_0\pi(1 - \omega^2/p_1^2)}\,. \tag{5.29}$$

Relation (5.29) makes it possible to calculate maximum admitted rotation velocity of the winder under the given geometrical and force parameters of the winding process

$$n \leq \frac{60}{L}\sqrt{\frac{P_0}{\gamma\pi D^2}\left(1 - \frac{2PC\sin\alpha}{\pi P_0[y]}\right)}\,. \tag{5.30}$$

The latter equation shows sense only in condition of positive radicand values, i.e.

$$1 - \frac{2PC\sin\alpha}{\pi P_0[y]} > 0\,. \tag{5.31}$$

Consequently, (5.31) can be employed for computation of C value

$$C < \frac{\pi P_0[y]}{2P\sin\alpha}\,. \tag{5.32}$$

Study of dynamics of the winding process shows that any combinations of parameters and operation regimes of the winder lying within the resonant region are intolerable, i.e. under $L > 350\,\mathrm{mm}$ and $n > 3600\,\mathrm{rpm}$ simultaneously. When there's a necessity to operate beyond the said limits of one of the parameters, then either the other parameter is reduced to a respective value or the resonant frequency of natural oscillations is shifted by increasing tension force of the main reinforcement. Limitations on oscillation amplitude within the rest range are imposed through the choice of corresponding values of C and P/P_0.

To reduce porosity of the material with a spirally reinforced filler one can use vacuum treatment. The questions of vacuum techniques have been most profoundly studied in glass plastics production. In this field the technology is used at the stage of forming and during filler preparation and impregnation to lessen the porosity of produced materials [232]. In [233] the use of a vacuum technique at the stage of material solidification has been analyzed. It has been established that mentioned technique produces materials with perfected physico-mechanical properties because air, vapor and solvent inclusions are removed. Variations of mechanical characteristics with vacuum

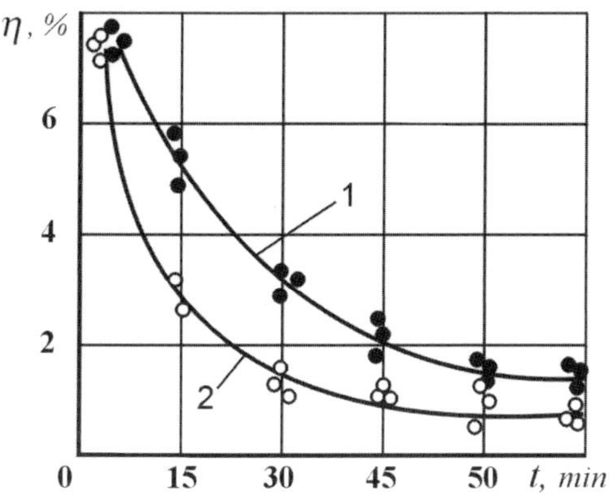

Fig. 5.6. Porosity dependence of the composite with the spirally reinforced filler on vacuumization time at different vacuum depths. 1 – 100 GPa; 2 – 10 GPa

depth are close to linear and coarse inclusions are removed already at 10 GPa pressure. Under lower pressures processes of partial destruction of the binder and isolation of dissolved in it gases are initiated. As a result, a tendency to saturation is observed in composites under such pressures and with pressure lowering the porosity remains practically constant. Analogous results were obtained in investigations of vacuum treatment effect on carbon plastic solidification processes. Besides, variation dynamics of porosity has been studied at vacuum treatment of materials with the spirally reinforced filler as dependent on vacuum depth and treatment duration. Samples were carbon plastic prepregs based on epoxy-phenolic binder UP-2220 with 4.6% residual content of volatile products. Experimental results are illustrated in Fig. 5.6. It's evident that vacuum treatment at 100–10 GPa is enough to lower porosity values. Treatment time prolongs proportionally to pressure rise in order to reach the same porosity values.

It's to be underlined that temperature, time of vacuumization start and decompression rate plays important roles during vacuum treatment. The temperature is to be high enough to ensure fluidity of the binder and free outlet of gaseous inclusions from the material well heated across thickness. Otherwise, a part of excessively viscous binder will be squeezed out onto the material surface together with removed gas inclusions. As a result, the inner structure of the material would be depleted of the binder and contain large amount of fine pores.

Analogous phenomenon is observed in case of late start of vacuumization when the binder thickens and becomes more viscous. In this case a structure with coarse pores is formed where the majority of pores are found between spirally reinforced elements and the material surface gets covered with bubbles and craters. Investigations have also proved that excessively high decompression rate brings about vigorous gas isolation accompanied by binder ejection from the material structure. Concrete parameters of vacuumization are determined during testing and depending on the binder grade and prod-

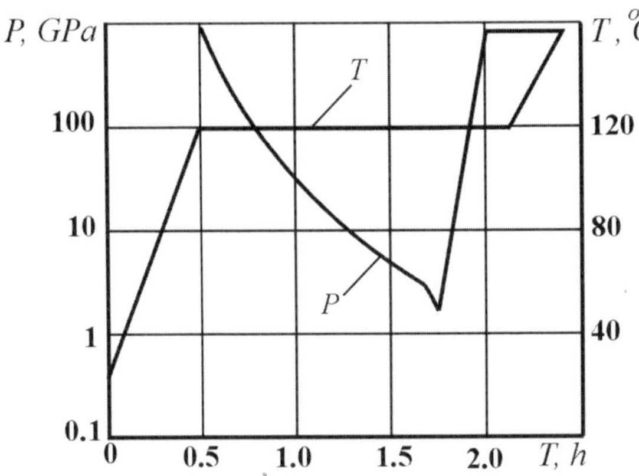

Fig. 5.7. Temperature and vacuumization regimes of materials with spirally reinforced fillers (binder – UP-2220)

uct dimensions. For instance, for binders EDT-10 and UP-2220 used in the present work it's recommended to start vacuumization as the article heats till, correspondingly, 90 and 120°C to ensure binder viscosity 5–10 Pa·s. The recommended decompression rate is 0.5 GPa/s. It can be adjusted by a throttling device mounted in a vacuum line that connects the thermal chamber with the vacuum system, or a leak-regulating complementary air inflow from the atmosphere.

At the beginning of gas isolation from the material structure pressure in the thermal chamber stabilizes and remains constant till the exhaust outdoes vapor and gas isolation rate, upon which pressure reduces (Fig. 5.7). Just in that instance vacuumization of the product can be finished as far as the main mass of gas and air inclusions has gone out.

To realize the process flow diagram, conventional equipment for manufacturing products from reinforced materials can be partially employed. However, additional stage of winding the main reinforcement with auxiliary reinforcement fibers necessitates a set of complementary parameters connected with achieving the spirally reinforced filler. Hence, some specific requirements are imposed on the equipment used.

5.1.3 Production Equipment for Materials with Spirally Reinforced Fillers

The device for stretching the main reinforcement should also provide for sufficient preliminary take up tension of the winding material as to avoid its twisting and distortion under the forces and moments arising at application of the spiral wrapping layer. While stretching the filler it's necessary to eliminate reinforcement slippage as it may violate the pitch and some other parameters of the material. With this aim the reinforcing material is commonly pulled either by light braking of spools with the reinforcement or by braking the reinforcement itself. The analysis of braking systems shows that they are acceptable only in conditions of light tension force values. This is because wound with tension reinforcement exerts pressure on the underlying strata, slips over them and engages with the previously laid coils. Fibers in the braid start to rupture, get fluffy, abraded, curly, and etc. Along with this, tension might be exercised by passing the braid through a system of immovable and rotating braking lightly rollers. Note that the braking force of each roller should be less than the friction force that ensures reinforcement motion together with the rollers and less the preliminary tension force created by braking of the spools with the reinforcement. When the prestress is low, e.g. when using carbon reinforcement, it's necessary to increase the contact angle, which means usage of larger number of bypass rollers and folding of the reinforcement in opposite directions. To eliminate the drawback, a new design of a total stretcher and that with continuous tape has been proposed. They ensure multicoil enveloping of the brake drum and exclude reinforcement slippage. To eliminate folding of the braid and preserve continuity of its fibers and threads its expedient to use pneumatic or hydrodynamic methods of stretching. The hydrodynamic procedure unites the processes of stretching and impregnation in one work cycle.

Currently used squeezing and forming devices for processing reinforced materials guarantee a stable perimeter and a reduction force of the filler. Most commonly employed today are spinneret-type devices. Their drawbacks are difficulties in filler charging and risk of damaging during squeezing or forming. Use of collapsible spinnerets heightens fault probabilities because of occasional ingress of fibers into the cutoff plane. To eliminate the drawbacks special designs are employed.

The main unit of the equipment for spiral winding of the reinforcing material is a winder which lays the auxiliary reinforcement on the filler surface. As the analysis has shown winders used in different industrial branches can be subdivided into two main groups – winders with tangential and those with axial reeling of fibers.

Systems with tangential reeling of fibers are simple in design and ensure high winding speeds. Besides, they are balanced and easily set the tension force of the winding fiber by braking lightly spools with the reinforcing material. These systems, however, restrict number of simultaneously installed spools with the winding material which hinders balanced state of the braid during winding. Furthermore, the spools have to be forcedly rotated to reel the fiber by the fiber. The latter circumstance forbids application of low-strength fine fibers under significant inertial loads. Besides, tension forces generated by high winding speeds appear to be so strong that may result in damaging of the main reinforcing material. Named facts were considered at manufacturing spirally reinforced fillers by the systems with axial reeling of the winding fiber. Such systems make provision for a needed number of spools with reinforcement mounted over the winder perimeter which guarantees steadiness of the wrapped braid and multiturn spiral winding. It's to

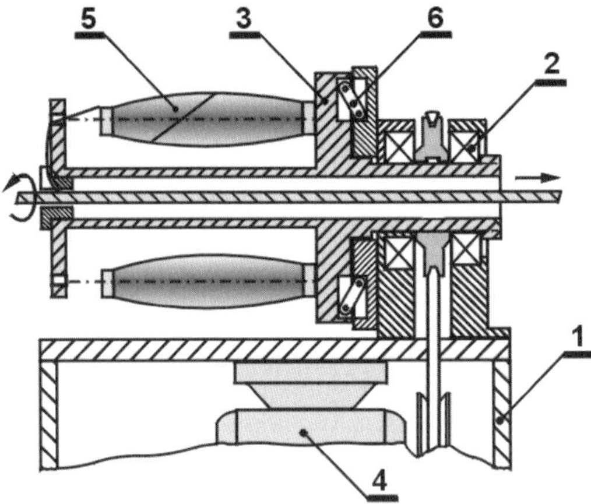

Fig. 5.8. A design of self-balancing winder. *1* – body; *2* – bearing unit; *3* – rotor; *4* – drive; *5* – bobbin with material; *6* – balancing device

be born in mind that winders of the type are large in size and, consequently, show low rotation velocities. Moreover, the winder rotor gets unbalanced as the auxiliary reinforcement exhausts and generates considerable vibrations. To reduce the phenomenon such designs of the winder are used that are able to automatically balance the rotor and stabilize the winding process. A schematic diagram of the setup is illustrated in Fig. 5.8, and its general view is shown in Fig. 5.9. The setup consists of a six-bobbin winder with a DC engine to ensure smooth handle of the winding pitch. Balls installed in an annular groove filled with a viscous fluid exercise automatic balancing in this case.

Considering that installation of stretchers on the rotor bounds the use of known braking methods, the friction procedure of tightening winding fibers by passing them through a system of stationary tension rollers or a braking sleeve with a helical guiding groove turns to be most acceptable. The tension force is smoothly regulated in the latter case by varying the contact angle of the braking sleeve by turning the fiber relatively the winder body. Experience has proved that there appears harmonic variation in the tension force at axial reeling of the fiber since a part of its path the fiber travels in direction of centrifugal forces and the other part oppositely. To avoid this and raise stability of the tension force a specific structure of the stretcher is used which compensates tension variations by turning the spring-backed braking sleeve.

Application of described designs and recommended devices for manufacturing samples and articles of the composite material based on the spirally reinforced filler has raised efficiency of the technological process and quality of the final product.

Fig. 5.9. General view of the winder

5.2 Choice of Testing Methods, Preparation of Samples and Processing of Experimental Results

A peculiarity of composites based on spirally reinforced fillers is that the materials contain a finite number of elements each of which is as if encapsulated into an individual shell. So, if a crack propagates inside a spirally reinforced element under loading, it can't go beyond it. This means that failure localizes and bearing capacity of the material doesn't exhaust. This's why testing methods and samples are selected so as to minimize structural damages which may result in violation of the auxiliary reinforcing layer during machining. Besides, composites, especially spirally reinforced ones depend strongly on the choice of technological regimes during molding since they are governing the form and relationship of the elements, their filling degree and, finally, affect the obtained material characteristics. Therefore, of importance is the choice of material samples size for testing. As it's been stated earlier, structure of composites with the spirally reinforced filler is composed of more coarse units as compared to elementary fibers of conventional reinforced plastics. Consequently, the scale factor effects significantly correctness of evaluated properties of the produced material and the necessity of lessening the effect of the sample rims.

Unidirectional samples have been tested during which their tensile and shear characteristics in corresponding directions have been evaluated. The testing methods have been elaborated on the base of standard requirements and recommendations. Derived data were compared to calculations described above.

To determine elastic modulus at tension in direction of the main reinforcement laying 220×25 mm in size pressed strips with 3 to 4 mm thickness corresponding to the initial sheet were used. Strain gauges were employed to register deformation. Characteristics of the samples were determined by thrice-repeated loading up to the value not exceeding 0.3 of the breaking one. Upon finding elastic modulus the samples were cut lengthwise into two parts having 220×12 dimensions and their tensile strength was determined. Veneer straps were glued onto the samples to avoid slippage in the testing machine clamps. Longitudinal elastic modulus of the winding material was estimated for tubular samples under the inner pressure effect.

Strength of winding composites with the spirally reinforced filler was determined by the method of two half-discs in direction of the main reinforcement using ring samples. On the working surfaces of the half-discs were installed rollers to ensure even stress distribution over the perimeter of the ring under testing.

Elastic and strength characteristics of 150×15 mm strip samples with glued veneer straps were recorded under transverse loading. Along with this, transversal tearing of tubular samples of the winding material was determined in order to eliminate rim effects completely. A distinctive feature of preparing such samples for testing consists in molding-on of each successive layer of the spirally reinforced filler upon laying on the mandrel. This guarantees high quality of the sample exterior, accurate geometrical parameters and needed structure. With reaching the required thickness of the sample, its end portions were additionally wound with preimpregnated glass braid to attain upon machining bearing areas for clutches. Thus-prepared sample is installed in a self-adjusting clutch which provides for high-accuracy centering of the sample relative to tension forces and eliminating any shift in the clutch. The employed tubular samples were of outer diameter $D_O = 36$ mm, inner diameter $D_I = 30$ mm and length of the working part -130 mm.

The loading regime with a constant strain rate $0.008\text{--}0.025\,\mathrm{s}^{-1}$ was maintained during testing. To determine the elastic modulus the samples were thrice-loaded till 10–20% of the breaking value. A special tensometer was used to avoid substantial deformations on loading perpendicularly to the main reinforcement direction.

Strength and elastic modulus at shear were determined for tubular samples using special wedge clamps evenly reducing the sample from the outside and inside. Characteristics of molded materials at shear were determined by warping the plates in a hinged quadric crank. Since materials with the spirally reinforced filler are like other unidirectional composites display anisotropy of mechanical properties, to register their shear characteristics the samples with incisions were used. A modified quadric crank was used with this aim to apply the load at some distance from the working surface and reducestress concentrator effect.

For the lengthwise and crosswise compression tests of the main reinforcement strips with 110×20 mm dimensions and tubes of above mentioned diameter were employed. One of the major requirements to be met tests is to maintain stable state of the sample on loading. For this purpose a special testing device has been developed where the samples are fixed with a cold-setting epoxy compound in cups with centering inserts. To preserve stability of the sample special supporting layings of foam plastics are installed from both sides and pressed against the sample.

The static bending tests for strip samples followed the three-point scheme. Their impact strength was estimated at impact bending of not incised samples of 55×15 size and 40 mm test basis. The samples were cut out of plates in direction of the main reinforcement.

Test results underwent static processing using known procedures. The mean values and mean-square deviations as well as variation factors and confidence intervals were evaluated.

5.3 The Effect of Major Structural and Technological Parameters on Mechanical Characteristics of Composites

It has been shown in 2 that technological and structural parameters of composites with spirally reinforced fillers are interrelated to a greater degree than in common reinforced materials. So, any variation of the technological process leads to corresponding changes at two structural levels. Therefore, when studying the effect of some structural parameter upon the material physico-mechanical properties we simultaneously get the dependencies of these characteristics on a technological parameter governing them, providing all the rest values are stable.

One of the main structural characteristics of reinforced plastics is the filling degree with a fibrous reinforcement. The characteristic of materials with the spirally reinforced filler is conditioned by the filling degree of elements, their form and location in the material bulk. Besides, the filling degree of the spirally reinforced elements is dependent on molding pressure q which is, in its turn, determined by tension force of the winding fiber, winding diameter and pitch.

In Fig. 5.10 experimental results are presented of the elements filling degree dependence with the main reinforcement upon molding pressure for different binders and packing types of the spiral reinforcement.

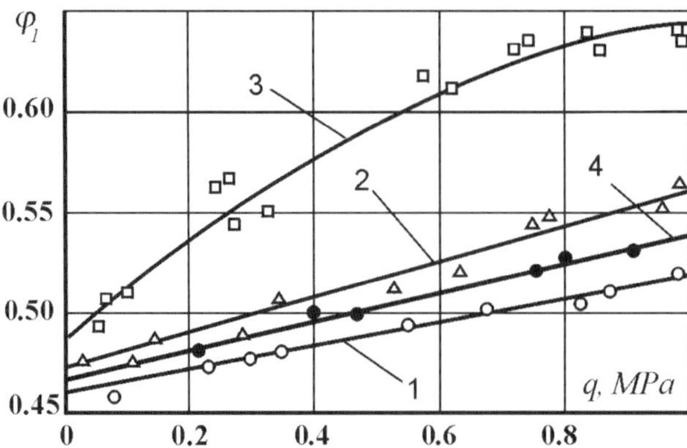

Fig. 5.10. Dependence of filling degree of elements by the main reinforcement versus molding pressure at winding. *1* – carbon-glass plastic, prepreg, UP-2220; *2* – organic-glass-reinforced plastic produced by wet method, EDT–10; *3* – the same, UP–2217; *4* – organic-glass-reinforced plastic, prepreg, UP-2217

It can be concluded that the type of preimpregnation of the main reinforcement exerts a significant effect on the filling degree. Thus, use of a prepreg from organic synthetic high-polymer (SHP) fibers based on UP-2217 binder yields $\varphi_1 = 0.53$ at a molding pressure $q = 0.6\,\text{MPa}$ and at wet method $-\varphi_1 = 0.62$. The analysis of presented dependencies shows that φ_1 to molding pressure q relationship is close to linear and can be described by the equation

$$\varphi_1 = Kq + \varphi_1^0, \tag{5.33}$$

where φ_1^0 – the initial filling degree of the braid conditioned by spinneret diameter; K – coefficient dependent on binder viscosity.

To reach above 75–80% filling at sufficiently high binder viscosity great tension forces of the auxiliary reinforcement are required. Their attainment is bounded by the winding fiber strength and possible loss of stability of the main reinforcing braid during winding. So far, about 80% of the filling degree can be assumed limiting for the spirally reinforced element core filling with the main reinforcement.

In the course of winding the elements of the spirally reinforced filler acquire round shape and are inclined to deformation which is dependent on their filing degree. Variation of percent reduction γ of materials having different initial packing of elements is shown in Fig. 5.11 as dependent on pressing force. The experiments followed sequentially the static analysis of microsections of the deformed structure under pressure augment. The presented data proves that the limiting percent reduction γ_{min} is reached under different molding pressure values and is dependent upon the initial packing type. Along with this, the percent reduction is conditioned by the initial filling degree of the

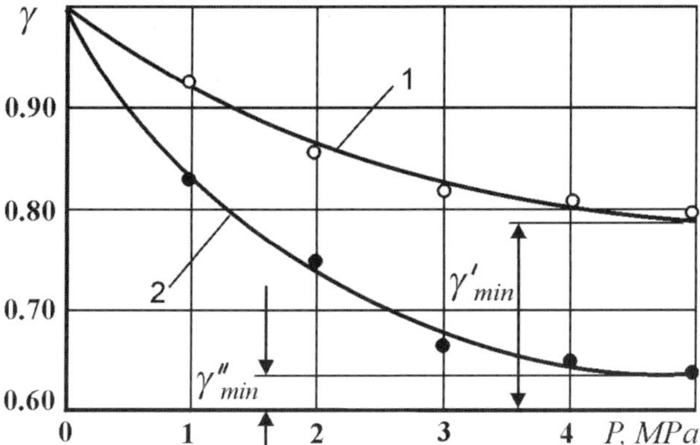

Fig. 5.11. Variation of percent reduction γ of the material structure depending on molding pressure. *1* – hexagonal packing; *2* – packing with clearance $h = 0.25R$

spirally reinforced element which is dependent on molding pressure by the winding fiber.

In Fig. 5.12, the effect of both parameters q and P on the percent reduction when the filler was laid with a clearance $h = 0.25R$. The analysis shows that with increasing q molding pressure P at which limiting deformation of the structure is reached rises.

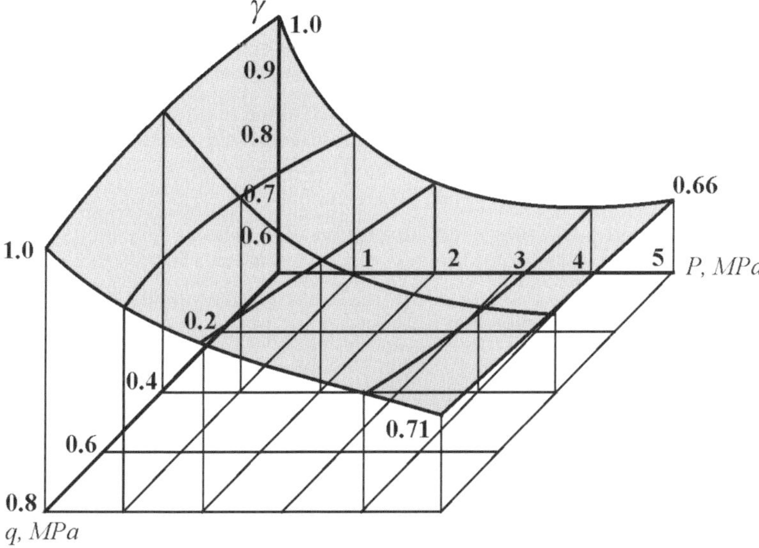

Fig. 5.12. Percent reduction of the material structure as a function of pressing force and reduction pressure by the main reinforcement

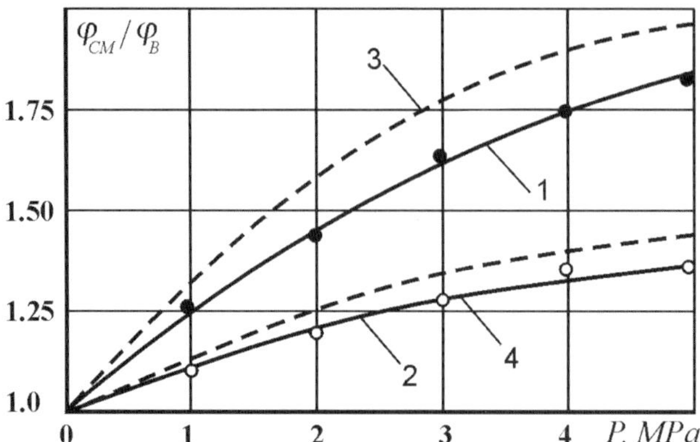

Fig. 5.13. Relative variation of material filling degree depending on molding pressure. *1* – organoplastic, UP– 2217; *2* – organoplastic, wet method, UP–2217; *3, 4* – theoretical computation based on dependencies from 3

Under high enough filling degree an inclination is observed in the element core to preserving its round shape. Notice that the material is formed by laying spirally reinforced braids in-between already-laid elements of the previous layer and thus produced structure approaches closely the hexagonal one. The final goal of the material filling is, in general, determined by packing density of the filler elements within the given space which is preset by the tension force of the spirally reinforced filler as you form winding articles, or by molding pressure if a molded material is produced. Experimental and theoretical evidences are illustrated in Fig. 5.13 for pressure effect on filling degree of different materials. With this aim, theoretical calculations made use of earlier-derived dependence $\gamma = f(P, q)$ presented in Fig. 5.12, using relations cited in Sect. 2. Variations of the filling degree in molded materials based on preliminary wound prepregs is more pronounced as compared to the filler obtained by the wet method. This is because in the latter case the initial filling degree with the main reinforcement φ_f is less.

As an example the dependence of the composite filling degree on molding pressure is given in Fig. 5.14 for different main reinforcement types and binders. It's clear that the filling degree variation is analogous to considered above one. Cited regularities are also true for composites with mutual engagement of the auxiliary reinforcement coils. Data obtained for such material structures are depicted in Fig. 5.15. Filling degree in these composites increments with rising molding pressure as in materials with a continuous layer. Naturally, their maximum filling values are also higher.

To analyze the dependence of mechanical characteristics of the materials on the degree of filling with the main reinforcement batches of samples have been manufactured from spirally reinforced organic-organic plastic differed in the number of elements laid in a similar space of the mould. Also, composites

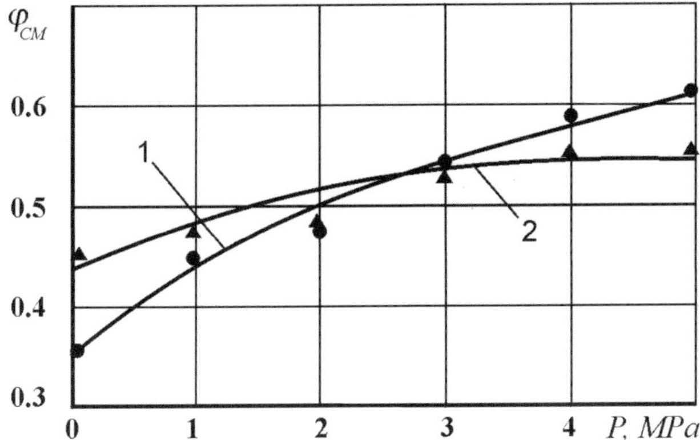

Fig. 5.14. Changes in filling degree of the composite with the spirally reinforced filler in response to pressure variation. *1* – carbon-glass-reinforced plastic (UP-2220); *2* – organic-glass-reinforced plastic (EDT-10)

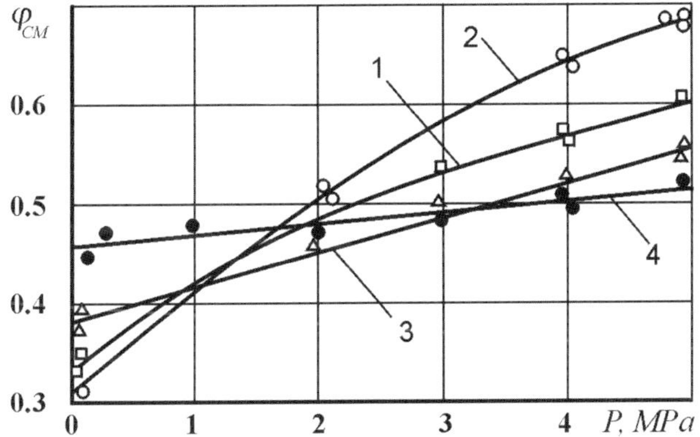

Fig. 5.15. Filling degree dependence of the material on molding pressure and type of auxiliary reinforcement. *1, 2* – UP-2220 (prepreg); *3, 4* – EDT-10 (wet braid); *1, 4* – continuous layer; *2, 3* – with coil engagement

based on fillers having either continuous or spaced winding that ensures mutual engagement of the coils by the winding fiber were examined. As the main reinforcement a braid of organic SHP fibers obtained by joining 5 fibers of 58.8 tex densities was employed. For the auxiliary reinforcement SHP fibers of 14.3 tex densities were used. For impregnation EDT-10 epoxy binder was applied. The winding fiber tension force was 1 N and 0.5 N at winding pitches of, correspondingly, 1 and 0.5 mm. As a result spirally reinforced elements with filling degree by the main reinforcement of about 65% were obtained. Test results are illustrated in Tables 5.1 and 5.2. For comparison test results of analogous unidirectional samples are also presented in the tables.

From analysis of the data obtained follows that application of the spirally reinforced filler improves characteristics of organic fibrous plastics under study. The highest properties are referred to composites whose auxiliary reinforcement is wound at 1 mm pitch, i.e. with mutual engagement of the winding fiber coils. Shear strength of mentioned materials increases from 48.0 MPa (unidirectional) till 72.1 MPa, i.e. by 50.2%. The rest characteristics vary similarly. Data presented in Tables 5.1 and 5.2 visualizes that some optimum filling degree value with the main reinforcement can be established just as for a continuous auxiliary reinforcement layer, so at coils engagement. In the latter case it's 57–59% and the amount of auxiliary reinforcement for such composites should be 6 to 7%.

Because of the low transversal strength of some materials used as the main reinforcement (e.g. carbon fibers) it's hard to reach high values of the initial filling degree through intensifying tension force. This's why structure of composites with the spirally reinforced filler can be changed using the filler which is wound by the auxiliary reinforcement fibers under light tension

Table 5.1. Mechanical Characteristics of Spirally Reinforced (Organoplastic). Composites at Shear and Compression in the Main Reinforcing Direction

Winding pitch, mm	Amount of fibers in an element	$\varphi_{MR}/\varphi_{AR}$, %	Shear τ_{XZ}, MPa	S, MPa	v, %	Compression σ_Z^-, MPa	S, MPa	v, %
Unidirectional		58.00	48.0	2.99	6.0	226	11.0	4.9
0.5	55	44.6/9.7	43.8	2.5	5.9	220	12.7	5.8
	63	51.1/11.2	61.7	1.5	2.4	268	5.0	1.9
	65	53.5/11.7	56.6	6.0	10.6	260	9.0	3.4
1.0	65	54.5/6.4	63.0	4.2	6.7	275	5.3	1.9
	70	57.8/6.8	72.1	2.1	2.9	287	8.6	3.0
	75	61.6/7.0	68.7	6.2	9.0	254	4.8	1.8

Table 5.2. Strength and Toughness of Organoplastics at Bending

Winding pitch, mm	Amount of fibers in an element	$\varphi_{MR}/\varphi_{AR}$, %	Shear σ_b, MPa	S, MPa	v, %	Compression $E_b \cdot 10^{-3}$, MPa	S, Mpa	v, %
Unidirectional		58.00	498	21	4.3	28.8	6.4	22.4
0.5	55	44.6/9.7	483	25	5.2	29.0	1.5	5.3
	63	51.1/11.2	591	13	2.2	38.2	2.2	8.5
	65	53.5/11.7	590	37	6.3	35.6	2.1	5.8
1.0	65	54.5/6.4	569	14	2.5	37.2	3.2	8.6
	70	57.8/6.8	615	13	2.1	44.5	1.4	3.1
	75	61.6/7.0	589	19	3.2	38.1	2.5	6.6

forces. Thus-obtained spirally reinforced element displays comparatively low filling degree and elevated deformability. Filling of the material grows in the course of manufacturing by way of lessening the cross-sectional area of the filler and fiber redistribution within the element core bounds. It's to be noted that in case dried prepregs are used, when wound they turn to be little filled with the main reinforcement and attain the needed filling value already during material forming. A drawback of the method is the probability of damage and distortion of the elementary fibers during their redistribution in the bulk of the spirally reinforced element, wherefrom, however, a chance appears of producing any in advance set structure of the material.

Experimental data in Table 5.3 depict mechanical characteristics of the composite with the spirally reinforced carbon plastic. The main reinforcement was a prepreg of carbon braid VMN-4 impregnated with EDT-10 binder and a glass fiber NSK-150/2 was used as a winding. Analysis of the results obtained show that the optimum filling degree with the main reinforcement for the carbon plastic based on spirally reinforced filler is 50–54% which is lower than the unidirectional carbon plastic has. The absolute values of mechanical properties are, nevertheless, higher, especially transversal and shear ones. So long as the varying tension force of the winding fiber brings about changes in the filling degree of the reinforced element and its deformability, it turns out that different structures and shapes of composite elements can be produced at a constant compacting pressure.

Table 5.3. Variations of Mechanical Characteristics of Carbon Plastic in Response to Filling Degree with the Main Reinforcement

Characteristic[*]	Unidirectional, $\varphi = 0.54$	Carbon plastic, $\varphi_{MR}/\varphi_{AR}$		
		0.47/0.06	0.52/0.07	0.56/0.07
$E_Z \cdot 10^{-3}$, MPa	$\dfrac{133.0}{169.7}$	$\dfrac{136.9}{142.8}$	$\dfrac{148.7}{157.8}$	$\dfrac{149.7}{169.7}$
υ_{XZ}	$\dfrac{0.34}{0.27}$	$\dfrac{0.39}{0.25}$	$\dfrac{0.36}{0.252}$	$\dfrac{0.38}{0.254}$
$E_X^+ \cdot 10^{-3}$, MPa	$\dfrac{6.69}{6.40}$	$\dfrac{8.3}{8.05}$	$\dfrac{9.36}{9.13}$	$\dfrac{8.7}{9.75}$
G_{XZ}, MPa	$\dfrac{4525}{2960}$	$\dfrac{5813}{5350}$	$\dfrac{5047}{5820}$	$\dfrac{4800}{6200}$
σ_Z^+, MPa	837	868	871	853
σ_Z^-, MPa	404	676	570	562
σ_X^+, MPa	25.1	39.3	36.4	35.5
σ_X^-, MPa	113	149	159	129
τ_{XZ}, MPa	32.5	56.3	61.2	42.4

[*] experimental values are in the numerator, theoretical from [132] and sections 3 and 4 of the present work are in the denominator

To establish the relationship of mechanical characteristics of the materials with the spirally reinforced filler upon above indicated parameter a carbon plastic was taken as an example. Carbon braid VMN-5 was used as the main reinforcement, the auxiliary reinforcement was based on organic SHP fibers (14.3 tex), and the binder was UP-2220. The choice of the winding material is explained by a higher strength of SHP fibers in contrast to glass filaments NSK 150/2 which produces a stronger tension force during winding. The molding pressure was taken constant for all batches and equaled to 3 MPa. Filling degree of the elements with the main reinforcement (carbon fibers) reached 65% and that of the material 0.48/0.07. Assuming that the element shape is a function of tension force of the winding fiber, then the derived results can be presented in double co-ordinates. Experimental evidences are illustrated in Table 5.3, where solid lines correspond to theoretical data derived in sections 3 and 4 of the present book.

From the analysis of results follows that the increasing tension force of the winding fiber results in reduced transversal characteristics σ_x, E_x, G_{xz} and some strength increment at a longitudinal shear τ_{xz}. Notice that experimental values of elastic properties of the material are in compliance with the theoretical ones.

It's been proved that a determining parameter of composites with the spirally reinforced filler is the auxiliary reinforcing layer thickness. Variations in the layer thickness ratio to the spirally reinforced element radius brings about variations in the filling degree with the main reinforcement and stress – strain state in it components. At least two methods of changing named parameter were found. They are the variation in geometrical parameters of

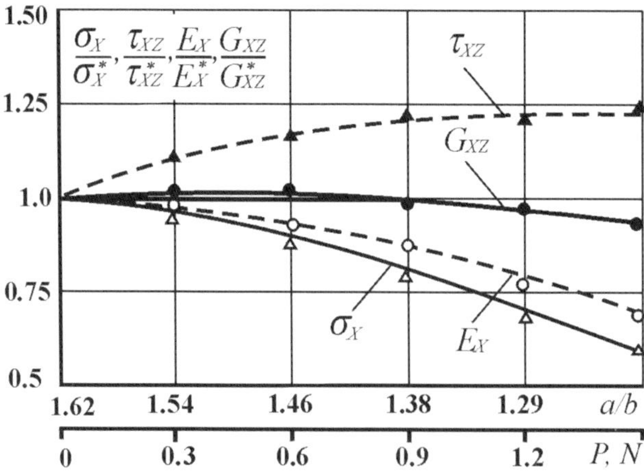

Fig. 5.16. Variations of mechanical characteristics of carbon-organoplastic depending on tension force of the winding fiber and the ratio of characteristic dimensions of the element (values of the characteristic at $P = 0$ are in the denominator)

the main reinforcement elements under preserved winding filament size, or enlarging the winding layer thickness, the amount of the main reinforcement being constant. Of great interest for experimental studies presents the latter variant as the scale factor effect arising at incrementing size of the spirally reinforced elements is eliminated in this case.

The glass fiber NSK-150/2 was used as the auxiliary reinforcement when preparing the samples for which winding layer thickness was varied by changing the amount of the original fibers in the winding reinforcement. The winding pitch was kept constant and equal to the original fiber width laid on the main reinforcement. This made provision for a linear increment of the winding layer thickness for different sample. Experimental results of mechanical characteristics of the material with the spirally reinforced filler as against to corresponding characteristics of a unidirectional material are shown in Figs. 5.17 and 5.18. It's evident that the increment in the winding layer thickness affects negatively the elastic and strength properties in reinforcement direction. In contrast, elastic characteristics in transverse direction under shearing and compressive strength perpendicularly to the main reinforcement increase perceptibly. From the above procedure follows that a complex of achieved data points to inexpedience of excessive augmenting relative thickness of the winding layer as it may lead to impaired bearing capacity of the material and efficiency of the spiral reinforcement procedure. Furthermore, the auxiliary reinforcement layer thickness should be estimated in terms of the requirements imposed on raising one or another characteristic of the structure under specific conditions and loading kind.

At the same time, an increase in the element size at a bounded material thickness results in violated homogeneity of the material stress field interre-

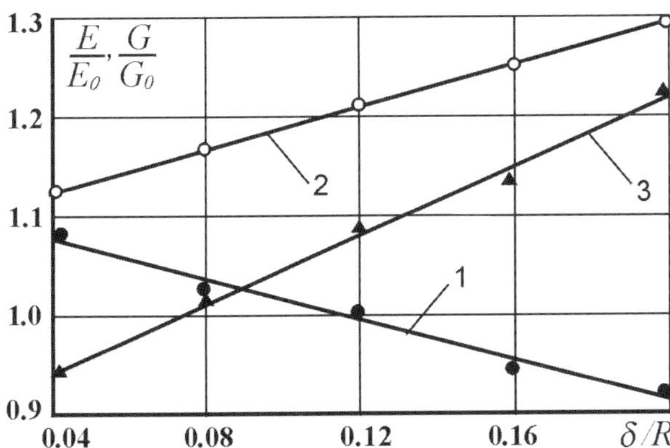

Fig. 5.17. Experimental dependencies of elastic characteristics of carbon-glass-reinforced plastic versus auxiliary reinforcement layer thickness (characteristics of unidirectional materials are in the denominator). *1 – E_Z/E_{Z0}; 2 – E_X/E_{X0}; 3 – G_{XZ}/G_{XZ0}*

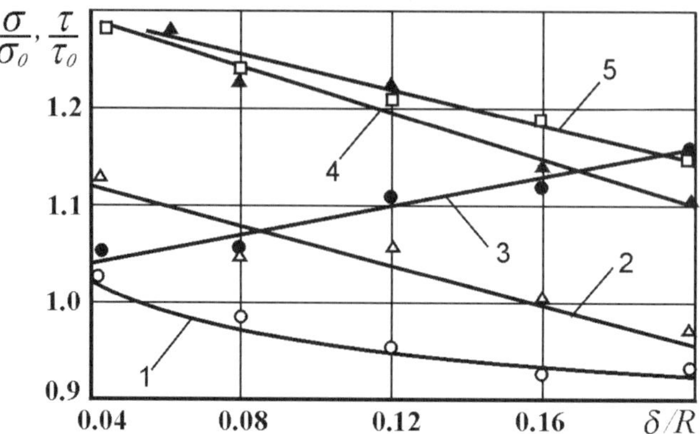

Fig. 5.18. Experimental dependencies of carbon-glass-reinforced plastic versus auxiliary reinforcement layer thickness (characteristics of unidirectional materials are in the denominator). $1 - \sigma_Z^+/\sigma_{Z0}^+$; $2 - \sigma_Z^-/\sigma_{Z0}^-$; $3 - \sigma_X^-/\sigma_{X0}^-$; $4 - \tau_{XZ}/\tau_{XZ0}$; $5 - \sigma_X^+/\sigma_{X0}^+$

lated with intensification of the edge effect. This suggests that the scale effect connected with variations in the cross-sectional area of the spirally reinforced element ratio to the sample thickness should be observed during testing. With this aim, samples with the spirally reinforced filler based on VMN-5 carbon braids and EDT-10 binder are manufactured. The binder is selected due to a necessity to evenly fill different size elements with the main reinforcement which is connected with application of significant tension forces to the auxiliary reinforcement if the prepregs are used for the process. Fibers based on organic SHP fibers of 1.3 tex densities were employed for the winding. Dimensions of the elements were set by introduction of different amounts of fibers into the main reinforcement which amount is necessary to achieve the given ratio of the element cross-sectional area to the sample thickness found within 3 to 4 mm.

From the analysis of experimental data presented in Fig. 5.19 can be made a conclusion that there exist some optimum correlations between geometrical parameters of spirally reinforced elements and sample thicknesses. Note that the scale factor effects stronger the transversal and shear characteristics of the materials.

At the same time, tensile stress variation in the main reinforcement direction is insignificant and doesn't surpass 5%. With increasing element size a considerable compressive strength increment is observed in direction of the main reinforcing. Presumably, this property may reach its maximum when the element dimensions are comparable with the sample thickness which is realized in coaxially reinforced elements. Notice that the compression strength increases in transversal direction as the element size diminishes, i.e. at a small-grain structure of the material. A combined analysis of the results ob-

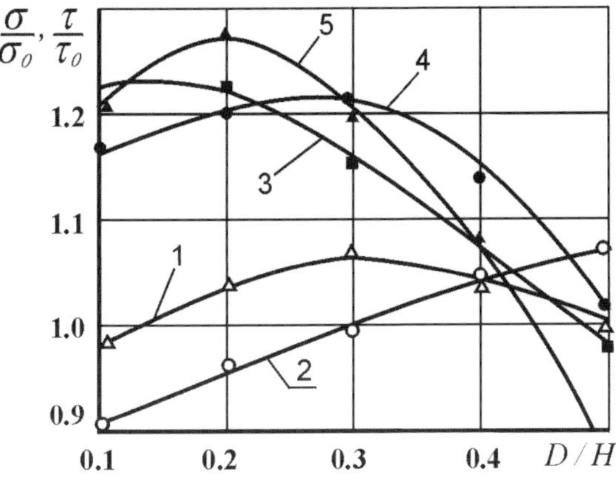

Fig. 5.19. Strength characteristics of carbon-organoplastic versus element cross-section ratio to sample thickness (notations correspond to Fig. 5.18)

tained has proved that most optimum for the material under study is the spirally reinforced element whose dimensions allow for the ratio of the element cross-sectional area to produced article thickness of the order of 0.2–0.3. Analogous results were derived for an organic plastic with spirally reinforced filler at bending tests. The test results are illustrated in Table 5.4. It's evident that the highest characteristics are displayed by the samples with spirally reinforced elements incorporating 5-folding of SHP-based fibers of 58.8 tex densities. They make provision for a 0.2 ratio of the element cross-sectional area versus the examined sample thickness.

Table 5.4. Mechanical Properties of Organoplastics at Bending

Material	E_b, MPa	S, MPa	v, %	σ_b, MPa	S, MPa	v, %
Unidirectional	$\dfrac{22720 - 3554}{28770}$	6440	22.4	$\dfrac{479 - 52}{498}$	21	4.3
Spirally reinforced, 3 braids	$\dfrac{28670 - 2978}{29180}$	560	1.9	$\dfrac{487 - 522}{501}$	19	3.7
Spirally reinforced, 5 braids	$\dfrac{40160 - 4660}{43830}$	3330	7.6	$\dfrac{602 - 658}{636}$	30	4.7
Spirally reinforced, 7 braids	$\dfrac{32230 - 3634}{34750}$	2210	6.4	$\dfrac{530 - 555}{545}$	13	2.4

* Main reinforcement – SHP fiber, 58.8 tex
Auxiliary reinforcement – SHP fiber, 14.3 tex
Binder – EDT-10.

5.4 Study of Mechanical Characteristics of Materials as Dependent on the Type of the Main and Auxiliary Reinforcement

The analysis conducted in the preceding chapters can make the base for recommending engineering and structural parameters of the composites with the spirally reinforced filling. Experimental studies of the materials have yielded perfect consistency between the theoretical and test data. Parameters shown in Table 5.3 are a confirmation that the forecasted elastic properties are close enough to experimental evidences, e.g. elastic modulus at tension in the main reinforcement direction deviates by 13.5%, the transversal elastic modulus by 12% and the transverse shear modulus by 29%. The comparison of experimental and theoretical data for transverse elastic modulus E_x presented in Fig. 5.20 suggests that with increasing filling degree till 0.55–0.6 the difference is minimal. As the filing grows the difference becomes more pronounced and is especially evident for the compression elastic modulus. As an example variations of elasticity moduli under tension and compression in transverse direction are depicted in Table 5.5 for the carbon-glass-reinforced plastic based on VMN-5 and UP-2220 and winding fiber NSK 150/2 in response to filling degree growth. At low filling degrees ($\varphi_{CM} < 0.5$) difference in elastic moduli is insignificant but increases with filling growth till 8–10%.

The agreement between theoretical and experimental data makes grounds for using earlier-proposed recommendations on the choice of structural and physico-mechanical properties of ingredients of the spirally reinforced composite. Later the materials structural characteristics maximum approaching optimum parameters were used for manufacturing test samples.

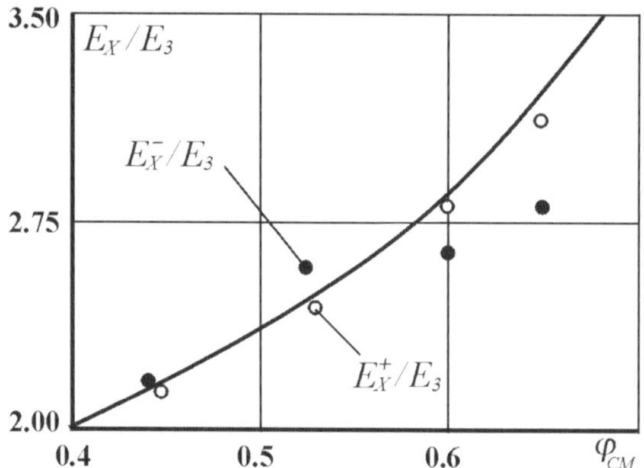

Fig. 5.20. The dependence of elastic modulus at tension and compression for a carbon-glass-reinforced plastic based on spirally reinforced filler (*Solid line* indicates calculated data obtained in 3)

Table 5.5. Investigation Results of Transverse Elastic modulus Dependence on Carbon-glass-reinforced Plastic Filling Degree with the Main Reinforcement

Filling degree, φ_{MR}	Total filling degree	Tensile elastic modulus			Compressive elastic modulus		
		E_X^+, MPa	S, MPa	v, %	E_X^-, MPa	S, MPa	v, %
0.4	0.47	7250	358	4.9	7300	390	5.4
0.46	0.53	8300	530	6.4	8640	230	2.7
0.53	0.60	9360	364	3.9	8490	125	1.3

Table 5.6. Variations of Strength and Elastic Properties of the Composite with the Spirally Reinforced Filler

Property[*]	Type of the main reinforcement[**]	
	Carbon braid of VMN-4	Glass fiber NSK 150/2
E_Z^+/E_{Z0}^+	0.97–1.12	0.965
E_Z^-/E_{Z0}^-	0.88–1.14	–
E_X^+/E_{X0}^+	1.2–1.4	1.3
E_X^-/E_{X0}^-	1.02–1.24	–
G_{XZ}/G_{XZ0}	1.28–1.06	1.18
σ_Z^+/σ_{Z0}^+	1.02–1.19	1.05
σ_Z^-/σ_{Z0}^-	1.02–1.44	1.10
σ_X^+/σ_{X0}^+	1.21–1.78	1.72
σ_X^-/σ_{X0}^-	1.14–1.41	1.16
τ_{XZ}/τ_{XZ0}	1.2–1.88	1.37
φ_{MR}	0.42–0.63	0.65

[*] Characteristics in the denominator correspond to unidirectional material.
[**] Auxiliary reinforcement – glass fiber NSK 150/2.

Investigations have visualized that spiral reinforcement helps to vary in fact all characteristics of the material. Table 5.6 presents strength and stiffness variations of carbon and glass fiber-based materials in provision of using glass auxiliary reinforcement only. It follows from the figure that transverse characteristics undergo greater variations, e.g. elastic modulus increases at transverse tension 1.2 to 1.4 times, shear modulus 1.6–1.28 times, strength at transverse separation 1.41–1.78 times strength at longitudinal shear 1.3–1.88 times. This behavior is affected to a greater extent by the material filling degree than by the main reinforcement type.

The considered characteristics of glass-glass-fiber-reinforced plastics are seen to be within the limits typical of the carbon-glass-reinforced plastic with different filling degrees. Strength at transverse tension and longitudinal shear are, as it's evident from the data, undergo most prominent changes. Variations of named characteristics in the carbon-glass-reinforced plastic constitute

Table 5.7. Strength and Elastic Characteristics of Carbon Plastics with the Spirally Reinforced Filler versus Type of Auxiliary Reinforcement

Characteristics	Carbon plastic VMN-5, UP-2220	
	Winding fiber 150/2	Winding fiber SHP (14.3 tex)
$E_Z \cdot 10^{-3}$, MPa	155.4	152.6
$E_X \cdot 10^{-3}$, MPa	9.63	11.8
$G_{XZ} \cdot 10^{-3}$, MPa	5.9	5.5
σ_Z^+, MPa	1024	997
σ_X^+, MPa	37.0	42.6
τ_{XZ}, MPa	58.0	64.7
$\gamma \cdot 10^{-4}$, N/m^3	1.55	1.49
$\varphi_{MR}/\varphi_{AR}$	0.54/0.07	0.52/0.07

1.78 and 1.88 in relation to unidirectional plastics, correspondingly. Similarly, for the glass-glass-fiber-reinforced plastic it's 1.72 and 1.37.

The carbon plastic with spirally reinforced filler was used to study the effect of auxiliary reinforcement on physico-mechanical characteristics of the composites. The main reinforcement was carbon braids VMN-5 and the binder – UP-2220.

Organic and glass fibers were used for winding. Investigation results illustrated in Table 5.7 indicate that the type of auxiliary reinforcement used exerts a negligible effect on characteristics in the main reinforcement direction.

The transverse elastic modulus augments by 26.9% as glass fibers are substituted by organic ones while strength increases on transverse separation just by 15.1%. All other characteristics vary but negligibly – 1.03–1.11 times. It's, however, to be born in mind that use of organic SHP fibers as the auxiliary reinforcement lowers the composite density and, consequently, its specific characteristics vary more noticeably. Thus, specific elastic modulus increases under transverse tension from 0.6×10^3 till 0.79×10^3 km when glass fibers are substituted by organic ones. Specific strength at shear varies similarly from 3.74 till 4.3 km and under transversal separation from 2.38 to 2.86 km. The rest specific characteristics of the material are roughly similar. Hence, it's possible to assert that carbon-organoplastics outdo carbon-glass-reinforced plastics by their specific characteristics.

Analysis of composites under study indicates that in spite of some loss in stiffness they gain in physico-mechanical characteristics in contrast to other analogous materials. From data presented in Table 5.8 a conclusion can be made that introduction of the spiral reinforcement reduces longitudinal elastic modulus from 161×10^3 down to 155.4×10^3 MPa, whereas transverse elastic modulus rises from 8.3×10^3 till 9.36×10^3 MPa and shear modulus from 5.0×10^3 till 5.9×10^3 MPa. It should be underlined that strength characteristics of the material with the spirally reinforced filler turn to be higher

Table 5.8. Mechanical Characteristics of Composites with Spirally Reinforced Fillers as Compared to Unidirectional Materials

Property	Carbon plastic VMN-5; UP-2220; NSK 150/2 fiber		Organic plastic SHP; EDT-10; SHP fiber		Glass plastic NSK; EDT-10; NSK 150/2 fiber	
	CP	SRCP	GP	SROP	GP	SRGP
$E_Z \times 10^{-3}$, MPa	161.0	155.4	73.3	75.7	38.5	37.0
$E_X \times 10^{-3}$, MPa	8.3	9.36	3.9	4.84	6.0	7.8
$E_b \times 10^{-3}$, MPa	124.2	131.3	28.8	44.5	–	–
$G_{XZ} \times 10^{-3}$, MPa	5.0	5.9	2.2	2.8	4.5	5.3
σ_Z^+, MPa	861.0	1024	1300	1420	1140	1200
σ_Z^-, MPa	461.0	637.0	226	287	490	540
σ_b, MPa	316.0	437.0	498	615	–	–
σ_X^+, MPa	29.0	37.0	14.1	20.6	25	43
σ_X^-, MPa	124.0	157.0	70.8	76.1	128	148
τ_{XZ}, MPa	31.0	58.0	48.0	72.1	38	52
$a \times 10^{-3}$, KJ/m^2	63.3	113.0	–	–	–	–
$\gamma \times 10^{-4}$, N/m^3	1.47	1.55	1.45	1.45	2.2	2.2
$\varphi_{MR}/\varphi_{AR}$	0.57/0	0.54/0.07	0.58/0	0.58/0.07	0.67/0	0.55/0.10

CP – carbon plastic;
SRCP – spirally reinforced carbon plastic;
SPOP – spirally reinforced organic plastic;
SP – glass plastic;
SRGP – spirally reinforced glass plastic

under all types of loading as compared to usual unidirectional composites even with less filling degree with the main reinforcement. As for instance, the strength in direction of the main reinforcement increases from 861 till 1024 MPa, although filling degree of the unidirectional material constitutes 0.57 and that with the spirally reinforced filler −0.54. This can be attributed to high-quality packing of fibers within the core of the spirally reinforced element along with eliminated fiber flexion and improved cohesion of fiber bunches at the interface.

Analysis of specific characteristics of the composite presented in Table 5.9 suggests that the spirally reinforced composites surpass unidirectional materials almost in all specific parameters. Figure 5.21, where a diagram of specific strength and stiffness at transverse loading is illustrated and Fig. 5.22 with specific strength variations at tension, compression and bending are futher proof to higher specific properties of the composites with the spirally reinforced filler in contrast to unidirectional reinforced ones. The comparison of anisotropic parameters of materials with the spirally reinforced filler (Table 5.10) and a hybrid one with a polyfibrous reinforcement shows that the composites are less anisotropic. For instance, the carbon plastic with

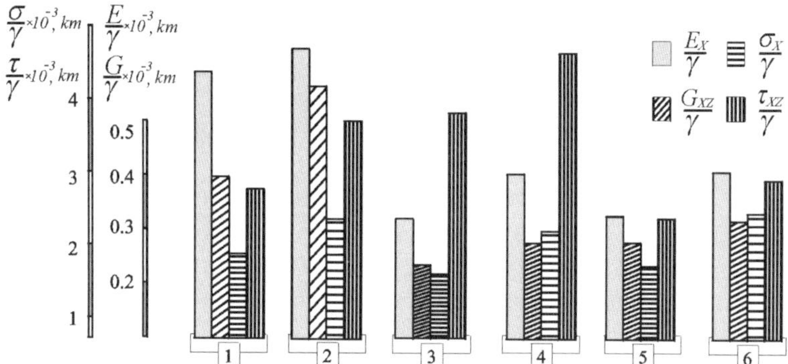

Fig. 5.21. Specific characteristics of composites under transversal tension and longitudinal shear. *1* – carbon plastic; *2* – carbon plastic with spirally reinforced filler; *3* – organic plastic; *4* – organic plastic with spirally reinforced filler; *5* – glass-reinforced plastic; *6* – glass-reinforced plastic with spirally reinforced filler

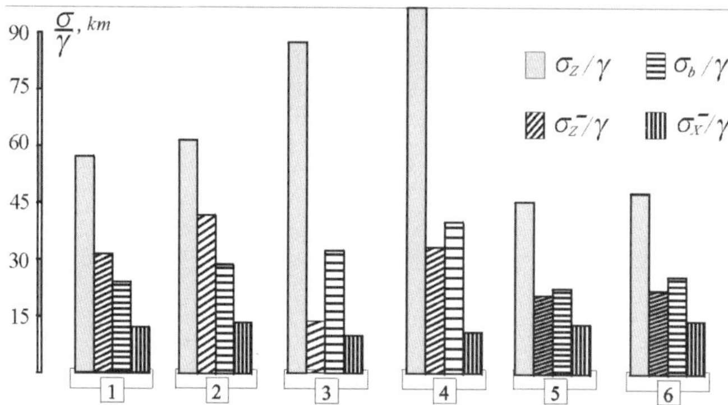

Fig. 5.22. Specific density of composites. Notations are similar to Fig. 5.21

the spirally reinforced filler displays $\sigma_Z^+/\tau_{XZ} = 14.2$–$17.5$, while the hybrid carbon-glass-reinforced plastic -30–35.

Of particular interest is the question of the spiral reinforcement effect on properties of the composites having organic or boron polyfibrous filler as the main reinforcement. In this connection, the effect of the amount of boron fibers in the main reinforcement on the main physico-mechanical properties of the material with the spirally reinforced filler has been studied. As the main reinforcement also organic SHP fibers of 14.3 tex linear densities were used.

The results presented in Fig. 5.23 visualize that shear strength doesn't depend essentially on the boron fiber content, whereas compression strength and elastic modulus at bending increase linearly with incrementing fiber amount. Bending strength values are strongly scattered, so far any statistically sig-

Table 5.9. Specific Properties of Composites

Specific properties, km	Carbon plastics VMN-5; UP-2220		Organic plastic SHP; EDT-10		Glass-reinforced plastic NSK; EDT-10	
	UD	SR 150/2 NSK 150/2	UD	SR SHP (14.3 tex)	UD 150/2	SR NSK
$E_Z^+/\gamma \times 10^{-3}$	10.9	10.4	5.05	5.22	1.75	1.68
$E_X^+/\gamma \times 10^{-3}$	0.56	0.60	0.27	0.33	0.27	0.35
$G_{XZ}/\gamma \times 10^{-3}$	0.34	0.38	0.15	0.19	0.20	0.24
σ_Z^+/γ	58.5	66.1	89.6	97.9	51.8	54.5
σ_X^+/γ	1.99	2.38	0.97	1.42	1.14	1.95
τ_{XZ}^+/γ	2.11	3.74	3.31	4.97	1.73	2.36

UD – unidirectional;
SR – with spirally reinforced filler.

Table 5.10. Parameters of Composites with the Spirally Reinforced Filler

Parameters/composite	E_Z^+/E_X^-	E_Z^+/G_{XZ}	σ_Z^+/σ_X^-	σ_Z^+/τ_{XZ}	σ_Z^+/σ_Z^-	σ_Z^+/σ_X^-
Carbon-glass-plastic (VMN-5, UP-2220)	16.5	26.5	28.8	17.5	1.6	6.5
Carbon-glass-reinforced plastic (VMN-4, EDT-10)	15.9	29.3	24.0	14.2	1.51	5.5
Carbon-organoplastic	12.9	27.6	23.3	15.4	–	–
Organo-organoplastic	15.6	27.0	68.6	19.7	4.95	18.6
Glass-glass-reinforced plastic	4.75	6.96	28.0	23.1	2.22	8.11
Organo-glass-reinforced plastic	10.4	13.7	49.0	22.4	2.6	12.3

nificant dependence of the characteristic on the boron fiber content isn't observed.

As it has been proved earlier [234], that the materials with the spirally reinforced filler display higher impact viscosity as compared to common composites. To establish impact viscosity relation to the auxiliary reinforcement content a carbon-glass-reinforced plastic has been tested. The derived results are illustrated in Fig. 5.24. It's evident that impact viscosity dependence tends to saturation as the auxiliary reinforcement content increases. Its limiting values reach the corresponding impact viscosity magnitude of the auxiliary reinforcement (glass-reinforced plastic in this case). An abrupt rise of the parameter is observed only within 10–12% of the auxiliary reinforcement content which indicates the expedience of introducing large amounts of the winding material.

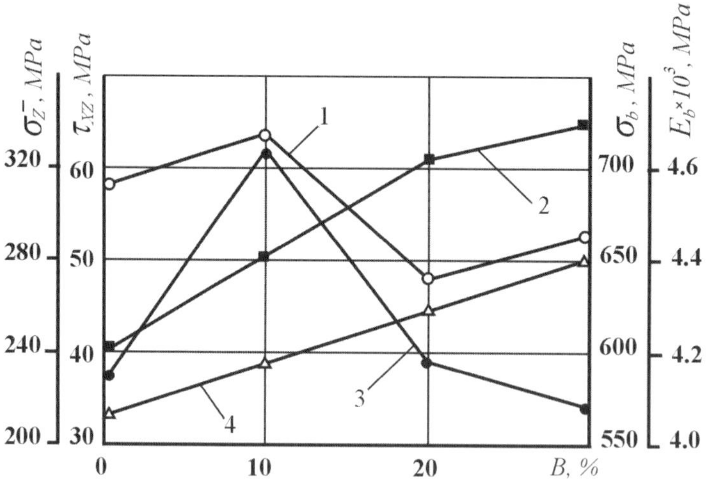

Fig. 5.23. The dependence of mechanical characteristics on boron fiber content in main reinforcement for organo-boron-reinforced plastics with the spirally reinforced filler. *1* – shear strength; *2* – compression strength; *3* – bending strength; *4* – elasticity modulus at bending

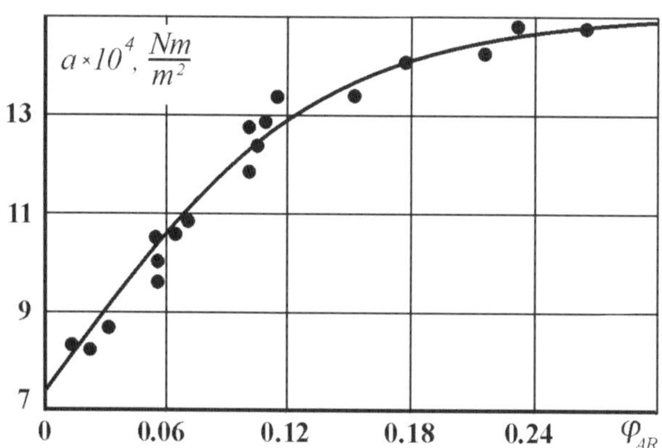

Fig. 5.24. The dependence of impact viscosity on filling degree of the carbon-glass-reinforced plastic with the auxiliary reinforcement

Above cited experimental results are consistent with theoretical findings that the developed material structure with a hybrid spatially reinforced filler yields composites with improved characteristics under transversal loading, shear and longitudinal compression.

5.5 The Analysis of Mechanic Characteristics of Coal-Plastics Based on Spirally Reinforced Fillers

While carrying out this research, the methodology of rational planning, based on the development of the casual balance method, has been used, with this the averaging of data happens due to a specially planned combination of factors, at which each of them appears only once in all variants. This is why fewer experiments are needed to average the data.

To simplify the calculation work while processing the results of the planned experiments the normalized parameters X_i have been introduced. They are:

$$X_i = \frac{Z_i - Z_{iH}}{\Delta Z_i}, \tag{5.34}$$

where: Z_{iH} – initial value of the factor; ΔZ_i – rate of variation.

Besides this, the introduction of normalized parameters allows better evaluation of the relative influence force of factors at their modification at a given variation interval. The transition to actual parameters is accomplished after obtaining the dependence $f(X_i)$ through substituting the values Z_i, defined from the formula (5.34). The intervals of factor variation, their initial values and variation rate, used while holding the planned experiment, are introduced in Table 5.11.

As far as the square of the cross-section of one carbon braid and one glass fiber have been used for lower parameters F_a^0 and F_a^b, the normalized parameters define, correspondingly, the number of braids in the element of the main reinforcement and the number of fibers of the armature winding.

In compliance with the plan of experiments for the analysis of the dependence of mechanic characteristics of coal-plastics based on spirally reinforced fillers from the complex of structural and technological parameters it is necessary to test 25 sets of samples. For this the following types of testing have been held.

1. Stretching in the direction of reinforcement.
2. Stretching in crosswise direction.
3. Compression in the direction of reinforcement.

Table 5.11. The Main Values of Varying Parameters

Factor	F_a^0, mm^2	F_a^b, mm^2	t, mm	T_H, H	t_y/D	P, MPa
The greatest value	1.018	0.059	5	0.5	1.4	2.5
The least value	0.204	0.012	1	0.2	1.0	0.5
Variation rate	0.204	0.012	1	0.075	0.1	0.5
Initial value	0	0	0	0.125	0.9	0
Normalized parameter	X_1	X_2	X_3	X_4	X_5	X_6

4. Compression in crosswise direction.
5. Shift in the bar placing plane.
6. Defining impact elasticity.

5.5.1 The Analysis of Elastic Characteristics

The results of the experiment have been grouped according to the meaning of each factor and to the received averaged values the graphs of dependencies of mechanical values and normalizeded values of structural and technological parameters have been built. To evaluate the utility of using spiral reinforcement, the dependencies are introduced as ratios to the corresponding characteristics of unidirectional coal plastic. For easier orientation in graphs of dependencies, the number of each curve in the picture corresponds to the number of the normalized parameter from Table 5.11. The light shapes define the points of averaged values on full curves of dependencies on the structural parameters, the shaded ones – on the dotted curves of technological factor dependencies.

To compare with the theoretical data, a calculation of elastic characteristics of coal-plastics based on spirally reinforced fillers at the structural parameter value, corresponding to the plan of experiment, and their similar grouping according to the value of normalized factors.

In Fig. 5.25 the graphs for experimental (a) and theoretical (b) dependencies of the averaged elastic moduli and technological parameters. The analysis of the point location on the curve shows that out of all varying parameters there is only one – the square of the winding thread cross-section (curve 2) that appears to be strong, and its elastic characteristics in the direction of reinforcement decrease linearly. In the given interval of varying this decrease can reach 15–13%. The comparison of experimental and calculation data shows that there should be another dependence on the winding rate (curve 3, Fig. 5.25a), but it does not appear to be so in practice, as far as the variation of the elastic moduli in the direction of reinforcement due to the rest of the factors stays within the limits of the experiment. It is worth

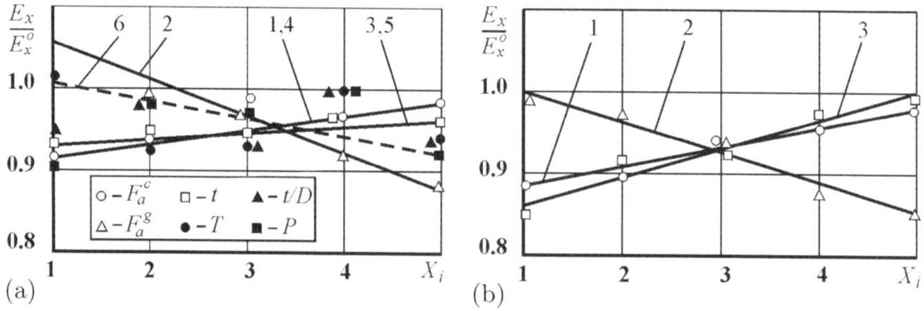

Fig. 5.25. The dependence of elastic moduli in the direction of reinforcement on the structural and technological parameters

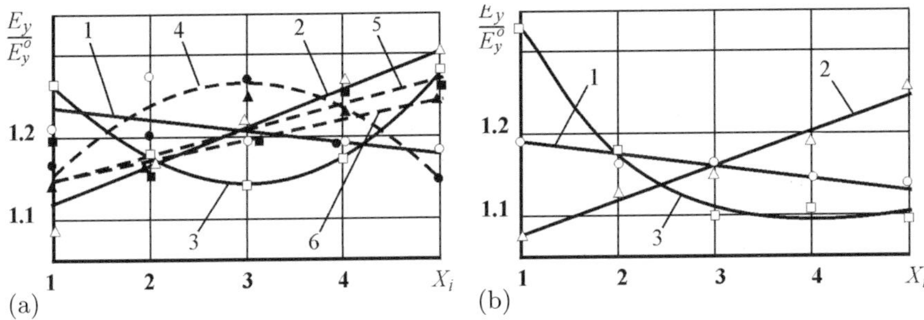

Fig. 5.26. The dependence of the cross modulus of elasticity on structural and technological parameters: notations are similar to Fig. 5.25

noting that the average values of the received experimental data with respect to the monodirected plastic are located higher than the theoretically calculated values, and while using a thin thread for winding they can equate to much more than one unit. The influence of variations in the given limits of technological parameters is rather weak and has no significant impact on the elastic moduli in the direction of reinforcement.

The analysis of the elastic characteristics in the direction, perpendicular to the reinforcement show that in this case there are contrary dependencies on structural parameters in comparison with the elastic moduli in the direction of reinforcement (Fig. 5.26).

Thus, the increase of the winding thread cross-section square contributes to the efficient increase of the cross elastic moduli (curve 2). At the same time, the increase of the main reinforcement square leads to its decrease (curve 1). The increase of the technological parameters value also causes the increase of the elastic characteristics in the cross direction (curves 5, 6), except the drawing power of the winding thread (curve 4), the best values of which are average. The received experimental data corresponds to the theoretical one (Fig. 5.26b). At this, both theoretical and experimental analyses of cross elastic moduli show that it is possible to increase this characteristic of coal-plastics based on spirally reinforced fillers in comparison with monodirected coal-plastics by 20–25%.

The analyses of the elastic characteristics while shifting in the plane of placing the bars show that their dependencies on structural parameters, both experimental and theoretical are of the same character with the elastic modulus in the cross-section (Fig. 5.27). However, the degree of influence of each of the parameters is stronger. Thus, the increase of the square of the main reinforcing braid leads to a decline of the shift moduli by 25% (curve 1), and an increase of winding thread square – to its similar rise (curve 2). The impact of technological parameters on the shift modulus does not exceed 10% (curves 4, 5, 6).

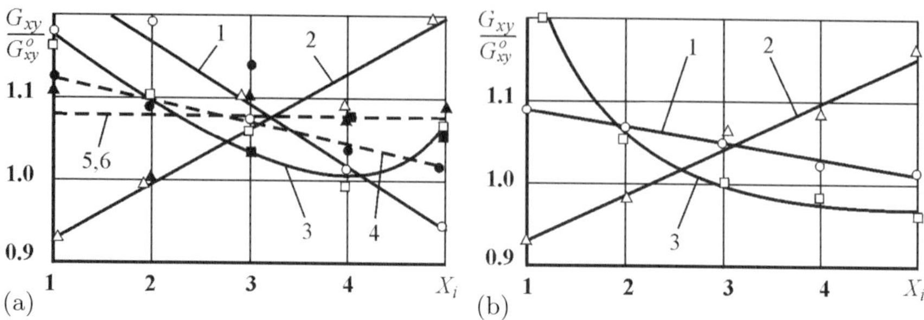

Fig. 5.27. The Dependence of longitudinal elastic moduli on structural and technological parameters: notations are similar to Fig. 5.25

5.5.2 The Analysis of Strengthening Characteristics

The processing of the results of the experiment by definition of the strengthening characteristics was held similarly to the analysis of the elastic characteristics. The dependence of durability of coal-plastics based on spirally reinforced fillers while stretching in the direction of reinforcement on structural and technological parameters is given in Fig. 5.28a.

As can be seen from the graphs, the structural parameters have the main impact on the strengthening characteristics. Moreover, with the increase of the cross-section square of the main and supplementary reinforcement, the durability decreases (curves 1 and 2), and with the increase of the winding rate it increases, approaching the durability of the monodirected plastic (curve 3). The location of the curves on the graphs shows that the average values of durability in the direction of reinforcement are by 5% lower than monodirected coal-plastics. The analysis of the strengthening characteristics while compressing (Fig. 5.28b) shows that the structural parameters appear to be the strongest (curves 1, 2, 3), and out of the technological ones – the drawing power of the winding thread (curve 4). The character of the curves points to the connection between the durability of coal-plastics based on spirally reinforced fillers while compressing and the possible initial bend of

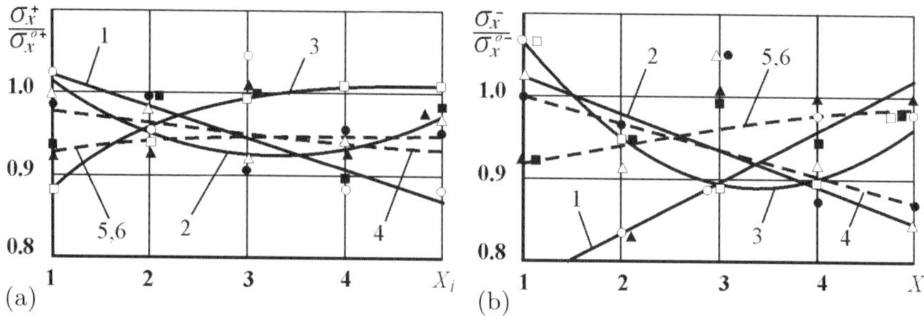

Fig. 5.28. Durability dependence on loading in the direction of reinforcement on structural and technological parameters: notations are similar to Fig. 5.25

fibers due to the loss of stability of the main reinforcement in the process of its spiral winding.

At the same time, the average values of coal-plastics durability based on spirally reinforced filler on compression exceeded the durability of a monodirected coal-plastic. This is connected with the fact that while compressing in the direction of reinforcement, the spiral winding with a low speed promotes the retention of fiber stability inside the structural elements, which in this case work as one entire fiber of a large diameter. At this, the values of critical loading increase and the factor of using the durability of the initial reinforcement rises. The impact of the given factors can interrupt the effect of decreasing the durability due to diminution of the main reinforcement filling factor, which leads to the increase of the absolute value of durability of this material in comparison with monodirected reinforced plastics.

The analysis of strengthening characteristics of coal-plastics based on spirally reinforced fillers at stretching and compressing in the direction, perpendicular to reinforcement (Fig. 5.29) reveals extremely strong and equal dependences on the cross-section square of the main reinforcement (curve 1).

The decrease of durability while increasing the diameter of the braid being wound in the given limits of variation is estimated at 30% on compression and 60% on stretching. This may be connected with a small thickness of the sample plates being tested, impeding the creation of a homogeneous tension field in the material with large structural elements. The influence of other factors, as well as while testing in the direction of reinforcement, is contrary, depending on the type of testing. The increase of thickness of the winding thread (curve 2) decreases the durability on stretching and increases on compression. The variations of technological factors to a greater extent affect the value of durability on compression (curves 5, 6)

Comparing the data of coal-plastics transversal durability based on spirally reinforced fillers and monodirected coal-plastic shows that all averaged values lay 10% higher when tested on compression and 20–25% higher at stretching. At the same time, using the method of choosing the necessary

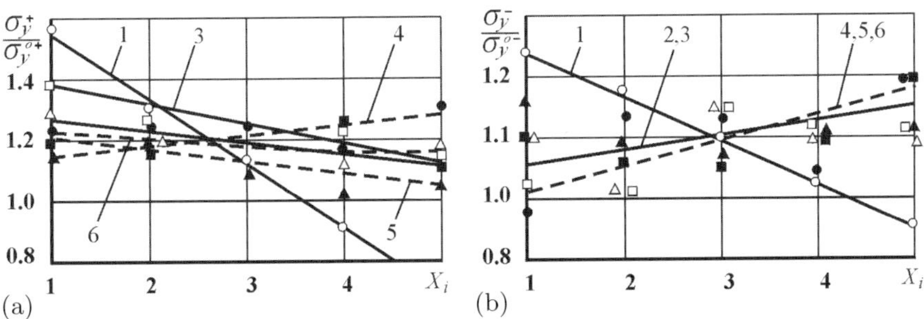

Fig. 5.29. The dependence of transversal durability on structural and technological parameters: notations are similar to Fig. 5.2

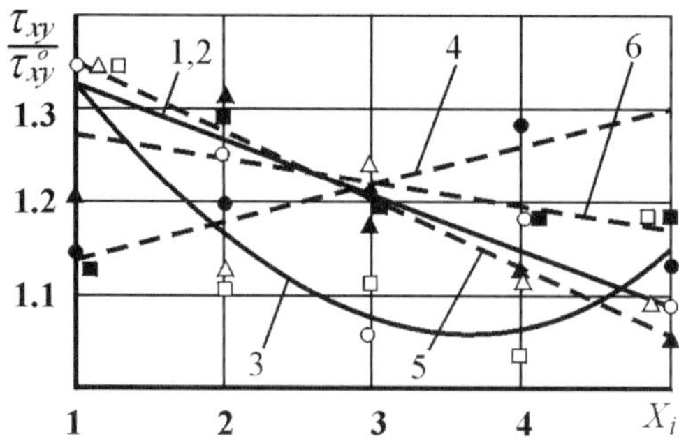

Fig. 5.30. The dependence of shift durability on structural and technological parameters: notations are similar to Fig. 5.25

parameter values there is an opportunity to increase them by 30% and 60% correspondingly.

The dependence of elasticity at shifting (Fig. 5.30) is of a character, similar to durability at stretching perpendicularly to reinforcement except the winding speed, whose impact at stretching insufficiently influences the value of strengthening characteristics, while the shift durability decrease with the increase of speed (curve 3). This is connected with the fact that at big speed values the mechanism of catching the winding wraps of neighboring elements is not working, which equips the thread of supplementary reinforcement at low winding speed.

The results of impact elasticity analysis (Fig. 5.31) show that it increases on average by 80–90% At this, the strongest are the factors, promoting the rise of supplementary reinforcement filling degree. Thus, increasing the thickness of the winding thread, the impact elasticity rises sharply (curve 2), and on increasing the speed – it falls (curve 3). A similar impact is made by the winding thread exertion (curve 4) and other technological factors, promoting the degree of reinforcement (curve 6).

The increase of impact elasticity on the cross-section square of the main reinforcement (curve 1) is connected with the peculiarity of defining the impact elasticity on impact bend. At this, the coal-plastics gets destroyed due to destruction of the main reinforcement fiber being compressed, and the dura-

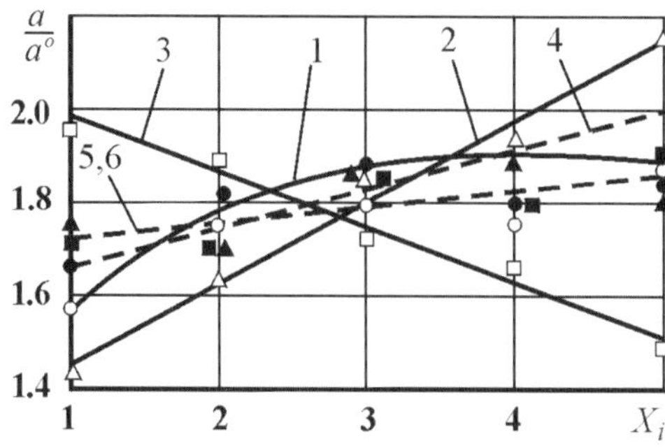

Fig. 5.31. The dependence of impact elasticity on parameters: notations are similar to Fig. 5.25

bility on compression, as is shown in Fig. 5.2b (curve 1), directly depends on the main reinforcing braid cross-section square.

5.5.3 The Analysis
of the Mechanic Characteristics Research Results

The results of the experimental and theoretical research show that structural and technological parameter variations of coal-plastics based on spirally reinforced fillers have a different impact on different mechanical characteristics. The complex of parameters, leading to the increase of one of the characteristics, may promote a decrease of other characteristics. To make the analysis of the combined influence of structural and technological factors on the complex of material properties more convenient, all data of its mechanical characteristics change at parameter variation are provided in Table 5.12. The sign "+" or "−" indicates the increase or decrease of this characteristic on variations of the parameter from the lower border to the upper one at a given variation interval. When changing a characteristic in the limits of 5–15% a figure of 1 is used, 20–25% – 2, more than 30% – 3.

It follows from the analysis of the table that the increase of the main reinforcement cross-section square leads to a drastic decrease of all strengthening characteristics, except the durability on compression in the direction of reinforcement and elastic characteristics, especially in the transversal direction and when shifted. The increase of the winding rate has a negative impact on practically all characteristics of the material. The overall balance of data of the winding thread drawing power and compacting pressume of approaching these parameters' value to the upper bound. At the same time, it suggests taking the speed close to the diameter of the structural element, while its increase leads to the decrease of durability at shift.

Table 5.12. The Influence of Structural-Technological Parameters on Mechanical Properties

Characteristics	Parameters					
	F_a^0	F_a^b	t	T_H	t_y/D	P
E_X	+1	−2	0	0	0	0
E_Y	−1	+2	−1	+1	+1	+1
G_{XY}	−2	+2	−2	−1	0	0
σ_X^+	−1	−1	+1	0	0	0
σ_X^-	+2	−1	−1	−1	0	+1
σ_Y^+	−3	−1	0	+1	−1	0
σ_Y^-	−3	+1	0	0	+1	+1
τ_{XY}	−2	−2	−2	+1	−2	−1
a	+2	+3	−3	+2	+1	+1

Table 5.13. The Reference Values of Structural and Technological Parameters for Standard Constructions

Construction type	Parameters					
	F_a^0, mm^2	F_a^b, mm^2	t, mm	T_H, H N	t_y/D	P, MPa
1. Thin shells of intrinsic pressure	0.2	0.012	2.5	0.5	1.1	1.0
2. Thin shells of external pressure	1.0	0.012	0.5	0.2	1.3	1.0
3. Thick wall shell external pressure	0.2	0.012	1.0	0.5	1.1	0.5
4. Rod constructions, working into stretching	1.0	0.012	2.5	0.5	1.1	1.0
5. Rod construction, working into compression and bend	0.6	0.012	0.5	0.5	1.0	1.0
6. General purpose plates	0.4	0.012	1.0	0.5	1.1	2.5
7. Impact-resistant constructions	1.0	0.059	0.5	0.5	1.1	2.5

Thus, the received results show that it is impossible to uniquely define the parameters of the optimal coal-plastic structure based on spirally reinforced fillers, and while developing a specific unit it is necessary to choose their values so that they will provide the increase of characteristics that are determining in this construction under certain conditions and types of loading.

The reference values of structural and technological parameters, recommended for the material types of some standard constructions are given in Table 5.13.

As can be seen from the analysis, optimizing all the parameters allows manufacturing coal-plastics based on spirally reinforced fillers with characteristics increasing the values of mechanical characteristics of monodirected materials by 40–60%. This leads to the rise of not only absolute but also specific characteristics of the material, which allows material consumption and the weight of constructions being manufactured to be decreased.

5.6 The Analysis of Fracture Processes in Composites with Spirally Reinforced Filler

The results of experimental investigations of strength characteristics of composites based on spirally reinforced fillers state that they are much higher than those of the unidirectional plastics under in fact all loading types. As far as heightened strength characteristics of materials based on the spirally reinforced filler transversely and on shearing have been substantiated theoretically, so their characteristics in the main reinforcement direction should, in principle, have lower values as the main reinforcement content in the material diminishes. Note that the winding layer effect on strength characteristics of

Table 5.14. Comparative Strength Characteristics of Fillers Based on Carbon Braid VMN-5

Filler type	Breaking tension, N	Variation factor, %
Original carbon braid	104.0	13.1
Carbon braid impregnated with EDT-10 binder	119.7	11.4
Dry carbon braid wound with glass fiber NSK 150/2	145.2	8.6
Carbon braid impregnated with binder and wound with glass fiber NSK 150/2	155.0	6.2

the reinforced plastic is perceived already at the stage of producing spirally reinforced elements. It follows that winding contributes to better technological effectiveness and material strength. This is attained through straightening of elementary fibers, compaction and reduction of the reinforcement braid to make them operate jointly on loading.

Comparative data on strength characteristics of fillers based on carbon braid VMN-5 and those of the spirally reinforced braid are illustrated in Table 5.14. The analysis of data indicates that winding with the auxiliary reinforcement improves filler strength by 40%.

In the course of manufacturing spirally reinforced fillers their main reinforcement twists to some angle under the torque induced by the winding fiber tension force. Its absolute value is dependent upon fiber disposition relative to the braid center and is found within 1–5° about the outer layer. It's indicated in [235, 236] that under named twist angles improved strength characteristics of composite materials are observed owing to the cable structure of the reinforcing braids.

The analysis of crack propagation and fracture in composites set forth in [118] has visualized that the increased diameter of reinforcing fibers augments the force with which they are pulled-out of the material structure. Since the technological effectiveness and strength of the fibers impairs with diameter increment there has to be a tradeoff consisting in fiber grouping into bunches which are sufficiently flexible at processing and behave as individual fibers on breakage. In the mentioned investigation resistance to cracking showed to improve in case fibers were grouped in bunches. This parameter underwent variations if mechanical properties of the bunch/matrix interface were varied. Finally, the bunches got broken in the crack plane. It was also shown in the work that stress distribution in fibers at transition to named fracture mode was more efficient as compared to pulling out by roughly an order of magnitude. Analogous results were obtained in [237, 238].

Introduction of the main reinforcement into materials with the spirally reinforced filler realizes the cable type of composite structure which allows for varying mechanical characteristics of the bunch interfaces. In Fig. 5.32 fracture modes of various samples based on carbon braids VMN-5 and UP-

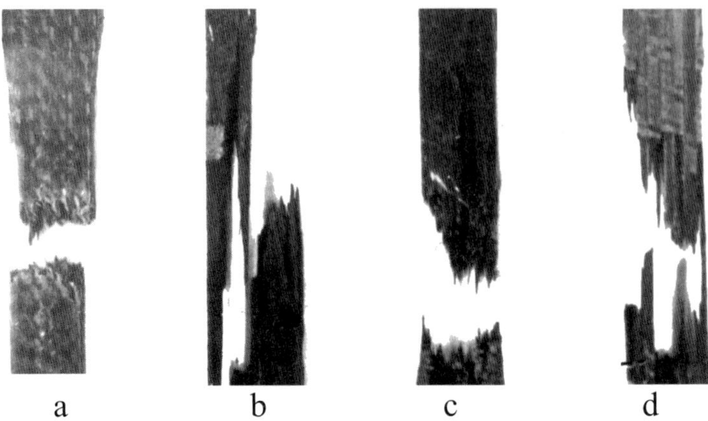

Fig. 5.32. Failure of samples at tensile tests in the main reinforcement direction

2220 binder are presented at tensile tests in direction of the main reinforcement packing. It's evident from 5.32, a that the materials based on spirally reinforced fillers acquire perfect bonds at the interfaces of fiber bunches of the main reinforcement (element diameter – 0.7 mm, glass fiber NSK 150/2 winding at 1 mm pitch with mutual engagement of coils), and their bearing capacity is preserved till certain concentration of broken fibers within a short portion, and their failure occurs without any delamination. These results corroborate well with conclusions made in [118].

Surface failure analysis of a carbon plastics based on the spirally reinforced filler has also shown (Fig. 5.33) that the number of pulled out fibers reduces considerably and their length is about 20–25 μ. As it's seen from Fig. 5.33, failure of the major part of fibers lies within the crack propagation plane. In contrast, failure of unidirectional carbon plastics is the result of static accumulation of damage in most weak and faulty areas of the sample followed by the lengthwise crack dissemination over fiber interfaces. Fracture is accompanied by pulling separate fibers, their groups or bunches out of the material structure (Fig. 5.32b).

Differences in fracture behavior of the samples with a continuous winding and engagement of coils in the main reinforcement can be judged from the comparison of Figs. 5.32a and b. In case of a small reinforced element size ($\varnothing$0.7) with continuous winding damages are accumulated within a longer portion of the sample which induces failure with rupture of individual elements on different portions followed by their pull out of the structure. In spirally reinforced elements with larger dimensions (till $\varnothing$1.6 mm) fiber bunches become insufficient for realizing failure mechanism typical of unidirectional materials. In this case, shear is observed at failure inside the element and at interfaces of the main and auxiliary reinforcements (Fig. 5.32d). Mentioned effects improve the use factor of the initial reinforcement strength in rein-

Fig. 5.33. View of a zone of the carbon plastic with the spirally reinforced filler at tension in direction of the main reinforcement packing ($\times 500$)

forcing direction and fully or in part overlap rated strength impairment by reducing filling degree with the main reinforcement.

Under compression in reinforcement direction the winding layer contributes to maintaining stability of fibers inside the spirally reinforced elements. They operate like an integral large-diameter fiber which augments, as is known [1], critical loading bringing about breakage of the material. In contrast, failure of reinforced in one direction materials on compression in reinforcement direction follows most often the delamination mechanism along fibers (Fig. 5.34b). Elimination or suppression of the delamination results naturally in raised bearing capacity of the reinforced polymer materials. It

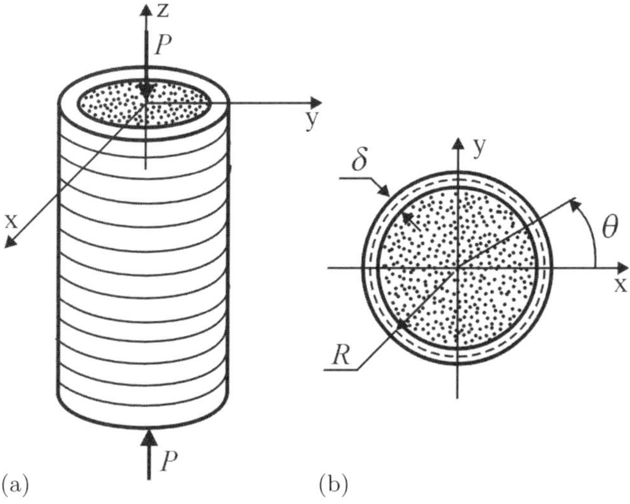

Fig. 5.34. On stress computation of a spirally reinforced element on compression

has been found out that the shear mode at failure of composites with the spirally reinforced filler is less probable since the winding layer exerts damping effect at tension transition to the interelement area. This phenomenon has been theoretically substantiated in preceding chapters. Spiral winding also promotes elevation of shear characteristics due to enlarging failure surface and mechanical engagement of winding coils of the neighboring spirally reinforced elements. The auxiliary reinforcement is involved in the phenomenon too together with the matrix and reinforcement/binder interface. As a result, the material with the spirally reinforced filler breaks on compression in reinforcement direction within the field of maximum stresses without any delamination in direction of the main reinforcement.

To validate the favorable effect of the auxiliary reinforcement let's consider the problem of a spirally reinforced element compression in direction of the main reinforcement, in case the element inside is a continuous transtropic cylinder (main reinforcement) and its outside is an orthotropic thin cylinder (auxiliary reinforcement layer) (Fig. 5.34).

In presumption of an ideal mechanical contact between both parts and smallness of the auxiliary reinforcement layer thickness, the equilibrium equation of displacements is considered to be valid on the median surface of the layer $(r = R)$. Substituting Lame's equation derivatives by their finite-difference analogs, we get

$$A_{11}(u_r^+ - u_r^-) + \frac{\delta}{2R}[A_{12}(u_r^+ - u_r^-) - R\sigma_r^-] + \frac{\delta}{2R}A_{13}R\left(\frac{\partial W^+}{\partial z} + \frac{\partial W^-}{\partial z}\right) = 0$$

(5.35)

$$(A_{11} - A_{12})(u_r^+ - u_r^-) - R\sigma_r^- + \frac{\delta}{2R}[(A_{12} - A_{22})(u_r^+ - u_r^-)]$$

$$+ (A_{13} - A_{33})R\left(\frac{\partial W^+}{\partial z} + \frac{\partial W^-}{\partial z}\right) = 0.$$

(5.36)

A_{jk} – are elasticity moduli of the auxiliary reinforcement layer; u_r^+, u_r^-, W^+, W^-, σ_r^- – displacements and stress on the layer inner surface (sign $-$) and outer surface (sign $+$).

We'll assume that $W^+ = W^- = a_1 z$; $u_r^{(1)} = -a_1 v_{zx}^{(1)} r + a_1 a_2 k r$, where

$$k = 2v_{zx}^{(1)} \frac{1 - 2v_{zx}^{(1)} v_{xz}^{(1)} - v_{xy}^{(1)}}{E_z^{(1)} v_{xz}^{(1)}},$$

and a_1, a_2 – unknown constants; $u_r^- = u_r^{(1)}(R_1)$; $u_r^{(1)}$ – travel of dots of the body inside; $R_1 = R - \delta/2$; $E_z^{(1)}$ – elastic modulus of the main reinforcement material in z axis direction.

Equilibrium equations for the inside are fulfilled identically.
Using Hook's law

$$\sigma_r^{(1)} = \sigma_r^- = 2a_1 a_2,$$

(5.37)

where $\sigma_r^{(1)}$ is the radial stress in the main reinforcement.

By substituting σ_r^- and u_r^- into (5.35) and (5.36), one gets two equations for determining constants a_r and u_r^+, wherefrom

$$a_2 = \frac{\delta}{2R}\,\frac{E_z^{(2)} v_{\theta z}^{(2)} \left(v_{zx}^{(2)} - v_{z\theta}^{(2)}\right)}{v_{z\theta}^{(2)}\left(1 - v_{\theta z}^{(2)} v_{z\theta}^{(2)}\right)}\,, \tag{5.38}$$

where index (2) denotes the auxiliary reinforcement layer characteristics.

To find constant a_1 (reduced stiffness on compression) we'll make use of boundary conditions on the element face [239]

$$\int_0^{2\pi R} \int_0^{+\delta/2} r\sigma_z\, dr\, d\theta = -P\,.$$

As a result,

$$a_1 = -\frac{P}{\pi R_1^2}\left[E_z^{(1)} + 2\frac{\delta}{R} E_z^{'(2)}\frac{1 - v_{zx}^{(1)} v_{z\theta}^{(2)}}{1 - v_{\theta z}^{(2)} v_{z\theta}^{(2)}} - 2v_{zx}^{(1)}\frac{\delta}{R}\frac{E_z^{(2)} v_{\theta z}^{(2)}\left(v_{zx}^{(1)} - v_{z\theta}^{(2)}\right)}{v_{z0}^{(1)}\left(1 - v_{\theta z}^{(2)} v_{z\theta}^{(2)}\right)}\right]^{-1} \tag{5.39}$$

As an example let's consider compression of a round spirally reinforced element whose inside is represented by carbon fibers wound by glass fibers at a pitch that ensures continuity of the winding layer.

By substituting geometrical and elastic parameters of the main reinforcement and the layer ($E_z^{(1)} = 133 \times 10^3$ MPa, $E_z^{(2)} = 8.5 \cdot 10^3$ MPa, $v_{zx}^{(1)} = 0.37$; $v_{\theta z}^{(2)} = 0.3$; $v_{z\theta}^{(2)} = 0.07$; $\delta = 0.06$; $R = 1.0$; $P = 1.0$) into (5.37) and (5.38) we find constants a_1 and a_2, and by (5.36) stress in the main reinforcement is calculated. Finally, we have $\sigma_r^{(1)} = -1.6 \times 10^{-2}$ MPa, wherefrom it can be stated that compressive loads are effective in the inner part of the body.

Calculations of the proposed model without an auxiliary reinforcement layer have proved that tensile stresses appear between fibers of the main reinforcement. The calculation results are in agreement with data obtained in 4 and [4]. Once the tensile stresses reach the adherence strength of the fibers to the matrix, a part of fibers exfoliate and the material breaks as it loses stability. This is what is taking place on the destruction surface of the considered unidirectional material (Fig. 5.35). In contrast, compressive stresses occurring in the main reinforcement of the spirally reinforced composite bring down, whatever their value is, the maximum tangential stress and hamper cracking thus avoiding delamination at material failure along the main reinforcement (Fig. 5.36).

Under the transversal compression cracks are originated in a weakest area in respect to shear strength of the material, namely in the element of the main reinforcement (Fig. 5.37).

Further intensification of loading leads to propagation of the trunk crack often passing through the element body of the main reinforcement, on the core contact with the winding fiber or neighboring elements (Fig. 5.38).

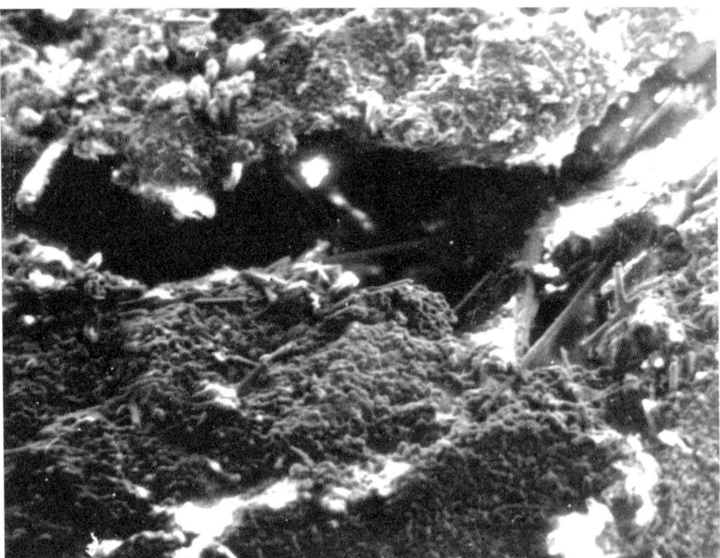

Fig. 5.35. Destruction surface image of a unidirectional composite on compression in reinforcing direction ($\times 200$)

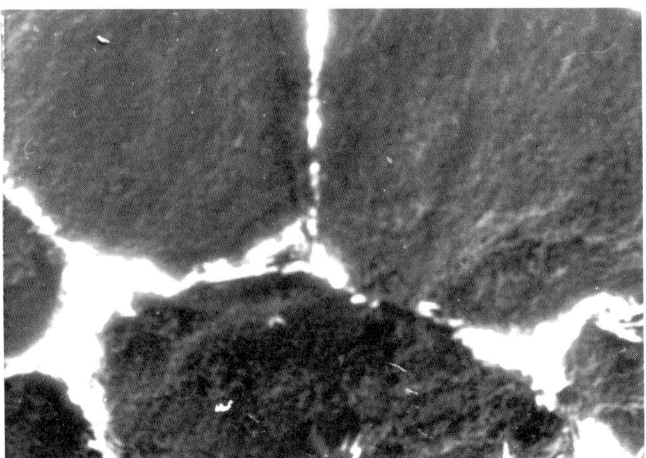

Fig. 5.36. Fracture surface of the composite with the spirally reinforced filler under compression in the main reinforcement direction ($\times 60$)

The winding fiber under transversal compression contributes to the joint provision of the element bearing capacity. In Fig. 3.39 one can observe a carbon plastic fracture surface. The auxiliary reinforcement fibers are seen to break perpendicularly to their packing direction. This means that interlayers hinder from deformation and relative shear of both spirally reinforced ele-

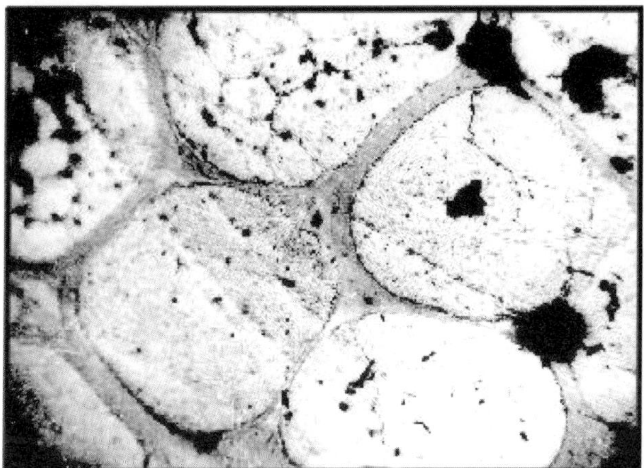

Fig. 5.37. Crack nucleation in the composite structure with the spirally reinforced filler at transverse compression ($\times 50$)

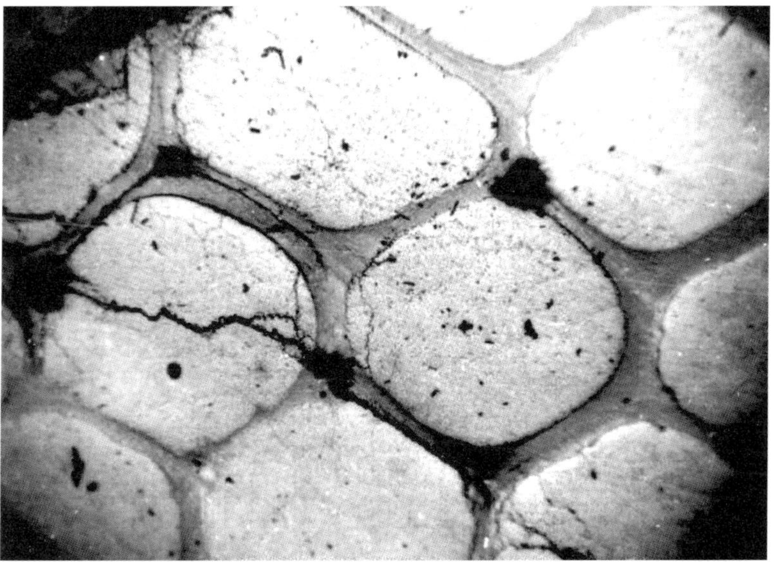

Fig. 5.38. Crack propagation in the composite stricture under transversal compression ($\times 50$)

ments and their chips even upon exfoliation. Since winding fibers are placed at an angle which mismatches the propagating crack surface, the destruction angle of the composite samples under study differs from that typical for unidirectional materials ($\sim 30°$) and varies with winding pitch, its percentage, shape and dimensions of the elements within 35–45° range (Figs. 5.40 and

Fig. 5.39. Destruction surface of a composite with the spirally reinforced filler under compression perpendicularly to the main reinforcement packing ($\times 150$)

Fig. 5.40. Sample failure under transversal compression ($\times 50$)

5.41). Noted destruction angle values are more characteristic of the spatially reinforced materials and constitutes for the 3D reinforced composites $\sim 45°$.

Under transverse tension the winding layer exerts damping effect at stress distribution between the matrix and spirally reinforced elements. In this case,

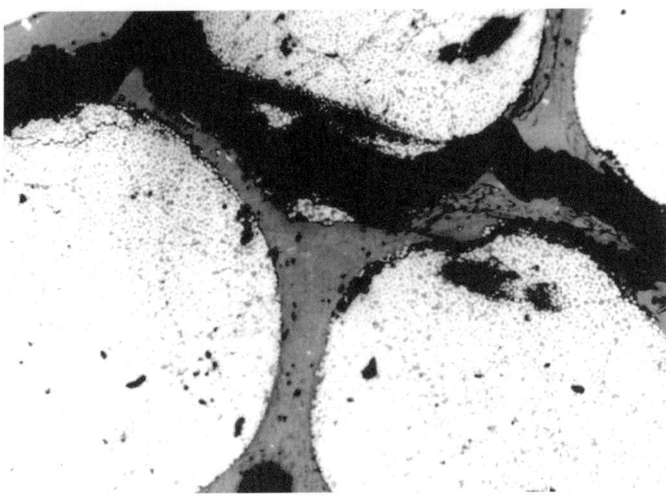

Fig. 5.41. Cracking in the material structure at transverse tension ($\times 90$)

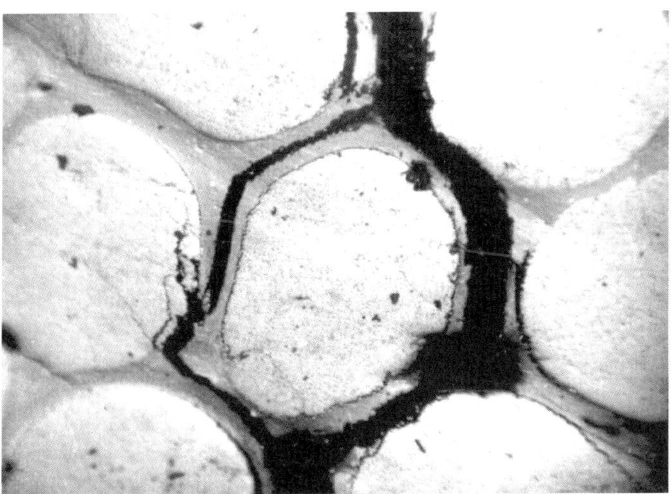

Fig. 5.42. Failure of the composite with the spirally reinforced filler under transversal tension ($\times 50$)

loading on the element core is transmitted through the matrix and winding layer found in tougher situation as they perceive bulk of the load. Therefore, destruction occurs, as a rule, in the interelement zone (Fig. 5.42), where of importance is roughness of the anticipated destruction zone, i.e. size of the spirally reinforced element, winding fiber and winding pitch.

Investigations of destruction zones of the studied composites under transversal tension have shown that destruction of the composites with the filler

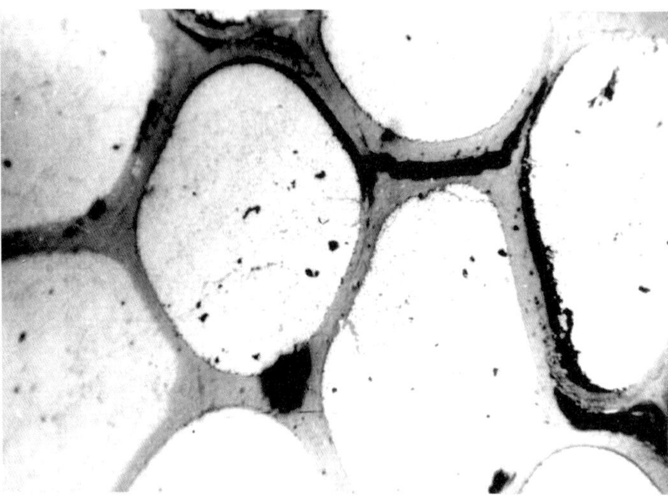

Fig. 5.43. Crack propagation under transversal shear ($\times 40$)

whose winding ensures mutual engagement of the neighboring elements coils is accompanied by tearing of the winding layer off the main reinforcement. In tests where the winding is continuous, the delamination isn't observed but the samples are less durable. So, in the former case the winding layer appears to operate more efficiently and it makes possible to partially transfer the deformation energy inside the spirally reinforced elements which is then scattered over to contribute into the failure in the contact zones between material components.

Failure behavior of the composites under transversal shear resembles that under transversal tension. An instant of a shear crack origination is depicted in Fig. 5.43 in a material with a hexagonal disposition of its elements. Destruction takes place at the interfaces between winding layers, the layer and the main reinforcement core. This partly involves the auxiliary reinforcement into the counteraction to these forces. In case of a tetragonal assembly of the spirally reinforced elements cracks are branching exclusively over the interfaces of the elements (Fig. 5.44). Roughness of the destruction surface thus reduces abruptly and the material shear strength declines.

The analysis of failure modes of the composites based on hybrid spirally reinforced fillers under various loading types is in qualitative agreement with theoretical computations of their stress-strain state. Introduction of the auxiliary reinforcement redistributes microstresses in the material structure and varies its failure mode. A conclusion has been made that the composite formulation, its elastic and strength characteristics can be optimized by varying structural parameters and physico-mechanical properties of the material components for different types of loading.

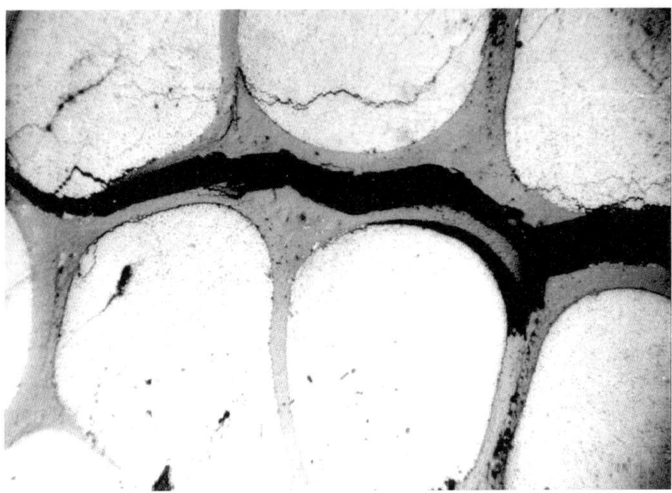

Fig. 5.44. Crack generation in the composite structure at transverse shear ($\times 40$)

5.7 Conclusions

1. Examination of the main reinforcement winding process with the auxiliary reinforcing fibers has specified a series of regularities observed at variations of structural and geometrical parameters of the spirally reinforced fillers. The regularities should be taken into account at designing the technological process of manufacturing the materials under study and equipment thereof.

2. Experimental data are consistent with theoretical investigations proving that the elaborated structure with hybrid spatial reinforced filler produces materials with perfected characteristics under transversal loading, shear and longitudinal compression.

3. Failure data analysis of the studied materials under different types of loading agrees with theoretical results on stress redistribution in the material structure as the auxiliary reinforcement is introduced whose dimensions, properties and shape exert an influence on the composite failure mode.

4. When studying physico-mechanical properties of the composite based on the spirally reinforced filler it has been established that it's impossible to give an unequivocal estimate of the optimum values of the filler structural and geometrical parameters. At designing certain products one is to select the parameters in terms of elevating particular characteristics of a given structure, set conditions and loading.

5. The results of experimental analysis suggest that the developed materials display higher characteristics if to insignificantly decrease their stiffness

in direction of the main reinforcement packing. In this case, practically all specific parameters of the composites outdo conventional unidirectional ones.

6 Design
of Composite Spirally Reinforced Rods

Rod elements made of fibrous composites are a frequent component of spatial trussed designs thanks to a set of specific advantageous properties, including low linear thermal expansion coefficient (LTEC), good radioparent features and some other. These advantages are however bounded by known drawbacks, most unfavorable among which is anisotropy of elastic and strength characteristics. High tensile properties of the material in reinforcement direction are in contradiction with poor transversal characteristics, low interlayer shear modulus, and etc. This, in its turn condition impaired strength of rods from unidirectional composites at a lengthwise compression and relatively low critical load values. Such rods start to break and loose their bearing capacity much earlier than the material strength and bearing capacity of its fibers are exhausted being dependent upon accumulation of microdamages in the bulk. Fracture of rods of unidirectional composites occurs as a result of delamination of their fibers arising from binder cracking and bulging of individual unstable fibers.

The critical force value under which the straight-line equilibrium state of the rod becomes unstable is determined as follows

$$P_C = EJ \left(\frac{\pi}{L}\right)^2 \frac{1}{1 + \dfrac{k}{GF} EJ \left(\dfrac{\pi}{L}\right)^2}, \tag{6.1}$$

where E – the efficient elastic modulus of the rod material; L – its length; J – cross-sectional inertia moment; k – coefficient dependent on cross-section geometrical form; G – longitudinal shear modulus; F – cross-sectional area.

The analysis of above dependence evidences that the critical load can be raised under given geometrical parameters of the structure through enhancing elastic constants of the material. As it has been indicated earlier, to increase the resistance to interlayer shear and transverse tearing it is worthwhile using hybrid spirally reinforced fillers and the spiral reinforcement procedure. Moreover, rather coarse structural units produced by spiral winding of one or a few types of fibrous reinforcements with fibers of some other type can be employed with this aim. As a result, a built-up cylinder is achieved in which coaxial longitudinal layers of the main reinforcement are alternating with thin stiffening layers. Hence, the spirally reinforced rod represents a multilayered hybrid composite.

The operation of composite rods under temperature difference requires high enough strength along with rigidity, length stability of the rod element, etc. Investigation results of a number of authors indicate that the integral LTEC of a hybrid unidirectional composite can be regulated by combining fibers with the positive (glass-reinforced plastic) and negative (carbon plastic) LTEC. An essential drawback of the method is introduction of majority of the low-modular filler in direction of the main reinforcement that reduces article rigidity. In a number of cases it's proposed to regulate the LTEC value of a thin-walled carbon plastic rod by various spatial reinforcement schemes to ensure high bearing capacity of the rod element that, unfortunately doesn't take into account the material temperature sensitivity.

In contrast, spiral reinforcement allows for diversity in choosing main and auxiliary materials, their mutual disposition, different geometrical parameters of stiffening layers, probability of varying in fact any properties of the produced rod, including the integral LTEC of the material. Thus, there arises a problem of predicting elastic and thermoelastic properties of spirally reinforced rods and their optimum designing with allowance for their performances. The studied rods represent inhomogeneous multilayered structures. The fundamentals of the theory of designs of the type are set forth in [227, 240] where relations for each element of the piecewise-homogeneous body are given and conditions of an ideal mechanical and thermomechanical contact between them are reported. Nevertheless, solutions of a number of practical problems by this method encounter perceptible analytical and computation difficulties and necessitate the development of new methods of solutions.

Numerical methods of computing inhomogeneous and multilayered structures have gained popularity lately. A most universal of them is considered the finite element method having numerous modifications. For example, in [241] the method has been applied to estimate efficient characteristics of structural elements under compression. The finite element method is applicable to practically any type of materials or designs but is bulky in realizing and requires profound preparatory stages.

More intricate in mathematical respect but simple in implementation is the method of finite elements, i.e. the method of finite integral equations [242]. It lowers dimensionality of the problems solved by substitution of stress functions or displacements by integrals inside the region over its boundary. The method is effective at solving 3D problems for bodies having boundaries with corner points and other peculiarities.

An important aspect in studying rod structures seems to be the problem of considering face loads on the elements. The problems of the kind are reduced to Saint-Venant's solutions, which are known only for certain classes of rods. Additional difficulties appear at solving mechanical problems in which, as is known the Saint-Venant principle is bounded due to anisotropy of composite properties. The analysis of requisite and suffice applicability conditions of the indicated principle for some particular types of inhomogeneity or anisotropy as well as solution of Saint-Venant's problem for above cases are considered in [243]. Notice that the issue of Saint-Venant's principle fairness and the general type of the named problem solution for inhomogeneous bodies with random anisotropy remains undecided.

In spite of significant steps forward in theoretical studies of elasticity and thermoelasticity problems of composite inhomogeneous bodies, they enable calculations of just piecewise-homogeneous bodies or those with some special types of inhomogeneity. More universal methods are extremely complex which hampers their adoption in computations and designing real structures.

By combining different reinforcing fibers in a hybrid spirally reinforced rod and varying geometrical parameters of the main and stiffening layers one can change in fact any properties of rod elements. Here arises a problem of defining elastic and strength characteristics of such rod elements and the issue of the choice of an optimum material structure for a set operation regime.

6.1 Selection of the Mathematical Model

As a design model of the hybrid spirally reinforced rod we choose a built-up cylinder of circular cross-section (Fig. 6.1) consisting of N thick main layers S_i rigidly connected to thin stiffening layers L_i $(i = 1, \ldots, N)$. One of the rod faces is considered as rigidly fastened. The coordinates are so that Z axis is parallel to the rod generatrix and its origin coincides with the gravity center of its idle face. In this case, plane $r\Theta$ is parallel to that of the cylinder cross-section.

It's been specified above that layers S_i, L_i $(i = 1, \ldots, N)$ are spirally anisotropic. The considered type of anisotropy relates to a case of a local

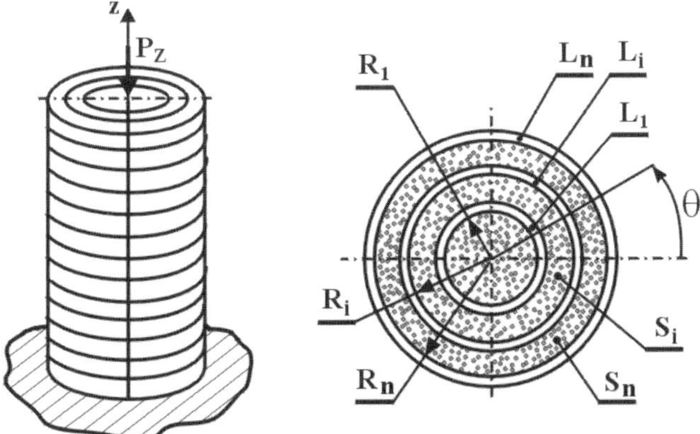

Fig. 6.1. The design model of a hybrid spirally reinforced rod

orthotropic body [195]. This means that there are 13 independent elastic constants in each point of the cylinder in the chosen coordinates $r\Theta z$.

Technical characteristics of layers S_i, L_i $(i = 1, \ldots, N)$ can be calculated on the basis of known formulas with account of the reinforcing fiber properties, that of the binder and filling degree of the layer. To compute elastic constants of layers S_i and L_i the local coordinates $r_1\Theta_1 z_1$ is chosen so that axis z_1 be aligned with the reinforcing layer direction, and axis r_1 be parallel to r axis. Transfer from the local coordinates $r_1\Theta_1 z_1$ to $r\Theta z$ is exercised by the pivot turn r by angle γ_i determined by twisting of layer S_i fibers under the action of the winding fiber, and by the packing angle β_i of layer L_i $(i = 1, \ldots, N)$. Dependencies thus obtained for elastic and thermoelastic properties of the layers are of the kind

Properties of the main material:

$$a_{11}^{(0)} = \frac{v_{z\Theta}}{v_{\Theta z} E_z}; \quad a_{12}^{(0)} = \frac{v_{r\Theta} v_{z\Theta}}{v_{\Theta r} E_z};$$

$$a_{13}^{(0)} = -\frac{v_{z\Theta}}{E_2}; \quad a_{33}^{(0)} = \frac{1}{E_z};$$

$$a_{44}^{(0)} = \frac{1}{G_{\Theta z}}; \quad a_{66}^{(0)} = \frac{2v_{z\Theta}(1 + v_{r\Theta})}{v_{\Theta z} E_z}, \tag{6.2}$$

where $E_z, G_{\Theta z}, v_{r\Theta}, v_{z\Theta}, v_{\Theta z}$ are technical characteristics of the layers.

The coefficients a_{ij} of Hook's law are found from

$$a_{ij} = a_{kL}^{(0)} q_{ij} q_{jL}, \tag{6.3}$$

where q_{ik}, q_{jL} are determined from Table 6.1.

For a unidirectional fibrous material having $E_z, G_{\Theta z}, v_{r\Theta}, v_{z\Theta}, v_{\Theta z}$ technical characteristics in coordinates $r_1\Theta_1 z_1$ the generalized Hook's law is of the form

$$\sigma_r = A_{11}^{(0)} \varepsilon_r + A_{12}^{(0)} \varepsilon_\Theta + A_{13}^{(0)} \varepsilon_z - \beta_{11}^{(0)} \Delta T; \quad \tau_{\Theta z} = A_{44}^{(0)} \gamma_{\Theta z};$$

$$\sigma_\Theta = A_{12}^{(0)} \varepsilon_r + A_{11}^{(0)} \varepsilon_\Theta + A_{13}^{(0)} \varepsilon_z - \beta_{11}^{(0)} \Delta T; \quad \tau_{rz} = A_{44}^{(0)} \gamma_{\Theta z};$$

$$\sigma_z = A_{13}^{(0)} \varepsilon_r + A_{13}^{(0)} \varepsilon_\Theta + A_{33}^{(0)} \varepsilon_z - \beta_{33}^{(0)} \varepsilon_z \Delta T; \quad \tau_{r\Theta} = A_{66}^{(0)} \gamma_{r\Theta}; \tag{6.4}$$

Table 6.1. Symbols q_{ij} to Transformation Formulas of Coefficients a_{ij}

		L					
		1	2	3	4	5	6
k	1	1	0	0	0	0	0
	2	0	1	γ^2	γ	0	0
	3	0	γ^2	1	$-\gamma$	0	0
	4	0	-2γ	2γ	$1 - \gamma^2$	0	0
	5	0	0	0	0	1	$-\gamma$
	6	0	0	0	0	γ	1

or

$$\varepsilon_r = a_{11}^{(0)}\sigma_r + a_{12}^{(0)}\sigma_\Theta + a_{13}^{(0)}\sigma_z + \alpha_r\Delta T; \quad \gamma_{\Theta r} = a_{44}^{(0)}\tau_{\Theta z};$$

$$\varepsilon_\Theta = a_{12}^{(0)}\sigma_r + a_{11}^{(0)}\sigma_\Theta + a_{13}^{(0)}\sigma_z + \alpha_r\Delta T; \quad \gamma_{rz} = a_{44}^{(0)}\tau_{rz};$$

$$\varepsilon_z = a_{13}^{(0)}\sigma_r + a_{13}^{(0)}\sigma_\Theta + a_{33}^{(0)}\sigma_z + \alpha_z\Delta T; \quad \gamma_{r\Theta} = a_{66}^{(0)}\tau_{r\Theta}; \tag{6.5}$$

where a_{ij} are given by (6.2).

Clearly,

$$A_{44} = G_{\Theta z}; \quad A'_{66} = \frac{v_{\Theta z}E}{2v_{z\Theta}(1 + v_{r\Theta})}.$$

By substituting (6.5) into (6.4) and equating coefficients of similar stresses, taking account of (6.2) we get:

$$A_{11}^{(0)} = \frac{1 - v_{\Theta z}v_{z\Theta}}{D}E_z; \quad A_{12}^{(0)} = \frac{v_{r\Theta} + v_{z\Theta}v_{\Theta z}}{D}E_z;$$

$$A_{13}^{(0)} = \frac{v_{z\Theta}(1 + v_{r\Theta})}{D}E_z; \quad A_{33}^{(0)} = \frac{v_{z\Theta}(1 + v_{r\Theta}^2)}{v_{\Theta z}D}E_z, \tag{6.6}$$

where

$$D = \frac{v_{z\Theta}(1 + v_{r\Theta})(1 - v_{r\Theta} - 2v_{z\Theta}v_{\Theta z})}{v_{\Theta z}},$$

$$\beta_{11}^{(0)} = (A_{11}^{(0)} + A_{12}^{(0)})\alpha_r + A_{13}^{(0)}\alpha_z;$$

$$\beta_{33}^{(0)} = 2A_{13}^{(0)}\alpha_r + A_{33}^{(0)}\alpha_z. \tag{6.7}$$

Elastic constants A_{ij} are calculated from $A_{ij}^{(0)}$ based on Table 6.2 and using the formula:

$$A_{ij} = A_{mn}^{(0)}q_{im}q_{jn}. \tag{6.8}$$

The Thermoelastic constants β_{ij} are found by the relations:

$$\beta_{11} = \beta_{11}^{(0)}; \quad \beta_{22} = \beta_{11}^{(0)}\cos^2\beta + \beta_{33}^{(0)}\sin^2\beta;$$

$$\beta_{33} = \beta_{11}^{(0)}\sin^2\beta + \beta_{33}^{(0)}\cos^2\beta;$$

$$\beta_{23} = \frac{1}{2}(\beta_{33}^{(0)} - \beta_{11}^{(0)})\sin 2\beta. \tag{6.9}$$

Table 6.2. Symbols q_{ij} to Transformation Formulas of Coefficients A_{ij}

		L					
		1	2	3	4	5	6
k	1	1	0	0	0	0	0
	2	0	$\cos^2\beta$	$\sin^2\beta$	$\sin 2\beta$	0	0
	3	0	$\sin^2\beta$	$\cos^2\beta$	$-\sin 2\beta$	0	0
	4	0	$-1/2\sin 2\beta$	$1/2\sin 2\beta$	$\cos^2\beta$	0	0
	5	0	0	0	0	$\cos\beta$	$-\sin\beta$
	6	0	0	0	0	$\sin\beta$	$\cos\beta$

6.2 Calculation of Loading Conditions

The analysis of the stress–strain state (SSS) of composite inhomogeneous rods under face loads presents certain difficulties. It has been reported in [203] that for rather long bodies it's possible to reduce the 3D problem to a 2D one using Saint-Venant's method and to simplify it significantly. The methods of designing homogeneous and isotropic rods based on the complex variable functions are well known [203]. As for the inhomogeneous and anisotropic rods only the Saint-Venant problem for the bodies with an arbitrary straight-line anisotropy [202] and the issues of rod twisting have been comprehensively studied so far [244].

What concerns rods under study, it's possible to avoid solution of Saint-Venant's problem for built-up spirally anisotropic bodies and the problem on a complex loading of a compound piecewise-homogeneous rod element by introduction of an assumption on smallness of the layer breadth as compared to contacting bulk bodies and the one that cross-sections of the rod remain plane at deformation. This will lead to the solution of 3 spatial problems of the theory of elasticity, namely on the rod loading by an axial force P_z, rod twisting by torque M_{TR} and bending moments M_1 and M_2 (Fig. 6.2).

Loading conditions in this case are accounted for as integral face conditions

$$\int_S \sigma_Z dS = P_Z ; \tag{6.10}$$

$$\int_S r\tau_{\theta Z} dS = M_{TR} ; \tag{6.11}$$

$$\int_S r\sin\Theta\sigma_Z dS = M_1; \quad \int_S r\cos\Theta\sigma_Z dS = M_2 ; \tag{6.12}$$

where

$$S = \left[\bigcup_N S_i\right] \cup \left[\bigcup_N L_i\right]; \quad dS = rdrd\Theta . \tag{6.13}$$

Layers S_i in the rod model considered are inhomogeneous and local-orthotropic. As it follows from [202], the integral face conditions (6.10)–(6.12) are true for each layer. Like in the case when layers S_i are homogeneous the assumption on smallness of layer breadths L_i enables consideration of face loads of the built-up spirally anisotropic rod using the integral face conditions.

Hence, the problem on a complex loading of a built-up spirally anisotropic rod is reduced to solution of three spatial problems: the problem on longitudinal loading, the twisting problem and that on bending with torques. The loading conditions are taken into account in the integral face conditions (6.10)–(6.12).

6.3 Rigid Contact Conditions

Let's consider massive bodies like layers S^+ and S^- that are rigidly linked with a thin anisotropic layer L of $2h$ width. Let's denote contact boundaries γ^+, γ^- of layer L with layers S^+ and S^-, and γ_O as the medial line of layer L (Fig. 6.2).

The chief principles of accounting for the layer thickness smallness have been expounded in [245]. From conditions of an ideal contact

$$U_r^{(L)} = U_r^{(S)}; \quad U_\Theta^{(L)} = U_\Theta^{(S)}; \quad W^{(L)} = W^{(S)} \tag{6.14}$$

$$\sigma_r^{(L)} = \sigma_r^{(S)}; \quad \tau_{r\Theta}^{(L)} = \tau_{r\Theta}^{(S)}; \quad \tau_{rz}^{(L)} = \tau_{rz}^{(S)} \tag{6.15}$$

on bounds γ^+, γ^- and suppositions that differential equilibrium equations

$$\frac{\partial \sigma_r}{\partial r} + \frac{1}{r}\frac{\partial \tau_{r\Theta}}{\partial \Theta} + \frac{\partial \tau_{rz}}{\partial z} + \frac{\sigma_r - \sigma_\Theta}{r} = 0;$$

$$\frac{\partial \tau_{r\Theta}}{\partial r} + \frac{1}{r}\frac{\partial \sigma_\Theta}{\partial \Theta} + \frac{\partial \tau_{\Theta z}}{\partial z} + 2\frac{\tau_{r\Theta}}{r} = 0;$$

$$\frac{\partial \tau_{r\Theta}}{\partial r} + \frac{1}{r}\frac{\partial \tau_{\Theta z}}{\partial \Theta} + \frac{\partial \sigma_{\Theta z}}{\partial z} + \frac{\tau_{rz}}{r} = 0 \tag{6.16}$$

are met in each point of layers S^-, S^+ and median γ_O of layer L, boundary conditions have been derived for the case when the layers and bulk bodies are isotropic.

In the rod model proposed layer L is local-orthotropic, i.e. stresses and displacements in layer L with allowance for dependencies

$$\varepsilon_r = \frac{\partial U_r}{\partial r}; \quad \varepsilon_z = \frac{\partial W}{\partial z}; \quad \varepsilon_\Theta = \frac{1}{r}\frac{\partial U_\Theta}{\partial \Theta} + \frac{U_r}{r};$$

$$\gamma_{r\Theta} = \frac{1}{r}\frac{\partial U_r}{\partial \Theta} + \frac{\partial U_\Theta}{\partial r} - \frac{U_r}{r};$$

$$\gamma_{\Theta r} = \frac{\partial U_\Theta}{\partial z} + \frac{1}{r}\frac{\partial W}{\partial \Theta}; \quad \gamma_{rz} = \frac{\partial W}{\partial r} + \frac{\partial U_r}{\partial z}, \tag{6.17}$$

are interrelated through Hook's law

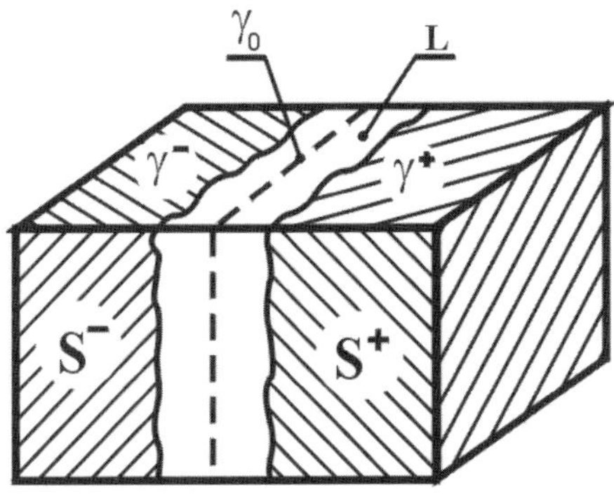

Fig. 6.2. To deriving contact conditions of bulk bodies with a thin spirally anisotropic layer

$$\sigma_r = A_{11}\frac{\partial U_r}{\partial r} + \frac{1}{r}A_{12}\left(\frac{\partial U_\theta}{\partial \Theta} + U_r\right) + A_{13}\frac{\partial W}{\partial z} + A_{14}\left(\frac{1}{r}\frac{\partial W}{\partial \Theta} + \frac{\partial U_\Theta}{\partial z}\right);$$

$$\sigma_\Theta = A_{12}\frac{\partial U_r}{\partial r} + \frac{1}{r}A_{22}\left(\frac{\partial U_\theta}{\partial \Theta} + U_r\right) + A_{23}\frac{\partial W}{\partial z} + A_{24}\left(\frac{1}{r}\frac{\partial W}{\partial \Theta} + \frac{\partial U_\Theta}{\partial z}\right);$$

$$\sigma_z = A_{13}\frac{\partial U_r}{\partial r} + \frac{1}{r}A_{23}\left(\frac{\partial U_\theta}{\partial \Theta} + U_r\right) + A_{33}\frac{\partial W}{\partial z} + A_{34}\left(\frac{1}{r}\frac{\partial W}{\partial \Theta} + \frac{\partial U_\Theta}{\partial z}\right);$$

$$\tau_{\Theta z} = A_{14}\frac{\partial U_r}{\partial r} + \frac{1}{r}A_{24}\left(\frac{\partial U_\theta}{\partial \Theta} + U_r\right) + A_{34}\frac{\partial W}{\partial z} + A_{44}\left(\frac{1}{r}\frac{\partial W}{\partial \Theta} + \frac{\partial U_\Theta}{\partial z}\right);$$

$$\tau_{rz} = A_{55}\left(\frac{\partial W}{\partial r} + \frac{\partial U_r}{\partial z}\right) + A_{56}\left(\frac{1}{r}\frac{\partial W}{\partial \Theta} + \frac{\partial U_\Theta}{\partial r} - \frac{U_\Theta}{r}\right);$$

$$\tau_{r\Theta} = A_{56}\left(\frac{\partial W}{\partial r} + \frac{\partial U_r}{\partial z}\right) + A_{66}\left(\frac{1}{r}\frac{\partial U_r}{\partial \Theta} + \frac{\partial U_\Theta}{\partial r} - \frac{U_\Theta}{r}\right), \tag{6.18}$$

where A_{ij} – elastic constants of layer L.

To calculate the first group of conjugation conditions let's make use of an ideal contact conditions (6.15). Let's differentiate with respect to variable r using non-difference formulas of numerical differentiation with nodes $R_O - h$, R_O, $R_O + h$ and considering ideal contact conditions (6.14).

From the first equation (6.18) we have

$$\sigma_r^+ = \frac{A_{11}}{2h}(3U_r^+ - 4U_r^O + U_r^-) + \frac{A_{12}}{R+h}\left(\frac{\partial U_\theta^-}{\partial \Theta} + U_r^+\right)$$
$$+A_{13}\frac{\partial W^+}{\partial z} + A_{14}\left(\frac{1}{R+h}\frac{\partial W^+}{\partial \Theta} + \frac{\partial U_\theta^+}{\partial z}\right) + \delta(h);$$

$$\sigma_r^- = -\frac{A_{11}}{2h}(3U_r^- - 4U_r^O + U_r^+) + \frac{A_{12}}{R-h}\left(\frac{\partial U_\theta^-}{\partial \Theta} + U_r^-\right)$$
$$+A_{13}\frac{\partial W^-}{\partial z} + A_{14}\left(\frac{1}{R-h}\frac{\partial W^-}{\partial \Theta} + \frac{\partial U_\theta^-}{\partial z}\right) + \delta(h); \tag{6.19}$$

where $\delta(h) \ll h$.

Here and later on $f^O = f(R)$, $f^+ = f(R+h)$, $f^- = f(R-h)$, f is one of the functions of stresses or displacements.

Taking

$$\frac{h}{R-h} = \frac{h}{R} + \delta(h); \quad \frac{h}{R+h} = \frac{h}{R} + \delta(h) \tag{6.20}$$

the first condition for a rigid contact is derived from (6.19)

$$A_{11}(U_r^+ - U_r^-) + \frac{h}{R}\left[A_{12}\left(\frac{\partial U_\Theta^+}{\partial \Theta} + \frac{\partial U_\Theta^-}{\partial \Theta} + U_r^+ + U_r^-\right)\right.$$
$$+RA_{13}\left(\frac{\partial W^+}{\partial z} + \frac{\partial W^-}{\partial z}\right) + A_{14}\left(\frac{\partial W^+}{\partial \Theta} + \frac{\partial W^-}{\partial \Theta} + R\left(\frac{\partial U_\Theta^+}{\partial \Theta} + \frac{\partial U_\Theta^-}{\partial \Theta}\right)\right)$$
$$\left. -R(\sigma_r^+ + \sigma_r^-)\right] = 0. \tag{6.21}$$

From (6.19) we also have the relation for U_r^O that is henceforth important

$$U_r^O = \frac{1}{2}(U_r^+ + U_r^-) + h\left[\frac{A_{13}}{4A_{11}}\left(\frac{\partial W^+}{\partial z} - \frac{\partial W^-}{\partial z}\right)\right.$$

$$+\frac{A_{14}}{4A_{11}}\left(\frac{1}{R}\left(\frac{\partial W^+}{\partial \Theta} - \frac{\partial W^-}{\partial \Theta}\right) + \frac{\partial U_\Theta^+}{\partial z} - \frac{\partial U_\Theta^-}{\partial z}\right)$$

$$\left.+\frac{A_{12}}{4RA_{11}}\left(\frac{\partial U_\Theta^+}{\partial \Theta} - \frac{\partial U_\Theta^-}{\partial \Theta} + U_r^+ - U_r^-\right) + \frac{1}{4A_{11}}(\sigma_r^+ - \sigma_r^-)\right]. \quad (6.22)$$

Similarly, from the 8th and 6th equations of (6.18) one has:

$$A_{55}\left[W^+ - W^- + \frac{h}{R}\left(\frac{\partial U_r^+}{\partial z} + \frac{\partial U_r^-}{\partial z}\right)\right]$$

$$+A_{56}\left[\frac{h}{R}\left(\frac{\partial U_r^+}{\partial \Theta} + \frac{\partial U_r^-}{\partial \Theta} - U_\Theta^+ - U_\Theta^-\right) + (U_\Theta^+ - U_\Theta^-)\right]$$

$$-h(\tau_{rz}^+ + \tau_{rz}^-) = 0; \quad (6.23)$$

$$A_{56}\left[W^+ - W^- + \frac{h}{R}\left(\frac{\partial U_r^+}{\partial z} + \frac{\partial U_r^-}{\partial z}\right)\right]$$

$$+A_{66}\left[\frac{h}{R}\left(\frac{\partial U_r^+}{\partial \Theta} + \frac{\partial U_r^-}{\partial \Theta} - U_\Theta^+ - U_\Theta^-\right) + (U_\Theta^+ - U_\Theta^-)\right]$$

$$-h(\tau_{r\theta}^+ - \tau_{r\theta}^-) = 0. \quad (6.24)$$

Further from the same equations we get

$$U_\Theta^0 = \frac{1}{2}(U_\Theta^+ + U_\Theta^-) + h\left[\frac{1}{4R}\left(\frac{\partial U_r^+}{\partial \Theta} - \frac{\partial U_r^-}{\partial \Theta} - U_\Theta^+ + U_\Theta^-\right)\right.$$

$$\left.+\frac{1}{A_{56}^2 - A_{55}A_{66}}\left(A_{56}(\tau_{rz}^+ - \tau_{rz}^-) + A_{55}(\tau_{r\Theta}^+ - \tau_{r\Theta}^-)\right)\right]; \quad (6.25)$$

$$W^0 = \frac{1}{2}(W^+ + W^-) + h\left[\frac{1}{4}\left(\frac{\partial U_r^+}{\partial z} - \frac{\partial U_r^-}{\partial z}\right)\right.$$

$$\left.+\frac{1}{A_{55}A_{66} - A_{56}^2}\left(-A_{66}(\tau_{rz}^+ - \tau_{rz}^-) + A_{56}(\tau_{r\Theta}^+ - \tau_{r\Theta}^-)\right)\right]. \quad (6.26)$$

On the median γ_O with account of (6.20) equilibrium equations (6.16) are of the kind

$$\frac{1}{2h}(\sigma_r^+ - \sigma_r^-) + \frac{1}{2Rh}A_{56}\left(\frac{\partial W^+}{\partial \Theta} - \frac{\partial W^-}{\partial \Theta}\right) + \frac{1}{R}A_{56}\frac{\partial^2 U_r^0}{\partial \Theta \partial z}$$

$$+\frac{1}{R^2}A_{66}\frac{\partial^2 U_r^0}{\partial \Theta^2} + \frac{1}{2Rh}A_{66}\left(\frac{\partial U_\Theta^+}{\partial \Theta} - \frac{\partial U_\Theta^-}{\partial \Theta}\right) - \frac{1}{R^2}A_{66}\frac{\partial U_\Theta^0}{\partial \Theta}$$

$$+\frac{1}{2h}A_{55}\left(\frac{\partial W^+}{\partial z} - \frac{\partial W^-}{\partial z}\right) + A_{55}\frac{\partial^2 U_r^0}{\partial z^2} + \frac{1}{R}A_{56}\frac{\partial^2 U_r^0}{\partial \Theta \partial z}$$

$$+\frac{1}{2h}A_{56}\left(\frac{\partial U_\Theta^+}{\partial z}-\frac{\partial U_\Theta^-}{\partial z}\right)-\frac{1}{R}A_{56}\frac{\partial U_\Theta^0}{\partial z}$$

$$+\frac{1}{2Rh}(A_{11}-A_{12})(U_r^+-U_r^-)+\frac{1}{R^2}(A_{12}-A_{22})\frac{\partial U_\Theta^0}{\partial\Theta}$$

$$+\frac{1}{R^2}(A_{12}-A_{22})U_r^0+\frac{1}{R}(A_{13}-A_{23})\frac{\partial W^0}{\partial z}$$

$$+\frac{1}{R^2}(A_{14}-A_{24})\frac{\partial W^0}{\partial\Theta}+\frac{1}{R}(A_{14}-A_{24})\frac{\partial U_\Theta^0}{\partial z}=0\,;\tag{6.27}$$

$$\frac{1}{2h}(\tau_{r\Theta}^+-\tau_{r\Theta}^-)+\frac{1}{2Rh}A_{12}\left(\frac{\partial U_r^+}{\partial\Theta}-\frac{\partial U_r^-}{\partial\Theta}\right)+\frac{1}{R^2}A_{22}\frac{\partial^2 U_r^0}{\partial\Theta^2}$$

$$+\frac{1}{R^2}A_{22}\frac{\partial U_r^0}{\partial\Theta}+\frac{1}{R}A_{23}\frac{\partial^2 W^0}{\partial\Theta\partial z}+\frac{1}{R^2}A_{24}\frac{\partial^2 W^0}{\partial\Theta^2}$$

$$+\frac{1}{R}A_{24}\frac{\partial^2 U_\Theta^0}{\partial\Theta\partial z}+\frac{1}{2h}A_{14}\left(\frac{\partial U_r^+}{\partial z}-\frac{\partial U_r^-}{\partial z}\right)+\frac{1}{R}A_{24}\frac{\partial^2 U_\Theta^0}{\partial\Theta\partial z}$$

$$+\frac{1}{R}A_{24}\frac{\partial U_r^0}{\partial z}+A_{34}\frac{\partial^2 W^0}{\partial z^2}+\frac{1}{R}A_{44}\frac{\partial^2 W^0}{\partial\Theta\partial z}+A_{44}\frac{\partial^2 U_\Theta^0}{\partial z^2}$$

$$+\frac{A_{56}}{Rh}(W^+-W^-)+\frac{2A_{56}}{R}\frac{\partial U_r^0}{\partial z}+\frac{2}{R^2}A_{66}\frac{\partial U_r^0}{\partial\Theta}$$

$$+\frac{A_{66}}{Rh}(U_\Theta^+-U_\Theta^-)-\frac{2A_{66}}{R^2}(U_\Theta^++U_\Theta^-)=0\,;\tag{6.28}$$

$$\frac{1}{2h}(\tau_{rz}^+-\tau_{rz}^-)+\frac{1}{2Rh}A_{14}\left(\frac{\partial U_r^+}{\partial\Theta}-\frac{\partial U_r^-}{\partial\Theta}\right)+\frac{1}{R^2}A_{24}\frac{\partial^2 U_\Theta^0}{\partial\Theta^2}$$

$$+\frac{1}{R^2}A_{24}\frac{\partial U_r^0}{\partial\Theta}+\frac{1}{R}A_{34}\frac{\partial^2 W^0}{\partial\Theta\partial z}+\frac{1}{R^2}A_{44}\frac{\partial^2 W^0}{\partial\Theta^2}$$

$$+\frac{1}{R}A_{44}\frac{\partial^2 U_\Theta^0}{\partial\Theta\partial z}+\frac{1}{2h}A_{13}\left(\frac{\partial U_r^+}{\partial z}-\frac{\partial U_r^-}{\partial z}\right)+\frac{1}{R}A_{23}\frac{\partial^2 U_\Theta^0}{\partial\Theta\partial z}$$

$$+\frac{1}{R}A_{23}\frac{\partial U_r^0}{\partial z}+A_{33}\frac{\partial^2 W^0}{\partial z^2}+\frac{1}{R}A_{34}\frac{\partial^2 W^0}{\partial\Theta\partial z}+A_{34}\frac{\partial^2 W^0}{\partial z^2}$$

$$+\frac{1}{2Rh}A_{55}(W^+-W^-)+\frac{1}{R}A_{55}\frac{\partial U_r^0}{\partial z}+\frac{1}{R^2}A_{56}\frac{\partial U_r^0}{\partial\Theta}$$

$$+\frac{1}{2Rh}A_{56}(U_\Theta^+-U_\Theta^-)-\frac{1}{R^2}A_{56}U_\Theta^0=0\,.\tag{6.29}$$

Thus, using (6.22), (6.25), and (6.26) the second group of conjunction conditions is obtained:

$$R(\sigma_r^+-\sigma_r^-)+A_{56}\left(\frac{\partial W^+}{\partial\Theta}-\frac{\partial W^-}{\partial\Theta}\right)+A_{66}\left(\frac{\partial U_\Theta^+}{\partial\Theta}-\frac{\partial U_\Theta^-}{\partial\Theta}\right)$$

$$+RA_{55}\left(\frac{\partial W^+}{\partial z}-\frac{\partial W^-}{\partial z}\right)+RA_{56}\left(\frac{\partial U_\Theta^+}{\partial z}-\frac{\partial U_\Theta^-}{\partial z}\right)$$

$$+(A_{11}-A_{12})(U_r^+-U_r^-)+\frac{h}{R}\left\{A_{66}\left(\frac{\partial^2 U_r^+}{\partial\Theta^2}+\frac{\partial^2 U_r^-}{\partial\Theta^2}\right)\right.$$

$$+ R^2 A_{55} \left(\frac{\partial^2 U_r^+}{\partial z^2} + \frac{\partial^2 U_r^-}{\partial z^2} \right) + 2RA_{56} \left(\frac{\partial^2 U_r^+}{\partial \Theta \partial z} + \frac{\partial^2 U_r^-}{\partial \Theta \partial z} \right)$$

$$+ (A_{12} - A_{22})(U_r^+ + U_r^-) + (A_{12} - A_{22} - A_{66}) \left(\frac{\partial U_\Theta^+}{\partial \Theta} - \frac{\partial U_\Theta^-}{\partial \Theta} \right)$$

$$+ R(A_{14} - A_{24} - A_{56}) \left(\frac{\partial U_\Theta^+}{\partial z} - \frac{\partial U_\Theta^-}{\partial z} \right) + R(A_{13} - A_{23}) \left(\frac{\partial W^+}{\partial z} + \frac{\partial W^-}{\partial z} \right)$$

$$+ (A_{14} - A_{24}) \left(\frac{\partial W^+}{\partial \Theta} + \frac{\partial W^-}{\partial \Theta} \right) \Bigg\} + \delta(h) = 0 \,; \tag{6.30}$$

$$R(\tau_{r\Theta}^+ - \tau_{r\Theta}^-) + A_{12} \left(\frac{\partial U_r^+}{\partial \Theta} - \frac{\partial U_r^-}{\partial \Theta} \right) + RA_{14} \left(\frac{\partial U_r^+}{\partial z} - \frac{\partial U_r^-}{\partial z} \right)$$

$$+ 2A_{66}(U_\Theta^+ - U_\Theta^-) + 2A_{56}(W^+ - W^-)$$

$$+ \frac{h}{R} \Bigg\{ (A_{22} + 2A_{66}) \left(\frac{\partial U_r^+}{\partial \Theta} + \frac{\partial U_r^-}{\partial \Theta} \right)$$

$$+ R(A_{24} + A_{56}) \left(\frac{\partial U_r^+}{\partial z} + \frac{\partial U_r^-}{\partial z} \right) + A_{22} \left(\frac{\partial^2 U_\Theta^+}{\partial \Theta^2} + \frac{\partial^2 U_\Theta^-}{\partial \Theta^2} \right)$$

$$+ 2RA_{24} \left(\frac{\partial^2 U_\Theta^+}{\partial \Theta \partial z} - \frac{\partial^2 U_\Theta^-}{\partial \Theta \partial z} \right) + R^2 A_{44} \left(\frac{\partial^2 U_\Theta^+}{\partial z^2} + \frac{\partial^2 U_\Theta^-}{\partial z^2} \right)$$

$$- 2A_{66}(U_\Theta^+ + U_\Theta^-) + R^2 A_{34} \left(\frac{\partial^2 W^+}{\partial z^2} + \frac{\partial^2 W^-}{\partial z^2} \right)$$

$$+ R(A_{23} + A_{44}) \left(\frac{\partial^2 W^+}{\partial \Theta \partial z} + \frac{\partial^2 W^-}{\partial \Theta \partial z} \right)$$

$$+ A_{24} \left(\frac{\partial^2 W^+}{\partial \Theta^2} + \frac{\partial^2 W^-}{\partial \Theta^2} \right) \Bigg\} + \delta(h) = 0 \,; \tag{6.31}$$

$$R(\tau_{rz}^+ - \tau_{rz}^-) + A_{14} \left(\frac{\partial U_r^+}{\partial \Theta} - \frac{\partial U_r^-}{\partial \Theta} \right) + RA_{13} \left(\frac{\partial U_r^+}{\partial z} - \frac{\partial U_r^-}{\partial z} \right)$$

$$+ A_{56}(U_\Theta^+ - U_\Theta^-) + A_{55}(W^+ - W^-)$$

$$+ \frac{h}{R} \Bigg\{ (A_{24} + A_{56}) \left(\frac{\partial U_r^+}{\partial \Theta} + \frac{\partial U_r^-}{\partial \Theta} \right)$$

$$+ R(A_{23} + A_{55}) \left(\frac{\partial U_r^+}{\partial z} + \frac{\partial U_r^-}{\partial z} \right) + A_{24} \left(\frac{\partial^2 U_\Theta^+}{\partial \Theta^2} + \frac{\partial^2 U_\Theta^-}{\partial \Theta^2} \right)$$

$$+ R(A_{44} + A_{23}) \left(\frac{\partial^2 U_\Theta^+}{\partial \Theta \partial z} + \frac{\partial^2 U_\Theta^-}{\partial \Theta \partial z} \right) - A_{56}(U_\Theta^+ + U_\Theta^-)$$

$$+ R^2 A_{34} \left(\frac{\partial^2 U_\Theta^+}{\partial z^2} + \frac{\partial^2 U_\Theta^-}{\partial z^2} \right) + 2RA_{34} \left(\frac{\partial^2 W^+}{\partial \Theta \partial z} + \frac{\partial^2 W^-}{\partial \Theta \partial z} \right)$$

$$+ A_{44} \left(\frac{\partial^2 W^+}{\partial \Theta^2} + \frac{\partial^2 W^-}{\partial \Theta^2} \right) + R^2 A_{33} \left(\frac{\partial^2 W^+}{\partial z^2} + \frac{\partial^2 W^-}{\partial z^2} \right) \Bigg\} + \delta(h) = 0 \,. \tag{6.32}$$

The derived equalities (6.21), (6.23), (6.24), (6.30)–(6.32) are boundary conditions of the rigid contact of a thin local-orthotropic layer L with massive bodies, in provision differential equations (6.16) are met only on the layer L median.

The generalized Hook's law under thermal shock is of the form

$$\sigma_r = \sigma_r^* - \beta_{11} dT \,;$$

$$\sigma_\Theta = \sigma_\Theta^* - \beta_{22} dT \,;$$

$$\sigma_z = \sigma_z^* - \beta_{33} dT \,;$$

$$\tau_{\Theta z} = \tau_{\Theta z}^* - \beta_{23} dT \,;$$

$$\tau_{rz} = \tau_{rz}^* \,;$$

$$\tau_{r\Theta} = \tau_{r\Theta}^* \,, \tag{6.33}$$

where dT – temperature variation, β_{ij} – thermoelastic coefficients for layer L, and σ^*, τ^* are given by (6.18).

It follows from (6.23) that contact conditions (6.23) and (6.24) under thermal loading remain invariable and from (6.21) we have:

$$A_{11}(U_r^+ - U_r^-) + \frac{h}{R}\left[A_{12}\left(\frac{\partial U_\Theta^+}{\partial \Theta} + \frac{\partial U_\Theta^-}{\partial \Theta} + U_r^+ + U_r^- \right) \right.$$

$$\left. + RA_{13}\left(\frac{\partial W^+}{\partial z} + \frac{\partial W^-}{\partial z} \right) + A_{14}\left(\frac{\partial W^+}{\partial \Theta} + \frac{\partial W^-}{\partial \Theta} + R\left(\frac{\partial U_\Theta^+}{\partial z} + \frac{\partial U_\Theta^-}{\partial z} \right) \right) \right]$$

$$- R(\sigma_r^+ + \sigma_r^-) = 2h\beta_{11} dT \,. \tag{6.34}$$

With account of (6.33) the first of differential equilibrium equations (6.16) is of the kind

$$\frac{\partial \sigma_r^*}{\partial r} + \frac{1}{r}\frac{\partial \tau_{r\Theta}^*}{\partial \Theta} + \frac{\partial \tau_{rz}^*}{\partial z} + \frac{\sigma_r^* - \sigma_\Theta^*}{r} + \left[-\frac{\partial}{\partial r}\beta_{11} - \frac{1}{r}(\beta_{11} - \beta_{22}) \right] dT = 0 \,.$$

$$\tag{6.35}$$

Wherefrom, we get conditions of a rigid contact

$$R(\sigma_r^+ - \sigma_r^-) + A_{56}\left(\frac{\partial W^+}{\partial \Theta} - \frac{\partial W^-}{\partial \Theta} \right) + A_{66}\left(\frac{\partial U_\Theta^+}{\partial \Theta} - \frac{\partial U_\Theta^-}{\partial \Theta} \right)$$

$$+ RA_{55}\left(\frac{\partial W^+}{\partial z} - \frac{\partial W^-}{\partial z} \right) + RA_{56}\left(\frac{\partial U_\Theta^+}{\partial z} - \frac{\partial U_\Theta^-}{\partial z} \right)$$

$$+ (A_{11} - A_{12})(U_r^+ - U_r^-) + \frac{h}{R}\left\{ A_{66}\left(\frac{\partial^2 U_r^+}{\partial \Theta^2} + \frac{\partial^2 U_r^-}{\partial \Theta^2} \right) \right.$$

$$+ R^2 A_{55}\left(\frac{\partial^2 U_r^+}{\partial z^2} + \frac{\partial^2 U_r^-}{\partial z^2} \right) + 2RA_{56}\left(\frac{\partial^2 U_r^+}{\partial \Theta \partial z} + \frac{\partial^2 U_r^-}{\partial \Theta \partial z} \right)$$

$$+ (A_{12} - A_{22})(U_r^+ + U_r^-) + (A_{12} - A_{22} - A_{66})\left(\frac{\partial U_\Theta^+}{\partial \Theta} + \frac{\partial U_\Theta^-}{\partial \Theta} \right)$$

$$+ R(A_{14} - A_{24} - A_{56})\left(\frac{\partial U_\Theta^+}{\partial z} + \frac{\partial U_\Theta^-}{\partial z} \right)$$

$$+R(A_{13} - A_{23})\left(\frac{\partial W^+}{\partial z} + \frac{\partial W^-}{\partial z}\right) + (A_{14} - A_{24})\left(\frac{\partial W^+}{\partial \Theta} + \frac{\partial W^-}{\partial \Theta}\right)\Big\}$$
$$= 4h(\beta_{11} - \beta_{22})dT \,. \tag{6.36}$$

The second and third equations of (6.16) and, consequently conditions (6.31) and (6.32) stay invariable under thermal loading.

Thus the supposition on smallness of the layer width in contrast to contacting massive bodies and the condition of meeting differential equations (6.16) only on layer L median have made grounds to obtain from conditions of an ideal contact (6.14) and (6.15) the conditions of a rigid contact of massive bodies S^+ and S^- with a local-orthotropic layer L under mechanical (6.21), (6.23), (6.24), (6.30)–(6.32) and temperature ((6.34), (6.23), (6.24), (6.31), (6.32), (6.36)) loading.

6.4 Solution of a Spatial Problem for an Inhomogeneous in Radial Direction Rod

The problems of the inhomogeneous theory of elasticity are difficult to solve in analytical respect, so far some simplified models of inhomogeneity of elastic materials are proposed. More often [202] a supposition is put forward that Poisson's ratio of the material is constant while elasticity and shear moduli vary following one or another law. Or otherwise, a material model whose shear modulus is taken constant and Poisson's ratio is randomly dependent upon three Cartesian coordinates is treated. Investigation results of elasticity and thermoelasticity of bodies with a continuous dependence of inhomogeneity of physico-mechanical characteristics on three coordinates have been reported elsewhere [204]. Solution of the 3D problem is reduced to nonlinear differential equations and is constrained by substantial computation difficulties.

For the rod element under study it's possible to avoid both model simplification and nonlinearity of the determinative system of equations thanks to the smallness of the angular value. Let's consider an axisymmetric loading of a spirally anisotropic rod S whose cross-section is a circular ring or circle. The elasto-equivalent directions of the rod material are disposed along the helical lines whose inclination to the generatrix is conditioned by fiber twisting under the action of the winding thread.

Since the elastic constants of the rod are independent of variable Θ, then under the axisymmetric loading

$$\tau_{rz} = \tau_{r\theta} = 0 \,. \tag{6.37}$$

For further transformations let's write coefficients a_{ij} of Hook's law as

$$a_{ij} = a_{ij}^{(0)} + \gamma_0^2 r^2 C_{ij}; \quad i,j = 1,2,3; \ i = j = 4,5,6$$
$$a_{i4} = \gamma_0 r C_{i4}; \quad i = 1,2,3; \quad a_{56} = \gamma_0 r C_{56} \,, \tag{6.38}$$

where $a_{ij}^{(O)}$ are elastic properties of rod S in reinforcement direction.

The dependencies for determining constants C_{ij} through $a_{ij}^{(O)}$ are of the form

$$C_{11} = 0; \qquad\qquad C_{14} = 2(a_{13}^{(0)} - a_{12}^{(0)});$$
$$C_{12} = a_{13}^{(0)}; \qquad\qquad C_{22} = 2a_{13}^{(0)} + a_{44}^{(0)};$$
$$C_{13} = a_{12}^{(0)}; \qquad\qquad C_{23} = a_{11}^{(0)} + a_{33}^{(0)} - a_{44}^{(0)};$$
$$c_{24} = -2a_{11}^{(0)} + 2a_{13}^{(0)} + a_{44}^{(0)}; \qquad C_{44} = 4a_{11}^{(0)} + 4a_{13}^{(0)} + 4a_{33}^{(0)} - 2a_{44}^{(0)};$$
$$C_{33} = 2a_{13}^{(0)} + a_{44}^{(0)}; \qquad\qquad C_{55} = a_{66}^{(0)}; \quad C_{56} = a_{44}^{(0)} - a_{66}^{(0)};$$
$$C_{34} = -a_{13}^{(0)} + a_{33}^{(0)}; \qquad\qquad C_{66} = a_{44}^{(0)}.$$

Since coefficients a_{ij} of Hook's law depend on small angle γ_0, it's natural to apply the perturbation theory for the problem solution. With this aim the stresses and strain will be written as

$$f = f^{(0)} + \gamma_0 f^{(1)} + \gamma_0^2 f^{(2)}, \tag{6.39}$$

taking that $\gamma_O^n \approx 0$ at $n \geq 3$.

Since the unstrained face of the rod is rigidly fixed, then

$$W = az, \tag{6.40}$$

$$U_\Theta = a_O r z, \tag{6.41}$$

where unknown constants a, $a^{(0)}$ are found from face conditions (6.10) and (6.11).

The problem is solved in three stages. The solution corresponding to γ_0^0, i.e. to transversely isotropic material is found at the first stage. As is known [202, 203], in this case the solution of the plane problem is of the form

$$\sigma_r^{(0)} = A_{-1} r^{-2} + A_1; \quad \sigma_\Theta^{(0)} = -A_{-1} r^{-2} + A_1, \tag{6.42}$$

where unknowns A_{-1}, A_1 are calculated from boundary conditions, while $A_{-1} = 0$ is for a solid cross-section. Stresses $\sigma_z^{(0)}$, $\tau_{\Theta Z}^{(0)}$ and displacements $U_r^{(0)}$ are easily found from Hook's law

$$\sigma_z^{(0)} = \frac{1}{a_{33}^{(0)}} a - 2\frac{a_{13}^{(0)}}{a_{33}^{(0)}} A_1;$$

$$\tau_{\Theta z}^{(0)} = \frac{1}{a_{44}} a_O r;$$

$$U_z^{(0)} = \frac{a}{a_{33}^{(0)}} r + A_{-1}(a_{12}^{(0)} - a_{11}^{(0)})r^{-1} + A_1 b_1 r, \tag{6.43}$$

where $b_1 = a_{12}^{(0)} + a_{11}^{(0)} - \dfrac{2(a_{13}^{(0)})^2}{a_{33}^{(0)}}$.

At the second stage the solution corresponding to γ_0^0 and γ_0^1 is sought. By equating coefficients of similar exponents of the small perimeter γ_0 from Hook's law

$$\frac{dU_r^{(1)}}{dr} = a_{11}^{(0)}\sigma_r^{(1)} + a_{12}^{(0)}\sigma_\Theta^{(1)} + a_{13}^{(0)}\sigma_z^{(1)} + rC_{14}\tau_{\Theta z}^{(0)} \,;$$

$$\frac{U_r^{(1)}}{r} = a_{12}^{(0)}\sigma_r^{(1)} + a_{11}^{(0)}\sigma_\Theta^{(0)} + a_{13}^{(0)}\sigma_z^{(1)} + rC_{24}\tau_{\Theta z}^{(0)} \,;$$

$$0 = a_{13}^{(0)}\sigma_r^{(1)} + a_{13}^{(0)}\sigma_\Theta^{(1)} + a_{33}^{(0)}\sigma_z^{(1)} + rC_{34}\tau_{\Theta z}^{(0)} \,;$$

$$0 = rC_{14}\sigma_r^{(0)} + rC_{24}\sigma_\Theta^{(0)} + rC_{34}\sigma_z^{(0)} + a_{44}^{(0)}\tau_{\Theta z}^{(1)} \,. \qquad (6.44)$$

From the last equation we get

$$\tau_{\Theta z}^{(1)} = -A_{-1}\frac{1}{a_{44}^{(0)}}(C_{14} - C_{24})r^{-1} - A_1\frac{1}{a_{44}^{(0)}}\left(C_{14} + C_{24} - \frac{2a_{13}^{(0)}C_{34}}{a_{33}^{(0)}}\right)r$$
$$-\frac{C_{34}}{a_{33}^{(0)}a_{44}^{(0)}}ar \,. \qquad (6.45)$$

From of (6.44) it follows that

$$\sigma_r^{(1)} = B_{-1}r^{-2} + B_1 + a_0 B_3 r^2 \,;$$

$$\sigma_\Theta^{(1)} = -B_{-1}r^{-2} + B_1 + 3a_0 B_3 r^2 \,;$$

$$\sigma_z^{(1)} = -2\frac{a_{13}^{(0)}}{a_{33}^{(0)}}B_1 + a_0 b_2 r^2 \,;$$

$$U_r^{(1)} = B_{-1}(a_{12}^{(0)} - a_{11}^{(0)})r^{-1} + B_1 b_1 r + a_0 b_3 r \,, \qquad (6.46)$$

where unknown constants B_{-1}, B_1 are determined from boundary conditions (for the circle $B_{-1} = 0$) and relations b_2, b_3, B_3 look like

$$b_2 = \frac{C_{34}}{a_{33}^{(0)}a_{44}^{(0)}} - \frac{4a_{13}^{(0)}}{a_{33}^{(0)}}B_3 \,;$$

$$b_3 = B_3(a_{12}^{(0)} + 3a_{11}^{(0)}) + \frac{C_{44}}{a_{44}^{(0)}} \,;$$

$$B_3 = \frac{2a_{13}^{(0)}C_{34} - a_{33}^{(0)}(3C_{24} - C_{14})}{8a_{44}^{(0)}(a_{11}^{(0)}a_{33}^{(0)} - (a_{13}^{(0)})^2)} \,.$$

At the third stage the terms corresponding to $\gamma_O^0, \gamma_O^1, \gamma_O^2$ are defined. From Hook's law we have

$$\frac{dU_r^{(2)}}{dr} = r^2\left(C_{11}\sigma_r^{(0)} + C_{12}\sigma_\Theta^{(0)} + C_{13}\sigma_z^{(0)}\right)$$
$$+rC_{14}\tau_{\Theta z}^{(1)} + a_{11}^{(1)}\sigma_r^{(2)} + a_{12}^{(1)}\sigma_\Theta^{(2)} + a_{13}^{(1)}\sigma_z^{(2)} \,;$$

$$\frac{U_r^{(2)}}{r} = r^2\left(C_{12}\sigma_r^{(0)} + C_{22}\sigma_\Theta^{(0)} + C_{23}\sigma_z^{(0)}\right)$$
$$+rC_{24}\tau_{\Theta z}^{(1)} + a_{12}^{(1)}\sigma_r^{(2)} + a_{11}^{(1)}\sigma_\Theta^{(2)} + a_{13}^{(1)}\sigma_z^{(2)} \,;$$

$$0 = r^2\left(C_{13}\sigma_r^{(0)} + C_{23}\sigma_\Theta^{(0)} + C_{33}\sigma_z^{(0)}\right)$$

$$+rC_{34}\tau_{\Theta z}^{(1)} + a_{13}^{(1)}\sigma_r^{(2)} + a_{13}^{(1)}\sigma_\Theta^{(2)} + a_{33}^{(1)}\sigma_z^{(2)};$$

$$0 = r^2 C_{44}\tau_{\Theta z}^{(0)} + a_{44}^{(1)}\tau_{\Theta z}^{(2)} + r(C_{14}\sigma_r^{(1)} + C_{24}\sigma_\Theta^{(1)} + C_{34}\sigma_z^{(1)}). \qquad (6.47)$$

From the first two equations in (6.47) follows that

$$\sigma_r^{(2)} = C_{-1}r^{-2} + C_1 + C_0 A_1 \ln r + C_{31}A_1 r^2 + C_{32}ar^2;$$

$$\sigma_\Theta^{(2)} = C_{-1}r^{-2} + C_1 + C_0 A_1(\ln r + 1) + 3r^2(C_{31}A_1 + C_{32}a), \qquad (6.48)$$

where unknown constants C_{-1}, C_1 ($C_{-1} = 0$ for the circle) are found from boundary conditions and C_O, C_{32}, C_{31} are of the kind

$$C_0 = \frac{a_{33}^{(0)}\left[2C_{12} - 2C_{11} + \dfrac{(C_{14} - C_{24})^2}{a_{44}^{(0)}}\right]}{2[a_{11}^{(0)}a_{33}^{(0)} - (a_{13}^{(0)})^2]};$$

$$C_{31} = \frac{1}{8[a_{33}^{(0)}a_{11}^{(0)} - (a_{13}^{(0)})^2]}\left[2(C_{11} + C_{12})a_{33}^{(0)} + 2a_{13}^{(0)}(C_{13} - 3C_{23})\right.$$

$$\left. + \frac{1}{a_{44}^{(0)}}(C_{14} - 3C_{24})(a_{33}^{(0)}(C_{14} + C_{24}) - 2a_{13}^{(0)}C_{34})\right]$$

$$C_{32} = \frac{1}{8[a_{33}^{(0)}a_{11}^{(0)} - (a_{13}^{(0)})^2]}\left[C_{13} - 3C_{23}\frac{C_{34}}{a_{44}^{(0)}}(C_{14} - 3C_{24})\right];$$

Hence, SSS of the spirally anisotropic rod is determined during axisymmetric loading by dependencies (6.40), (6.41) and

$$\sigma_r = (A_{-1} + \gamma_0 B_{-1} + \gamma_0^2 C_{-1})r^{-2} + \gamma_0 A_{-1}C_0 \ln r + A_1 + \gamma_0 B_1 + \gamma_0^2 C_1$$

$$+(\gamma_0 a_0 B_3 + \gamma_0^2 A_1 C_{31} + \gamma_0^2 aC_{32})r^2;$$

$$\sigma_\Theta = -(A_{-1} + \gamma_0 B_{-1} + \gamma_0^2 C_{-1})r^{-2} + \gamma_0^2 A_{-1}C_0 \ln r + \gamma_0^2 C_0 A_{-1} + A_1$$

$$+\gamma_0 B_1 + \gamma_0^2 C_1 + (3\gamma_0 a_0 B_3 + 3\gamma_0^2 A_1 C_{31} + 3\gamma_0^2 aC_{32})r^2;$$

$$\sigma_z = -2\frac{a_{13}^{(0)}}{a_{33}^{(0)}}(A_1 + \gamma_0 B_1 + \gamma_0^2 C_1 + \gamma_0^2 A_{-1}b_4 + \gamma_0^2 A_{-1}C_0 \ln r) + \frac{a}{a_{33}^{(0)}}$$

$$+(\gamma_0 b_2 a_0 + \gamma_0^2 A_1 b_5 + a\gamma_0^2 b_6)r^2;$$

$$\tau_{\Theta z} = \frac{C_{24} - C_{14}}{a_{44}^{(0)}}r^{-1}(\gamma_0 A_{-1} + \gamma_0^2 B_{-1})$$

$$+\left[a_0 + b_7(\gamma_0 A_1 + \gamma_0^2 B_1) - \gamma_0\frac{C_{34}}{a_{33}^{(0)}a_{44}^{(0)}}a\right]r + b_8 a_0;$$

$$U_r = (a_{12}^{(0)} - a_{11}^{(0)})(A_{-1} + \gamma_0 B_{-1} + \gamma_0^2 C_{-1})r^{-1}$$

$$+\left[b_1(A_1 + \gamma_0 B_1 + \gamma_0^2 C_{-1}) + a\frac{1}{a_{33}^{(0)}} + A_{-1}b_{11}\right]r$$

$$+[\gamma_0 a_0 b_3 + \gamma_0^2 A_1 b_9 + \gamma_0^2 ab_{10}]r^3 + \gamma_0^2 A_{-1}C_0(a_{11} + a_{12})r \ln r, \qquad (6.49)$$

where unknown constants $A_{-1}, A_1, B_{-1}, B_1, C_{-1}, C_1, a, a_0$ are found from the boundary and face conditions. Coefficients b_i are calculated from

$$b_1 = a_{12}^{(0)} + a_{11}^{(0)} - \frac{2(a_{13}^{(0)})^2}{a_{33}^{(0)}} \; ;$$

$$b_4 = \frac{1}{2a_{13}^{(0)}} \left[C_{13} - C_{23} - \frac{C_{14} - C_{24}}{a_{44}^{(0)}} C_{34} + a_{13}^{(0)} C_0 \right] \; ;$$

$$b_5 = -\frac{1}{a_{33}^{(0)}} \left[C_{13} + C_{23} - \frac{2a_{13}^{(0)} C_{13}}{a_{33}^{(0)}} - \frac{C_{34}}{a_{44}^{(0)}} \left(C_{14} + C_{24} - \frac{2a_{13}^{(0)} C_{34}}{a_{33}^{(0)}} \right) \right.$$
$$\left. +4C_{31} a_{13}^{(0)} \right] \; ;$$

$$b_6 = -\frac{1}{a_{33}^{(0)}} \left(4a_{13}^{(0)} C_{32} - \frac{C_{34}^{(0)}}{a_{33}^{(0)} a_{44}^{(0)}} \right) \; ;$$

$$b_7 = \frac{1}{a_{44}^{(0)}} \left(-C_{14} - C_{24} + \frac{2a_{13}^{(0)} C_{34}}{a_{33}^{(0)}} \right) \; ;$$

$$b_8 = -\frac{1}{(a_{44}^{(0)})^2} \left(C_{44} + \frac{C_{34}}{a_{33}^{(0)}} \right) - \frac{1}{a_{44}^{(0)}} B_3 \left(C_{14} + 3C_{24} - \frac{4C_{34} a_{13}^{(0)}}{a_{33}^{(0)}} \right) \; ;$$

$$b_9 = C_{12} + C_{22} - 2C_{23} \frac{a_{13}^{(0)}}{a_{33}^{(0)}} + \frac{C_{24}}{a_{44}^{(0)}} \left(\frac{2a_{13}^{(0)} C_{34}}{a_{33}^{(0)}} - C_{14} - C_{24} \right)$$
$$+C_{31}(a_{12}^{(0)} + 3a_{11}^{(0)}) \; ;$$

$$b_{10} = \frac{C_{23}}{a_{33}^{(0)}} - \frac{C_{24} C_{34}}{a_{33}^{(0)} a_{44}^{(0)}} + C_{32}(a_{12}^{(0)} + 3a_{11}^{(0)}) \; ;$$

$$b_{11} = C_{12} - C_{22} + \frac{C_{24}(C_{24} - C_{14})}{a_{44}^{(0)}} + C_0 a_{12}^{(0)} \; ;$$

$$b_{12} = \frac{a_{33}^{(0)}}{8a_{44}^{(0)}} \left[C_{14} - 3C_{24} + 2\frac{a_{13}^{(0)} C_{34}}{a_{44}^{(0)}} \right] \frac{1}{a_{11}^{(0)} a_{33}^{(0)} - (a_{13}^{(0)})^2} \; .$$

Let's consider thermal loading of rod S as its temperature varies under thermal shock by a value ΔT. Within the interval Δt both elastic and thermoelastic properties of the rod are taken constant. Similarly to axisymmetric loading we presume [187] that equalities (6.40) and (6.41) are satisfied. The first stage of solution yields that

$$\sigma_{r,T}^{(0)} = \sigma_r^{(0)}; \; \sigma_{\Theta,T}^{(0)} = \sigma_\Theta^{(0)}; \; \tau_{\Theta z,T}^{(0)} = \tau_{\Theta z}^{(0)} , \tag{6.50}$$

$$\sigma_{z,T}^{(0)} = \sigma_z^{(0)} - \frac{1}{a_{33}^{(0)}} \alpha_z^{(0)} \Delta T , \tag{6.51}$$

$$U_{r,T}^{(0)} = U_r^{(0)} + r\alpha_r^{(0)} \Delta T . \tag{6.52}$$

Here and further $f_{-,T}$ is either stress or displacement at thermal loading.

We'll denote for the second stage

$$f^{(T)} = f^{(1)}_{-,T} - f^{(1)} \,. \tag{6.53}$$

With account of notations (6.53) from Hook's law we obtain

$$\frac{dU^{(T)}}{dr} = a^{(0)}_{11}\sigma^{(T)}_r + a^{(0)}_{12}\sigma^{(T)}_\Theta + a^{(0)}_{13}\sigma^{(T)}_z \,;$$

$$\frac{U^{(T)}_r}{r} = a^{(0)}_{12}\sigma^{(T)}_r + a^{(0)}_{11}\sigma^{(T)}_\Theta + a^{(0)}_{13}\sigma^{(T)}_z - \alpha^{(0)}_z r\Delta T \,;$$

$$0 = a^{(0)}_{13}\sigma^{(T)}_r + a^{(0)}_{13}\sigma^{(T)}_\Theta + a^{(0)}_{33}\sigma^{(T)}_z + \alpha^{(0)}_r r\Delta T \,;$$

$$0 = a^{(0)}_{44}\tau^{(T)}_{\Theta z} + \left[-rc_{34}\frac{1}{a^{(0)}_{33}}\alpha^{(0)}_z + \alpha^{(0)}_r - \alpha^{(0)}_z \right]\Delta T \,. \tag{6.54}$$

From the last equation in (6.54) follows

$$\tau^{(1)}_{\Theta z,T} = \tau^{(1)}_{\Theta z} + r\left[\alpha^{(0)}_z\left(1 + \frac{C_{34}}{a^{(0)}_{33}}\right) - \alpha_r\right]\frac{1}{a^{(0)}_{44}}\Delta T \,. \tag{6.55}$$

From (6.54) one gets

$$\sigma^{(T)}_z = \beta_{13}\sigma^{(T)}_r + \beta_{13}\sigma^{(T)}_\Theta - \frac{\alpha^{(0)}_r}{a^{(0)}_{33}}r\Delta T \,; \tag{6.56}$$

$$\varepsilon^{(T)}_r = \beta_{11}\sigma^{(T)}_r + \beta_{12}\sigma^{(T)}_\Theta + \beta_{13}\alpha^{(0)}_r r\Delta T \,; \tag{6.57}$$

$$\varepsilon^{(T)}_\Theta = \beta_{12}\sigma^{(T)}_r + \beta_{11}\sigma^{(T)}_\Theta + (\beta_{13}\alpha_r - \alpha_z)r\Delta T \,; \tag{6.58}$$

where [202]:

$$\beta_{ij} = a^{(0)}_{ij} - \frac{a^{(0)}_{13}a^{(0)}_{j3}}{a^{(0)}_{33}}, \quad i,j = 1,2; \ \beta_{13} = -\frac{a^{(0)}_{13}}{a^{(0)}_{33}} \,. \tag{6.59}$$

Let's introduce stress function $F(r)$ so that

$$\sigma^{(T)}_r = \frac{1}{r}F(r); \quad \sigma^{(T)}_\Theta = F'(r) \,. \tag{6.60}$$

From (6.57) and (6.58) the differential equation follows

$$rF'' + F' - \frac{1}{r}F = \frac{1}{\beta_{11}}(2\alpha^{(0)}_z - \alpha^{(0)}_r\beta_{13})r\Delta T \,. \tag{6.61}$$

Solution of the homogeneous equation corresponds to the differential one (6.61) which is $\sigma^{(1)}_r, \sigma^{(1)}_\theta$. To find $\sigma^{(T)}_r, \sigma^{(T)}_\theta$ just the local solution of differential equation (6.61) is necessary, which is

$$F = \frac{1}{3\beta_{11}}(2\alpha^{(0)}_z - \alpha^{(0)}_r\beta_{13})r^2\Delta T \,. \tag{6.62}$$

From (6.60), (6.62) and (6.69) we have

$$\sigma_r^{(T)} = \frac{2a_{33}^{(0)}\alpha_z + a_{13}^{(0)}\alpha_r}{3[a_{11}^{(0)}a_{33}^{(0)} - (a_{13}^{(0)})^2]} r\Delta T\,;$$

$$\sigma_\Theta^{(T)} = 2\frac{2a_{33}^{(0)}\alpha_z + a_{13}^{(0)}\alpha_r}{3[a_{11}^{(0)}a_{33}^{(0)} - (a_{13}^{(0)})^2]} r\Delta T\,. \tag{6.63}$$

From (6.56)

$$\sigma_z^{(T)} = -\left[\frac{a_{13}^{(0)}}{a_{33}^{(0)}} \frac{2a_{33}^{(0)}\alpha_z^{(0)} + a_{13}^{(0)}\alpha_r^{(0)}}{(a_{13}^{(0)})^2 - a_{11}^{(0)}a_{33}^{(0)}} - \frac{\alpha_r^0}{a_{33}^{(0)}}\right] r\Delta T\,. \tag{6.64}$$

and from (6.58)

$$U_r^{(T)} = \left[\left(a_{11}^{(0)} + 2u_{12}^{(0)}\ 3\frac{(a_{13}^{(0)})^2}{a_{33}^{(0)}}\right)\frac{2a_{33}^{(0)}\alpha_z^{(0)} + a_{13}^{(0)}\alpha_r^{(0)}}{3[a_{11}^{(0)}a_{33}^{(0)} - (u_{13}^{(0)})^2]}\right.$$
$$\left. - \left(\frac{a_{13}^{(0)}}{a_{33}^{(0)}}\alpha_r^{(0)} + \alpha_z^{(0)}\right)\right] r^2\Delta T\,. \tag{6.65}$$

By reasoning analogously we have at the third stage

$$\tau_{\Theta z,T}^{(2)} = \tau_{\Theta z}^{(2)} - \frac{1}{a_{44}^{(0)}}\left[\left(C_{14} + 2C_{24} - 3\frac{a_{13}^{(0)}}{a_{33}^{(0)}}C_{34}\right)\right.$$
$$\left. \times \left(\frac{2a_{33}^{(0)}\alpha_z^{(0)} + a_{13}^{(0)}\alpha_r^{(0)}}{a_{11}^{(0)}a_{33}^{(0)} - (a_{13}^{(0)})^2} - C_{34}\frac{\alpha_r^{(0)}}{a_{33}^{(0)}}\right)\right] r^2\Delta T\,;$$

$$\sigma_{r,T}^{(2)} = \sigma_r^{(2)} + b_{12}(\alpha_z - \alpha_r)r^3\Delta T\,;$$

$$\sigma_{\Theta,T}^{(2)} = \sigma_\Theta^{(2)} + 3b_{12}(\alpha_z - \alpha_r)r^2\Delta T\,;$$

$$\alpha_{z,T}^{(2)} = \sigma_z^{(2)} + \frac{1}{a_{33}^{(0)}}(2\alpha_r^{(0)} - \alpha_z^{(0)})\left(4a_{13}^{(0)}b_{12} + \frac{C_{34}}{a_{44}^{(0)}}\right)r^2\Delta T\,;$$

$$U_{r,T}^{(2)} = U_r^{(2)} + \left[(3a_{11}^{(0)} + a_{12}^{(0)} - 4a_{13}^{(0)})b_{12} + \frac{C_{24} + C_{34}}{a_{44}^{(0)}}\right](\alpha_z^{(0)} - \alpha_r^{(0)})r^3\Delta T\,.$$
$$\tag{6.66}$$

Hence, solution of the thermoelasticity problem for spirally anisotropic rods S is

$$\sigma_{r,T} = \sigma_r + \left[\gamma_0 r\frac{2a_{33}^{(0)}\alpha_z^{(0)} + a_{13}^{(0)}\alpha_r^{(0)}}{3[a_{11}^{(0)}a_{33}^{(0)} - (a_{13}^{(0)})^2]} + \gamma_0^2 r^2 b_{12}(\alpha_z^{(0)} - \alpha_r^{(0)})\right]\Delta T\,;$$

$$\sigma_{\Theta,T} = \sigma_\Theta + \left[2\gamma_0 r\frac{2a_{33}^{(0)}\alpha_z^{(0)} + a_{13}^{(0)}\alpha_r^{(0)}}{3[a_{11}^{(0)}a_{33}^{(0)} - (a_{13}^{(0)})^2]} + 3\gamma_0^2 r^2 b_{12}(\alpha_z^{(0)} - \alpha_r^{(0)})\right]\Delta T\,;$$

$$\sigma_{z,T} = \sigma_z + \left[\gamma_0 r \left(\frac{a_{13}^{(0)}}{a_{33}^{(0)}} \frac{2a_{33}^{(0)} \alpha_z^{(0)} + a_{13}^{(0)} \alpha_r^{(0)}}{(a_{13}^{(0)})^2 - a_{11}^{(0)} a_{33}^{(0)}} - \frac{\alpha_r^{(0)}}{a_{33}^{(0)}} \right) \right.$$

$$\left. + \gamma_0^2 r^2 \frac{1}{a_{33}^{(0)}} \left(4a_{13}^{(0)} b_{12} + \frac{c_{34}}{a_{44}^{(0)}} \right) (\alpha_r^{(0)} - \alpha_z^{(0)}) \right] \Delta T \,;$$

$$\tau_{\Theta z,T} = \tau_z + \left[\gamma_0 r \frac{\alpha_z^{(0)} - \alpha_r^{(0)}}{a_{44}^{(0)}} - \gamma_0^2 r^2 \frac{1}{a_{44}^{(0)}} \left(\left(C_{14} + 2C_{24} - 3 \frac{a_{13}^{(0)}}{a_{33}^{(0)}} C_{34} \right) \right. \right.$$

$$\left. \left. \times \left(\frac{2a_{33}^{(0)} \alpha_z^{(0)} + a_{13}^{(0)} \alpha_r^{(0)}}{a_{11}^{(0)} a_{33}^{(0)} - (a_{13}^{(0)})^2} - \frac{C_{34} \alpha_r^{(0)}}{a_{33}^{(0)}} \right) \right) \right] \Delta T \,;$$

$$U_{r,T} = U_r + \left\{ \alpha_r^{(0)} r + \gamma_0 r^2 \left[\left(a_{11}^{(0)} + 2a_{12}^{(0)} - 3 \frac{(a_{13}^{(0)})^2}{a_{33}^{(0)}} \right) \right. \right.$$

$$\times \frac{2a_{33}^{(0)} \alpha_z^{(0)} + a_{13}^{(0)} \alpha_r^{(0)}}{3[a_{11}^{(0)} a_{33}^{(0)} - (a_{13}^{(0)})^2]} - \left(\frac{a_{13}^{(0)}}{a_{33}^{(0)}} \alpha_r^{(0)} + \alpha_z^{(0)} \right) \bigg]$$

$$\left. + \gamma_0^2 r^3 \left[(3a_{11}^{(0)} + a_{12}^{(0)} - 4a_{13}^{(0)}) b_{12} + \frac{C_{24} + C_{34}}{a_{44}^{(0)}} \right] \right.$$

$$\left. \times (\alpha_z^{(0)} - \alpha_r^{(0)}) \right\} \Delta T \,, \tag{6.67}$$

where $\sigma_r, \sigma_\theta, \tau_{\theta z}, U_r$ are found from (6.49).

By the perturbation theory method solutions for the elastic (6.49) and thermoelastic (6.67) problems are found for a spirally anisotropic inhomogeneous rod S. The solution contains 6 unknown constants defined from boundary conditions and 2 unknown constants a, a_0 found from integral face conditions (6.10), (6.11).

6.5 Reduction of the Elastic Theory Problem to Solution of a System of Algebraic Equations

Under mechanical loading with a longitudinal force P_Z, torque M_{TR} and thermal loading of the rod described in Sect. 6.2 SSS in each main layer S_i $(i = 2, \ldots, N)$ and layer S_1 (in provision $E_Z^{(1)} \neq 0$) is described by (6.37), (6.40), (6.41) and (6.67) where $A_{-1}^{(1)} = B_{-1}^{(1)} = C_{-1}^{(1)} = 0$.

Assuming that the outer, inner (at $E_Z^{(1)} = 0$) and side surfaces of the rod are unloaded, i.e. taking

$$\sigma_r(R_1) = \sigma_r(R_N + 2h_N) = 0 \,, \tag{6.68}$$

the additional unknown constants $D_k^{(L)}$ are introduced, where $k = 0, 1, 2$; $L = 1, N$ such that

$$U_r(R_1) = D_0^{(1)} + \gamma_0 D_1^{(1)} + \gamma_0^2 D_2^{(1)} ; \tag{6.69}$$

$$U_r(R_N + 2h_N) = D_0^{(N)} + \gamma_0 D_1^{(N)} + \gamma_0^2 D_2^{(N)} . \tag{6.70}$$

Boundary conditions (6.23), (6.24), (6.31) and (6.32) on the outer side surface and interfaces of layers S_i, S_{i+1} ($i = 1, \ldots, N-1$) with layer L_i are, with account of (6.37), (6.40), (6.41), (6.67)–(6.70) identical, and from (6.21) and (6.30), and dependencies

$$\gamma_0^{(i)} = K_i \gamma_0 , \tag{6.71}$$

where $i = 1, \ldots, N$; $K_N = 1$, by equating coefficients at similar exponents γ_0 we get $6N$ linear algebraic equations with unknowns X_j, $j = 1, \ldots, M$, where

$$M = 6N + 2 \tag{6.72}$$

$$
\begin{aligned}
x_1 &= A_1^{(1)}\delta + D_0^{(1)}(1-\delta) ; \\
x_2 &= B_1^{(1)}\delta + D_1^{(1)}(1-\delta) ; \\
x_3 &= C_1^{(1)}\delta + D_2^{(1)}(1-\delta) ; \\
x_k &= A_{-1}^{(i)}; \quad x_{k+1} = B_{-1}^{(i)}; \quad x_{k+2} = C_{-1}^{(i)} ; \\
x_{k+3} &= A_1^{(i)}; \quad x_{k+4} = B_1^{(i)}; \quad x_{k+5} = C_1^{(i)}; \quad k = 6i - 8; \; i \geq 2 ; \\
x_{M-4} &= D_0^{(N)}; \quad X_{M-3} = D_1^{(N)}; \quad X_{M-2} = D_2^{(N)} ; \\
x_{M-1} &= a; \quad x_M = a_0 ,
\end{aligned}
\tag{6.73}
$$

$$
\delta = \begin{cases} 1, \Rightarrow E_z^{(1)} \neq 0; \\ 0, \Rightarrow E_z^{(1)} = 0. \end{cases}
\tag{6.74}
$$

Loading conditions of the rod element are allowed for by integral face conditions (6.10) and (6.11). Two more linear algebraic equations are obtained relatively unknowns X_j, $j = 1, \ldots, M$.

When determining integrals of the type

$$\int_{R_i}^{R_i + 2h_i} r f(r) dr$$

smallness of layer L_i was allowed for

$$\int_{R_i}^{R_i + 2h_i} r f(r) dr = 2R_i h_i f(R_i + h_i) . \tag{6.75}$$

It appears that the problem on a joint action of the mechanical (axial force and torque) and temperature loading of a built-up spirally anisotropic rod is reduced with attraction of the rigid contact conditions (6.21), (6.23), (6.24), (6.30)–(6.32) to solution $6N + 2$ of a system of algebraic equations.

For $i = 2, \ldots, N - 1$ and $L = 1, 2$

$$\sum_{j=1}^{2}(Q_{1,0}^{(L)} X_k + Q_{2,0}^{(L)} X_{k+3} + Q_{3,0}^{(L)} X_{M-1} + Q_{4,0}^{(L)} X_M) = B_{1i}^{(L)} \, ;$$

$$\sum_{j=1}^{2}(Q_{5,i}^{(L)} X_{k+1} + K_{i+j=1} Q_{2,i}^{(L)} X_{k+4} + Q_{6,i}^{(L)} X_M) = B_{2i}^{(L)} \, ;$$

$$\sum_{j=1}^{2}(Q_{7,i}^{(L)} X_k + Q_{8,i}^{(L)} X_{k+2} + Q_{9,i}^{(L)} X_{k+3} + Q_{10,i}^{(L)} X_{k+5} + Q_{11,i}^{(L)} X_{M-1}) = B_{3i}^{(L)} \, ;$$

$$Q_2^{(L)}(1,1)X_1 + Q_1^{(L)}(1,2)X_4 + Q_2^{(L)}(1,2)X_7$$
$$+ \sum_{j=1}^{2}(Q_3^{(L)} X_{M-1} + Q_4^{(L)} X_M) = B_{11}^{(L)} \, ;$$

$$K_1 Q_2^{(L)}(1,1)X_2 + Q_5^{(L)}(1,2)X_5 + K_2 Q_2^{(L)}(1,2)X_8 + \sum_{j=1}^{2} Q_6^{(L)} X_M = B_{21}^{(L)} \, ;$$

$$Q_9^{(L)}(1,1)X_1 + Q_{10}^{(L)}(1,1)X_3 + Q_7^{(L)}(1,2)X_4 + Q_8^{(L)}(1,2)X_6$$
$$+ Q_9^{(L)}(1,2)X_7 + Q_{10}^{(L)}(1,2)X_9 + \sum_{j=1}^{2} Q_{10}^{(L)}(i,j)X_{M-1} = B_{31}^{(L)} \, ;$$

$$Q_1^{(L)}(N,1)X_{M-10} + Q_2^{(L)}(N,1)X_{M-7}$$
$$+ \sum_{j=1}^{2}(Q_3^{(L)}(N,j)X_{M-1} + Q_4^{(L)}(N,j)X_M) + Q_2^{(L)}(N,2)X_{M-4} = B_{1N}^{(L)} \, ;$$

$$Q_5^{(L)}(N,1)X_{M-9} + K_2 Q_2^{(L)}(N,1)X_{M-7} + Q_2^{(L)}(N,2)X_{M-3}$$
$$+ \sum_{j=1}^{2} Q_6^{(L)}(N,j)X_M = B_{2N}^{(L)} \, ;$$

$$Q_7^{(L)}(N,1)X_{M-10} + Q_8^{(L)}(N,1)X_{M-8} + Q_9(N,1)X_{M-7}$$
$$+ Q_{10}^{(L)}(N,1)X_{M-5} + Q_9^{(L)}(N,2)X_{M-4} + Q_{10}^{(L)}(N,2)X_{M-2}$$
$$+ \sum_{j=1}^{2} Q_{10}^{(L)}(N,j)X_{M-1} = B_{3N}^{(L)} \, ; \tag{6.76}$$

$$\sum_{L=1}^{M} D_{jL} X_L = B_{M-2+n}, \quad n = 1, 2 \, . \tag{6.77}$$

Constants $Q_n^{(L)}$ are found from (a_{mn} – properties of layer S_{i+j-1}, A_{mn} – properties of layer L_i.):

$$Q_1(i,j) = ((-1)^j A_{11} + h_i/R_i A_{11})[(a_{12} - a_{11})/R + b_{11}R] - h_i/R_i^2\,;$$
$$Q_2(i,j) = ((-1)^j A_{11} + h_i/R_i A_{12})b_1 R - h_i\,;$$
$$Q_3(i,j) = ((-1)^j A_{11} + h_i/R_i A_{12})/a_{33}R + h_i A_{13}\,;$$
$$Q_4(i,j) = K_i((-1)^j A_{11} + h_i/R_i A_{12})[(a_{12} - a_{11})/R - h_i R_i]\,;$$
$$Q_5(i,j) = K_i((-1)^j A_{11} + h_i/R_i A_{12})b_3 R^3 - h_i R_i^2 B_3]\,;$$
$$Q_6(i,j) = -K_i^2 h_i C_0 \ln R\,;$$
$$Q_7(i,j) = K_i^2\{((-1)^j A_{11} + h_i/R_i A_{12})[(a_{12} - a_{11})/R + b_9 R^3 - h_i/R_i^2]\}\,;$$
$$Q_8(i,j) = -K_i^2 C_{13} h_i R_i^2\,;$$
$$Q_9(i,j) = ((-1)^j A_{11} + h_i/R_i A_{12})b_1 R - h_i\,;$$
$$Q_{10}(i,j) = K_i^2\{((-1)^j A_{11} + h_i/R_i A_{12})[b_{10}R^3 + C_0(a_{11} + a_{12})R\ln R]$$
$$-h_i R_i^2 C_{32}\}\,.$$

In this case, $R = R_i + 2(j-1)h_i$.

6.6 Estimation of Effective Characteristics

Solution of the system of equations (6.76) contributes to calculation of SSS using relations (6.67) in any point of the rod. The value of a_0 conditions the relative twist angle of the rod under loading.

From (6.40) it follows that at

$$\sigma_N = \frac{P_Z}{D(R_N + 2h_N)^2} = -1 \quad (M_{TR} = 0;\ \Delta T = 0)\,,$$

value $E = -1/a$ (6.77) equals the effective longitudinal elasticity modulus of the rod element under compression and at $P_Z = M_{TR} = 0$

$$\alpha = \frac{a}{\Delta T}\,, \tag{6.78}$$

corresponds to its effective longitudinal LTEC.

6.7 Conclusions

1. A new modification of the rigid contact conditions for massive bodies with a thin local-orthotropic layer has been devised.
2. The elasticity and thermoelasticity problems for an inhomogeneous spirally anisotropic cylinder have been solved.
3. The solution for the elasticity and thermoelasticity problems for the main material layers enables to reduce estimation of SSS, the efficient longitudinal elastic modulus and effective longitudinal LTEC to solution of the algebraic system of equations.

7 The Effect of Structural and Geometrical Parameters on Rod Properties

7.1 Principles of Experimental Investigations of Rod Elements

Rod elements were tested to study the threshold of delamination and strength of rod elements in response to variations in their structural and geometrical parameters. Test results were compared to theoretical conclusions derived by the method described in Chap. 6.

7.1.1 Preparation and Manufacture of the Samples

The process of manufacturing spirally reinforced rods includes their molding from a fibrous material by pultrusion followed by packing of the spiral layers. The needed filling degree and dimensions of each intermediate layer and the article as a whole are reached at every stage of forming with the help of spinnerets of the required open flow area. The reinforcing layers are packed immediately after leaving the spinneret using special winders. Solid rod elements are produced 20 mm in diameter and tubular ones have 25 mm inner and 35 mm outer diameters, and their longitudinal to transversal layer relation is 1:0 till 2:3. There are 2 coaxial layers (main reinforcement) and 3 layers of the auxiliary reinforcement. Winding pitch of the spiral reinforcement is 1 mm. The initial materials were glass braids, carbon braids and organic fibers of 29.4 tex, boron fibers and epoxide compound-based binders. Properties of the rod element ingredients are given in Table 7.1. The winding fiber tension force is chosen within 5–8 N/braid. The obtained braids are cured in a polymerizing oven following a standard time–temperature regime. As far as inner stresses might appear during cooling of the rod element a part of the samples were formed so as to exclude the effect of SSS during testing. With this aim each layer was cured immediately after molding and the reinforcing layer was laid on the cold-set binder after polymerization and cooling of the main layer.

The formed rod elements were cut into 50 mm long specimens and endured less than 20 ± 2 gC and 65% humidity not less than for 16 h.

Table 7.1. Properties of Reinforcing Fibers

Parameters	Boron	Organic	Glass	Carbon
Breaking tensile stress, MPa, $\times 10^2$	25–38	22	35	20–22
Elastic modulus, MPa, 10^2				
under tension	4000	1100	950	2700
under shear	1600	15	40	75
Relative elongation at rupture, %	0.8	3	4	1.2
LTEC 10^6 K^{-1}	2.4	3	5	−0.5
Poisson's ratio	0.2	0.3	0.3	0.22

7.1.2 Testing Procedure

To register the breaking strength value and delamination threshold under compression the highest loading has been recorded under which the material started to exfoliate and the first cracks appeared t longitudinal compression of the sample. The strain rate was chosen within 1–5% per minute which corresponds to approaching velocity of two bedplates of the testing machine 0.2–1.0 mm/min. Initially the sample is loaded by a force of about 10–20% of the anticipated short-term static strength that was then diminished to $0.05\,P_B$. This state was taken for the initial value.

To obtain the real strength values of the rods under study under compression various types of face closure has been studied:

1. The faces are simply supported, polished, and assist conditions of free cross-deformation of face areas.
2. The faces are inserted in metal shells to simulate conditions of constrained cross-deformation of outer areas of the rod with some limitations of the inner ones.
3. Faces are inserted into metal shells and filled up with a cold-set compound to simulate conditions of constrained deformation of all face areas of the rod.

The unidirectional and spirally reinforced glass-plastic rods underwent testing. The test results are presented in Table 7.2.

Proceeding from the above, the closure type of rod faces effects noticeably their strength during compression testing. The samples containing spiral layers are more sensitive to this factor as compared to unidirectional ones. Further investigations have given evidences that the samples incorporating a spiral reinforcing layer doesn't exhaust their bearing capacity but breakdown as a result of stability loss of the sample. To eliminate the phenomenon a tubular guide has been mounted for the shells to move with filled-in faces aimed at preserving relative coaxiality. As a result, trustworthy values of strength characteristics of the spirally reinforced rods under compression have been obtained.

Table 7.2. The Compressive Strength Dependence of Rod Elements on Face Closure Type

Type of closure	Unidirectional			Wrapped		
	σ_P, MPa	S, MPa	W, %	σ_P, MPa	S, MPa	W, %
Simple supported, polished	232	31	13.4	201	25	12.5
Inserted in metal shells	295	20	6.9	249	23	9.2
Filled up in metal shells	382	6	1.6	389	12	6.2
Filled up in metal shells and movable at loading along guides	396	17	4.3	416	22	5.4

7.2 The Analysis of the Auxiliary Effect on SSS of the Rod Under Compression

The dependence of SSS of the spirally reinforced rods on elastic properties and geometrical parameters of the spiral layer has been studied by way of numerical analysis of solid monofibrous rods with an external auxiliary layer. Two types of rod designs have been examined:

– rods of a high-modular material (carbon plastic) having a spiral reinforcing layer of low-modular (glass plastic) winding;
– rods of a low-modular (glass plastic) material with a spiral reinforcing layer of high-modular (organoplastic) winding.

It has been presumed that there's no twisting in the main material or residual stresses.

The calculations have shown that the interaction between the min and auxiliary layer changes substantially deformation behavior of the compressed rod. In Fig. 7.1 one can see the dependencies of the relative twisting angle of the rod material under compression. Here and further in Figs. 7.2–7.4 curve 1 corresponds to a carbon plastic rod with a thin glass plastic layer of $2h = 0.03D$ width. Curve 2 is the same rod but of $2h = 0.07D$ width, curves 3 and 4 present a glass plastic rod with a reinforcing layer of organoplastic $2h = 0.03D$ and $2h = 0.07D$ wide, respectively.

The calculation results show that along with deformations typical of a unidirectional rod there occurs twisting of the rod whose value is conditioned by the spiral layer packing angle, its relative thickness and the elastic characteristics ratio of the main to the reinforcing layers. The twisting direction is a function of the winding layer packing angle only. Besides, the main material experiences the effect of transversal forces from the winding side whose sign and value are also dependent on above named parameters. The forces bring about transversal stresses (Fig. 7.2) and in spite of their relatively insignificant value in contrast to the limiting stresses in longitudinal direction, they exert a significant influence on the relative transverse strain magnitude

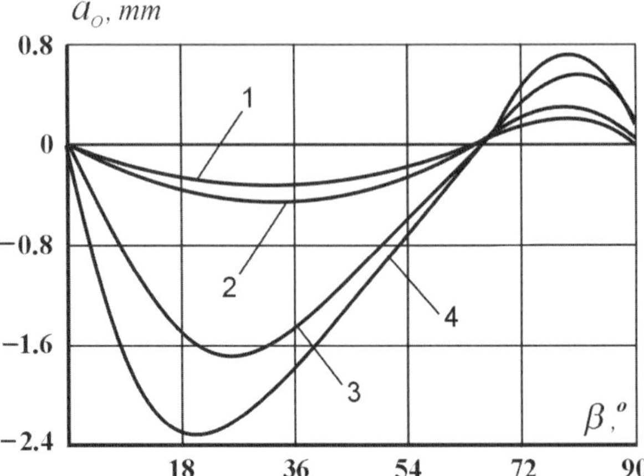

Fig. 7.1. The effect of packing angle of the spiral layer on rod twisting under compression

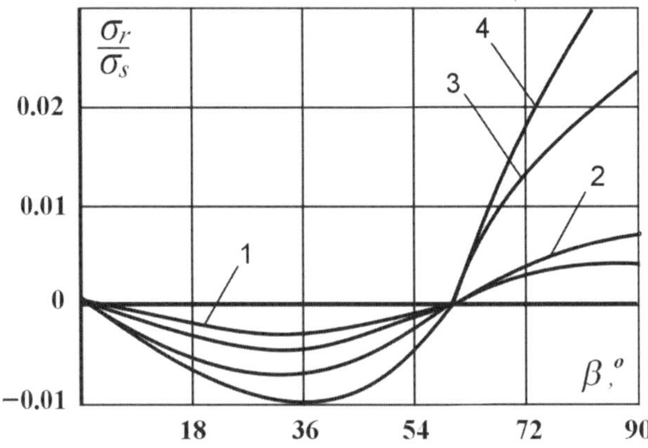

Fig. 7.2. Variations of radial stresses in the rod under compression as a function of the spiral layer packing angle

(Fig. 7.3). As a result, there appear longitudinal cracks in the main material leading to breakage of the formed strips owing to violated stability. Notice that the probability of delamination is similar for the solid rods in both circumferential and radial directions.

From Figures 7.2 and 7.3 follows that there exists some critical value of the packing angle of the spiral layer $\beta_C \approx 63°$ at which cross-stresses in the main material in reinforcement direction changes their sign. Thus at $\beta < \beta_C$ transverse strains in the main material are higher than those of a unidirectional rod. This means that introduced at $\beta < \beta_C$ angle reinforcing layer isn't a reinforcing but even assists in delamination of the rod under

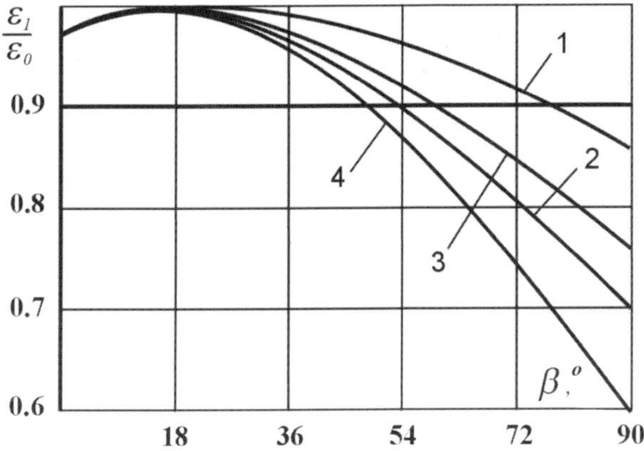

Fig. 7.3. Dependence of cross-strains in the main material on the spiral layer packing angle under compression

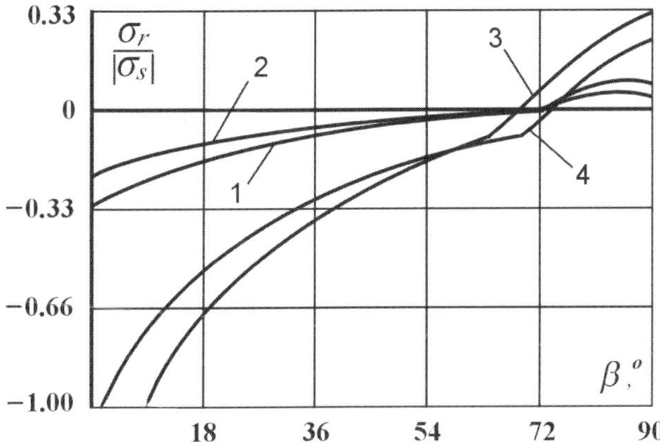

Fig. 7.4. Stress dependence in a spiral layer upon its packing angle under compression

compressive loads in longitudinal direction. The longitudinal compression of the rod at $\beta > \beta_C$ packing angle brings about tensile stresses in the reinforcing layer (Fig. 7.4) that are determined by the longitudinal elastic modulus of the layer and its width. Proceeding from Figs. 7.3 and 7.4, the minimum cross-strains and maximum tensile stress in the reinforcing layer in its reinforcement direction correspond to the layer packing angle $\beta = 90°$. As a consequence, one of the next reasons can lead to the rod material failure:

1. Transverse strains of the main material reach their limiting values at which the breaking stress

$$\sigma_S = \sigma_S^{(0)} \cdot \frac{\varepsilon_M}{\varepsilon_1}, \tag{7.1}$$

where $\sigma_S^{(0)}$ – strength threshold of the unidirectional rod under compression; ε_1 – transverse strain in the main material of the spirally reinforced rod calculated at a pressure equal to $\sigma_S^{(0)}$; ε_M – its limiting value.

2. Tensile stress σ_L in the spiral layer reaches its breaking strength at tension $\sigma_P^{(L)}$, and the breaking stress is here

$$\sigma_S = \sigma_S^{(0)} \cdot \frac{\sigma_P^{(L)}}{\sigma_L}, \tag{7.2}$$

where σ_L – stress in the spiral layer at stresses $\sigma_S^{(0)}$.

3. Stresses $\sigma_Z^{(S)}$ in the reinforcement direction of the main material reach their breaking strength at σ_{Zn}, so

$$\sigma_S = \sigma_{Zn} \cdot \frac{F}{F_S}, \tag{7.3}$$

where F – rod area, F_S – area of the main material.

From (7.1)–(7.3) follows that the breaking stress of the rod with an external spiral layer can be determined from

$$\sigma_S = \sigma_S^{(0)} \cdot \min \left\{ \frac{\varepsilon_M}{\varepsilon_1}, \frac{\sigma_P^{(L)}}{\sigma_L}, \frac{\sigma_{Zn} F}{\sigma_S^{(0)} F_S} \right\}. \tag{7.4}$$

To study experimentally the dependence of the rod element strength and delamination threshold upon the winding layer thickness and to verify relation (7.4) fairness the glass plastic rods of 20 mm diameter having outer organoplastic spiral layer of $2h = 0.1$–0.4 mm thickness and with 0.3 mm thick carbon plastic layer have been tested. The experimental results and those of theoretical calculations are presented in Table 7.3 and show good compliance. It's apparent that when the winding layer thickness is 0.1–0.2 mm the sample breaks as it reaches the breaking strength under tension. When the winding is made of $2h = 0.3$ mm organoplastic layer the continuity of the main material starts at stresses less than σ_S. Breaking stress is in this case somewhat higher than the rated one since the winding impedes loss of stability of the spalled fragments of the main material and the bearing capacity is lost only after the winding layer has broken provoking breakage of the delaminated main material.

Adding thickness to the winding layer till 0.4 mm doesn't result in any σ_S rise as it reduces the main material area and, consequently augments stress σ_Z in it. The sample fractures in the plane of the maximum tangential stresses under the 49° angle to the rod axis. If to use for the winding material

Table 7.3. Strength of Glass-reinforced Plastic Samples under compression as Dependent on Material Kind and Spiral Layer Thickness

2h, mm	Theory		Experiment			
	σ, MPa	Fracture mode	σ, MPa	S, MPa	w, %	Fracture mode
		Without winding				
–	–	Delamination	396	17	4.3	Delamination
		Organoplastic as the spiral layer material				
0.1	409	Winding breakage	408	21	5.1	Winding breakage
0.2	422	Winding breakage	427	13	3.0	Winding breakage
0.3	422	Delamination of the main material	458	26	5.6	Compression of the main material
0,4	458	Delamination of the main material	453	34	7.5	Material compression
		Carbon plastic as the spiral layer material				
0.3	465.6	Compression of the main material	468	27	5.8	Compression of the main material

the carbon plastic whose elastic modulus is twice as high as that of the organoplastic, the breaking stress of the rod element can be elevated much.

The conducted theoretical and experimental investigations have proved that introduction of a spiral layer changes SSS and fracture behavior of the rods made of fibrous CM under compression. The effectiveness of the layer is governed by its geometrical parameters and the relationship of the elastic moduli of the main and auxiliary reinforcements. Proposed earlier (Chap. 6) calculation method enables accurate forecasting of the delamination threshold and fracture behavior of the rod element.

7.3 The Effect of Technological Factors on Properties of the Rod Elements

The manufacturing process of spirally reinforced elements is distinguished from production of unidirectional rods by the presence of the reinforcing layer winding stage. Along with the layer parameters, properties of thus-obtained rods can exert influence on the manufacturing regimes of applying the winding layer, among which are tension of the winding fiber, twisting of the main material being the result of interaction between the main and auxiliary reinforcement fibers, and curing parameters and associated phenomena.

7.3.1 The Analysis of the Winding Fiber Tension Force Effect

Aimed at estimating tension force of the winding fiber a batch of glass plastic samples has been manufactured with the external reinforcing layer separated from the main material with an antitack agent. The winding layer and the antitack agent were removed upon setting to avoid the residual thermal stress effect and another spiral layer was applied instead using a cold-setting binder. The test results are given in Fig. 7.5.

It's evident that variation of the tension force T doesn't exert any essential influence on the rod strength except for the section $T = 0$–$3.0\,\mathrm{N}$. This can be attributed to relaxation of the original tension of the fiber during binder polymerization. At the same time, the increasing strength of the rod within $T = 0$–$3.0\,\mathrm{N}$ is apparently due to the winding fiber straightening and greater contribution to work.

Based on above, to raise strength of the spirally reinforced rod elements it's recommended to select the winding fiber tension force to within $T = 0.1$–$3.0\,\mathrm{P}_B$, where P_B is the fiber-breaking tension.

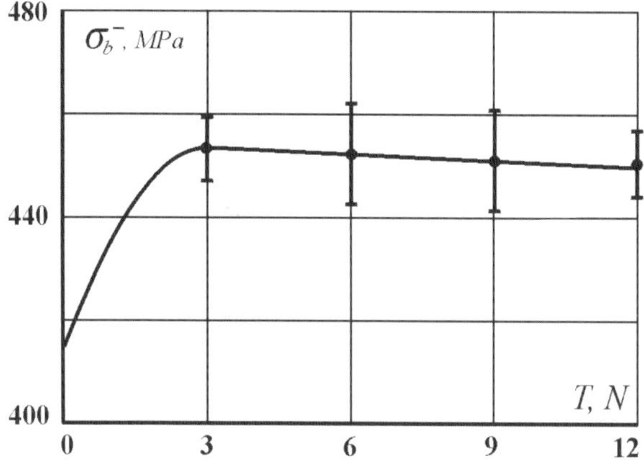

Fig. 7.5. Compressive strength of a spirally reinforced rod versus winding fiber tension force

7.3.2 Residual Temperature-induced Stresses in the Rod Element

As it has been stated in Sect. 6.1, there arise stresses and strains during thermal treatment of the rod with a spiral-reinforcing layer generated by differences in elastic properties of the layers. To study the dependence of mentioned parameters on the spiral layer properties a unidirectional solid rod with an external reinforcing layer has been used as an example. The rods whose main material has acquired negative (carbon plastic) and positive (glass plastic)

Table 7.4. LTEC and Elastic Moduli for Materials Used in Rod Elements

T, °K	Organic fiber $\alpha\times10^6$, K^{-1}	$E\times10^3$, MPa	Glass fiber $\alpha\times10^6$, K^{-1}	$E\times10^3$, MPa	Carbon fiber $\alpha\times10^6$ K^{-1}	$E\times10^3$ MPa	Boron fiber $\alpha\times10^6$, K^{-1}	$E\times10^3$, MPa	binder $\alpha\times10^6$, K^{-1}	$E\times10^3$, MPa
200	−5.6	99	2.5	64	−0.91	270	2.04	400	36	5.39
250	−5.9	94	3.1	64	−1.0	270	2.32	400	42	4.41
300	−6.2	89	3.4	64	−0.95	270	2.60	400	48	3.43
350	−6.4	84	3.8	64	−0.80	270	2.87	310	60	2.70
400	−6.6	79	4.2	64	−0.56	270	3.15	280	73	1.96

LTEC in longitudinal direction with a thin glass and organoplastic layer have been analyzed in numerical terms. To take account of thermal sensitivity of the rod material the temperature range (T_1, T_2) of the rod element cooling after polymerization from the polymerization oven temperature $(T_1 = 120°C)$ down to room temperature $(T_2 = 20°C)$ has been divided into intervals (T in which properties of components have been taken constant. The dependencies of elastic moduli and LTEC of rod element components on temperature are presented in Table 7.4 according to data of [246]. The calculations have been made for the spiral layer packing angles $\beta = 0$–$90°$ at layer thicknesses $2h = 0.03D$ and $2h = 0.07D$, where D is the rod outer diameter.

The analysis of results presented in Fig. 7.6 has proved that compressive stresses arise in the spiral layer during cooling of the rod whose value is mainly dependent upon stiffness, LTEC and geometrical parameters of the layer. As the winding broadens and its packing angle lessens the stresses go down.

Data in Table 7.5 shows that the compressive stresses in the spiral layer with close to $\pi/2$ packing angle may reach the material limiting strength and thus destabilize and make part of the reinforcement coils bulge.

The analysis of stresses in transverse direction in the main material (Fig. 7.7) has given evidences that their value is conditioned by the properties of the main and spiral layers and also by geometrical parameters of the latter. The residual tensile stresses in the main material can lead to its early cracking and, as a result to abrupt impairment of the bearing capacity of the rod. When the stresses at the contact boundaries exceed the interlayer strength, thermal shrinkage results in exfoliation of the spiral layer and spalling of separate strips of the main material which degrades its performance.

Table 7.5. Stresses in the Winding Layer at Small Packing Angles

Main material	Carbon plastic				Glass plastic			
Auxiliary material	Glass plastic		Organic plastic		Glass plastic		Organic plastic	
Layer thickness	0.03	0.07	0.03	0.07	0.03	0.07	0.03	0.07
σ_S, MPa	−274	−235	−368	−286	−169	−156	−228	−183

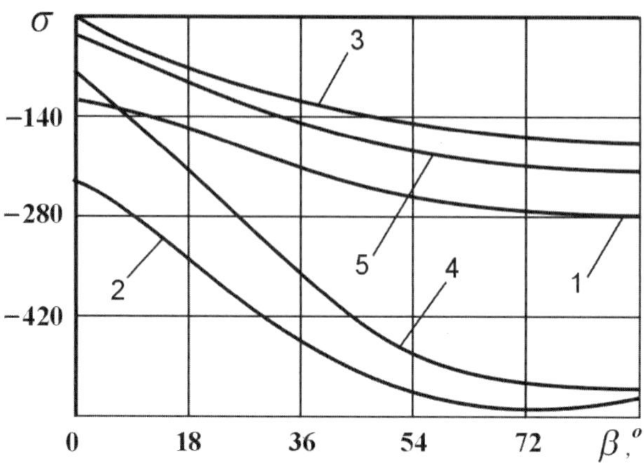

Fig. 7.6. Packing angle effect of the spiral layer on residual thermal stresses in it. *1, 2, 3* – the main material – glass plastic; *4, 5* – the main material – carbon plastic; *1, 2, 4* – spiral layer – organoplastic; *3, 5* – spiral-layer – glass-reinforced plastic; *1, 3, 4, 5* - $2h/D = 0.03$; *2* – $2h/D = 0.07$

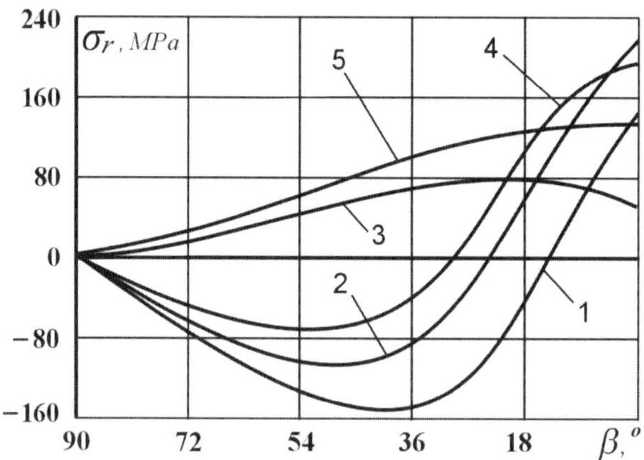

Fig. 7.7. Transversal stresses in the main material as a function of the spiral layer packing angle on thermal shrinkage: Notations correspond to Fig. 7.6

To evaluate the effect of residual temperature stresses on compressive strength glass-reinforced plastic rods with an external reinforcing glass-plastic layer of various thicknesses have been studied. The results obtained are given in Fig. 7.8. Subjected to thermal treatment rods are seen to have scatter in strength properties (curve 1) without any evident reason. Although scaling of the values with respect to the longitudinal reinforcement area operating in direction of applied forces has confirmed its efficiency rise, it hasn't revealed any definite dependence.

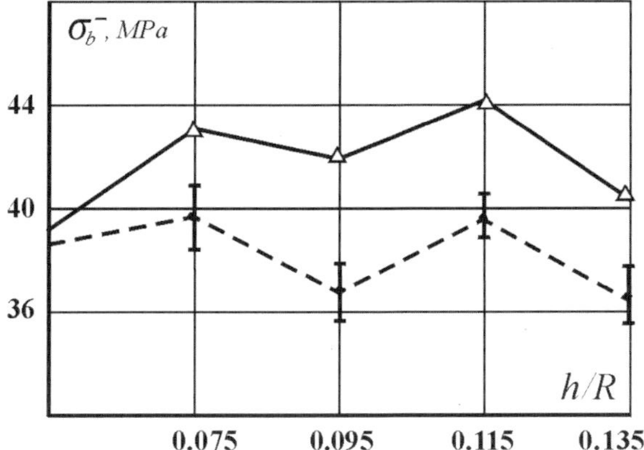

Fig. 7.8. Effect of residual thermal stresses on compressive strength of the rod

The analysis of the failure process dynamics has visualized that the thermally treated spirally reinforced rods undergo destruction with exfoliation during which the reinforcing layer remains intact. Further separate strips of the material get destabilized and undergo deformation under longitudinal bending and constrained strain in transverse directions. This is because the value of the residual thermal stresses arising in the main material during interaction approaches closely that of the compressive stresses induced by compression forces close to their threshold. This results in mutual compensation of indicated stresses and the reinforcing layer gets into gear only after the main material of the rod has delaminated.

This suggests that the reinforcing layer generates residual thermal stresses within the rod element bulk which markedly change the conditions of the main material work loading. When the spiral later packing angle is close to $\pi/2$, then the residual stress can exceed the strength limit of the main and reinforcing layers and that of the interlayer strength and bring to the rod material destruction as a result of thermal treatment.

7.3.3 The Effect of the Main Material Twisting on SSS and Warping of Rod Elements

As the analysis has shown, the main reason of the main material twisting in reinforced rods is the torsion moment arising during interaction of the main and auxiliary reinforcements (see Chap. 5). It has been established experimentally that the relative torsion angle of the external fibers of the main material is $\gamma_0 = 0.00178$–$0.087\,\text{rad/mm}$ in the rod elements 20 mm in diameter and more. As a result of thermal treatment the value increases somewhat. Simultaneously, the residual stresses in the main and auxiliary materials augment too. The stresses differ, however, only by 1–2% from residual stresses at $\gamma_0 = 0$.

Some positive effect of twisting is observed on mechanical loading. According to [247] strength improvement is observed under tension along the rod element axis due to the effect of cable structure of the main material fibers. Numerical calculations of SSS under compression of a carbon plastic rod of 20 mm diameter with an organoplastic outer layer has proved that twisting alleviates transverse strain of the main material by 5–7% and simultaneously lessens stresses in the auxiliary reinforcement direction by 20–26% thus raising its efficiency.

Along with hybrid spirally reinforced elements of 20 mm diameter and more carbon and silicon-based rods 1.0–1.5 mm in diameter with an external layer of spiral reinforcement (aimed at machine assembly of multidirectional structures) have been studied. Since only 1 to 3 braids of the main reinforcement are used to produce small-diameter rods, one of the main reasons of appearing unbalanced technological stresses is considered unevenness of braid tension. As soon as the rod hardens the bending moment able to bend its axis is formed in the rod in the absence of twisting. The curvature radius of resultant line is

$$\rho = \frac{E_Z \cdot J}{\Delta p \cdot C},$$

where

$$C = \frac{2n}{3\pi} \sqrt{\frac{F}{\pi}} \sin \frac{\pi}{n},$$

n – number of braids; F – cross-sectional area of the rod; J – its inertia moment; Δp – difference in tension forces of braids.

Using the small-diameter rods it's necessary to limit their deviation from linearity. When there's no twisting in the main material, the maximum bending value of a rod 400 mm long and 1.2 mm in diameter should be lower than 2 mm and the tension force of the braids is to be a controlled given at a high accuracy parameter.

In a general case, when there's present the material twisting the bending moment plane locates upon thermal treatment along the rod axis in a helical line.

The moment projections on YOZ and XOZ planes are defined by the relations

$$M_y = \Delta pC \sin \frac{2\pi}{L} z; \quad M_x = \Delta pC \cos \frac{2\pi}{L} z;$$

where L – lay of strand of the braids. Consequently, twisting of the small-diameter rods can lead to lengthwise inhomogeneities in the main material as the braids are differently pulled described by the dependence [202]

$$E_z(z) = \frac{E_z}{\Delta pC \sin \dfrac{2\pi}{L} z + \Delta pC \cos \dfrac{2\pi}{L} z}.$$

This changes notably SSS of the rod under axial loading.

Theoretical analysis of the joint effect of the main material fibers twisting and difference in their tension on the quality of the rods has made grounds for elaboration of requirements on the manufacturing process. Investigations in the technological factors effect on SSS of rod elements has visualized the necessity of ensuring the required tension force of the winding filament and elimination of difference in tension of braids of the main reinforcement to reach a high-quality of rod specimens. When designing rod elements the residual temperature stresses are to be allowed for so as not to impair their strength.

7.4 The Effect of Rod Element Structure on the Stress–strain Field Under Compression

The results of numerical computations of a carbon plastic rod of 20 mm diameter have been compared to three variants of arranging 0.6 mm thick organoplastic layer to estimate the effect of the mount and disposition of spiral layers on the rod element SSS:

1. The spiral layer is located on the outer surface (numerical results are cited in Sect. 7.2).
2. The spiral layer is placed inside the rod (to avoid the scale factor the winding layer halves the main material area $R_1 = 7$ mm).
3. There are two spiral layers inside ($R_1 = 7$ mm) and on the external surface of the rod.

Thermal stresses were presumed to be absent in the rod.

Introduction of a spiral layer in the rod material brings about transversal stresses σ_r and σ_θ in layers S_1 and S_2 (Fig. 7.9).

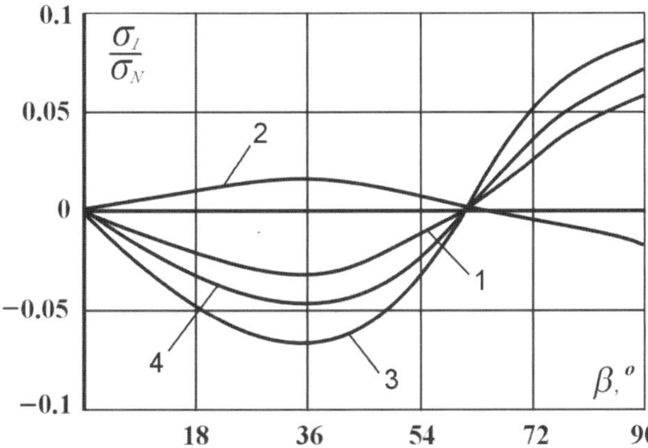

Fig. 7.9. Stresses in the carbon plastic rod with an internal spiral layer. $1 - \sigma_r(R_1 + 2h_1)$; $2 - \sigma_\Theta(R_1 + 2h_1)$; $3 - \sigma_r(R_1) = \sigma_\Theta(R_1)$; $4 - \sigma_\Theta(R_2)$

Table 7.6. Transverse Strains of Carbon Plastic Rods as a Function of Disposition of a Spiral Organoplastic Layer under Compression (ε_0 – deformation of a unidirectional rod)

Location of spiral layer	$\dfrac{\varepsilon_r(R_1)}{\varepsilon_0}$	$\dfrac{\varepsilon_r(R_1 + 2h_1)}{\varepsilon_0}$	$\dfrac{\varepsilon_r(R_2)}{\varepsilon_0}$	$\dfrac{\varepsilon_\theta(R_1 + 2h_1)}{\varepsilon_0}$	$\dfrac{\varepsilon_\theta(R_2)}{\varepsilon_0}$
Internal	0.68	1.33	1.24	0.81	0.93
External	0.76	0.76	0.76	0.76	0.76
External and internal	0.71	1.01	0.92	0.70	0.77

Table 7.7. Design Types of Rod Hybridization

Hybrid type	Internal main layer	External main layer
I	Intralayered hybrid	Intralayered hybrid
II	High-modular	Low-modular
III	Low-modular	High-modular

In contrast to the external layer these stresses are of different signs independently of packing angles and result not in reduction but an increase of the transverse strains (Fig. 7.6). Application of two winding layers (internal and external) doesn't lead to raising the delamination threshold either. Hence, introduction of a spiral layer inside a homogeneous material in inexpedient since decreases its delamination threshold. The efficiency of interlayer hybridization has been evaluated on the grounds of numerical calculation results of solid hybrid carbon-glass-reinforced plastic and carbon-boron plastic rods of three types (Table 7.7) with different spiral layers arrangements. Variations in transverse strains have been estimated as opposed to a corresponding unidirectional intralayered hybrid.

Presented in Tables 7.8 and 7.9 calculation results display that similarly to a homogeneous material introduction of only the inner spiral layer in the hybrid (interlayer) rod promotes its delamination at the layer interfaces under compression. If only an external spiral layer is used in the hybrid rod element, this lessens transverse strains like in homogeneous elements, though less efficiently.

One can trace here the difference between probable failure reasons of the solid (internal) and circular (external) layers of the main material. In the solid layer stresses are $\sigma_r = \sigma_\theta$ and, consequently strains $\varepsilon_r = \varepsilon_\theta$. Therefore, radial and circular cracks are equiprobable in this layer. As calculations have shown in the layer whose cross-section is a circle with radius r_2 ($r_1 < r_2$) the equality $\varepsilon_r(r_1) > \varepsilon_2(r_2)$ is met. Hence, radial strains in this layer won't lead to its delamination. Violation of the circular layer continuity may occur as the strains reach their limiting values and generate radial cracks.

Table 7.8. Strain Dependence of Hybrid Carbon Plastic Rods on their Design at Axial Compression. (ε_0 – strain of the 1st type unidirectional rod)

Hybridization type	Arrangement of reinforcing layer	$\dfrac{\varepsilon_r(R_1)}{\varepsilon_0}$	$\dfrac{\varepsilon_r(R_1+2h_1)}{\varepsilon_0}$	$\dfrac{\varepsilon_\theta(R_1+2h)}{\varepsilon_0}$	$\dfrac{\varepsilon_r(R_2)}{\varepsilon_0}$	$\dfrac{\varepsilon_\theta(R_2)}{\varepsilon_0}$
I	External	0.84	0.84	0.84	0.84	0.84
	Internal	0.71	1.23	0.88	1.13	0.95
	External and internal	0.76	1.03	0.72	0.96	0.80
II	Without	0.95	1.18	0.95	1.12	1.00
	External	0.67	1.02	0.76	0.93	0.76
	Internal	0.73	1.46	0.57	1.30	0.91
	External and internal	0.55	1.26	0.56	1.10	0.72
III	Without	1.07	0.82	1.07	0.88	1.00
	External	0.84	0.47	0.85	0.57	0.76
	Internal	0.87	1.05	0.86	1.00	0.90
	External and internal	0.72	0.71	0.69	0.71	0.70

Table 7.9. The Dependence of Transverse Strains of Hybrid Carbon Plastic Rods on their Design at Axial Compression (ε_0 – strain of a unidirectional 1st type rod)

Hybridization type	Reinforcing layer location	Strain relations				
		$\dfrac{\varepsilon_r(R_1)}{\varepsilon_0}$	$\dfrac{\varepsilon_r(R_1+2h_1)}{\varepsilon_0}$	$\dfrac{\varepsilon_\theta(R_1+2h_1)}{\varepsilon_0}$	$\dfrac{\varepsilon_r(R_2)}{\varepsilon_0}$	$\dfrac{\varepsilon_0(R_2)}{\varepsilon_0}$
I	Internal	0.89	1.21	0.87	1.14	0.96
	External	0.83	0.83	0.83	0.83	0.83
	Internal and external	1.04	0.72	0.97	0.97	0.80
II	Without	1.02	0.98	1.02	0.99	1.02
	Internal	0.96	1.27	0.86	1.18	0.96
	External	0.90	0.95	0.78	0.91	0.83
	External and internal	0.68	1.21	0.63	1.08	0.80
II	Without	1.00	1.07	1.00	1.06	1.02
	Internal	0.92	1.02	0.90	1.16	0.98
	External and internal	0.82	0.82	0.78	0.81	0.79

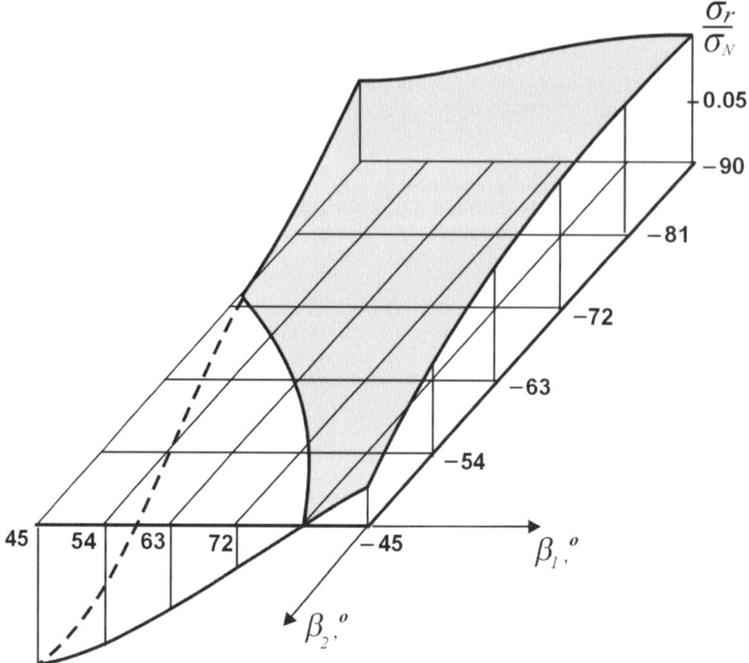

Fig. 7.10. Variations of stress $\sigma_r(R_1)$ in the 3rd type rod with two reinforcing layers

The dependence of transverse stresses on packing angle of spiral layers for the 3rd type rods with two spiral layers is presented in Table 7.10.

It's evident from the diagram that by changing packing angle of only one winding layer one can vary considerably stress values. This is because together with stresses σ_θ in layer S_2 they condition the level of limiting strains of the rod and introduction of the inner spiral layer is helpful in even redistribution of stresses to diminish maximum strains.

Calculation results demonstrate that to avert circular exfoliation of the rod material at interfaces of the layers of different materials it's required that the following equality be met for layers S_i, S_{i+1} of the spirally reinforced rod element

$$\frac{v_{zr}^{(i)}}{E_Z^{(i)}} > \frac{v_{zr}^{(i+1)}}{E_Z^{(i+1)}},\qquad(7.5)$$

where $v_{zr}^{(j)}$, $E_Z^{(j)}$ are Poison's ratio and elastic moduli of layer S_j, $j = i,\, i+1$.

Hence, the elastic moduli of the main layers of the rod element should rise as they approach the outer surface.

Data characterizing variation of the relative values of transverse strains in the 3rd type rods are presented in Table 7.10 as a function of the ratio

Table 7.10. Strain Dependence of the 3rd Type Hybrid Rods on Varied Filler Component Ratio

Inner radius	Winding thickness H_1	H_2	$\dfrac{\varepsilon_r(R_1)}{\varepsilon_0}$	$\dfrac{\varepsilon_r(R_1+2h_1)}{\varepsilon_0}$	$\dfrac{\varepsilon_0(R_1+2h_1)}{\varepsilon_0}$	$\dfrac{\varepsilon_r(R_2)}{\varepsilon_0}$	$\dfrac{\varepsilon_\theta(R_2)}{\varepsilon_0}$
0.4	–	–	1.23	1.31	1.23	1.28	1.26
	–	0.015	1.06	0.75	1.06	0.89	0.94
	–	0.03	0.96	0.43	0.96	0.66	0.74
	0.015	–	0.98	1.58	0.94	1.33	1.21
	0.015	0.015	0.86	1.03	0.81	0.94	0.90
	0.015	0.03	0.78	0.70	0.74	0.72	0.72
	0.03	–	0.84	1.77	0.70	1.37	1.17
	0.03	0.015	0.74	1.23	0.61	1.00	0.89
	0.03	0.03	0.69	0.90	0.56	0.77	0.71
0.6	–	–	1.09	1.16	1.09	1.13	1.11
	–	0.03	0.86	0.37	0.86	0.52	0.72
	0.015	–	0.93	1.32	0.92	1.19	1.04
	0.015	0.015	0.82	0.86	0.80	0.83	0.82
	0.015	0.03	0.75	0.56	0.73	0.61	0.67
	0.03	–	0.83	1.44	0.76	1.25	0.98
	0.03	0.015	0.74	1.00	0.67	0.91	0.78
	0.03	0.03	0.68	0.71	0.62	0.68	0.65
0.8	–	–	0.94	0.99	0.94	0.98	0.94
	–	0.015	0.82	0.59	0.82	0.63	0.78
	–	0.03	0.74	0.30	0.74	0.38	0.66
	0.015	–	0.83	1.09	0.83	1.05	0.87
	0.015	0.015	0.74	0.71	0.73	0.71	0.72
	0.015	0.03	0.67	0.44	0.66	0.47	0.62
	0.03	–	0.76	1.17	0.73	1.11	0.80
	0.03	0.015	0.68	0.83	0.65	0.79	0.68
	0.03	0.03	0.63	0.56	0.59	0.57	0.58

of the hybrid composite ingredients (change in radius R_1) to the spiral layer thickness. Low quantities of the high-modular filler (e.g. boron fibers) in the main material necessitate a thicker internal spiral layer and vice versa. Apparently, the choice of the main components should meet an optimum correlation between the inner and outer winding layer thicknesses at a given total cross-sectional area.

The analysis of SSS of solid hybrid rods has visualized its difference from that of a unidirectional intralayered hybrid. Note that to raise the rod material delamination threshold the equality (7.5) should be satisfied, which means that: (i) the low-modular layer of the main material must be always inside the high-modular one; (ii) the spiral layers can be introduced only at the interfaces of the layers of different materials. So far, introduction of

Table 7.11. The Design Effect of Solid Carbon-boron Plastic Rods on their Delamination and Rod Strength under compression (Spiral Layers – Carbon Braid, Layer Thickness – 0.15 mm, Packing Angle – 90°)

Internal spiral layer	External spiral layer	Violation of continuity			Destruction		
		σ, MPa	S, MPa	W, %	σ, MPa	S, MPa	W, %
Carbon-boron plastic as the main internal layer Carbon-boron plastic as the main external layer							
Without	Without	747	19.3	2.00	769	49	6.30
With	Without	613	68	11.1	690	14	2.00
Without	With	1016.7	51.9	58.1	1063	50.1	4.70
With	With	732	63.8	8.7	739	58	7.80
Boron plastic as the main internal layer Carbon plastic as the main external layer							
Without	Without	476	82	17.3	542	66	12.0
With	Without	413	64	15.5	500	45	9.00
Without	With	452	32	7.00	505	110	21.7
With	With	329	82	24.9	373	71	13.0
Carbon plastic as the main internal layer Boron plastic as the main external layer							
Without	Without	551	13	2.40	816	118	14.4
With	Without	530	125	23.5	745	178	24.0
Without	With	741	213	28.7	1027	63	6.20
With	With	776	81.9	10.5	88	58	65.9

spiral layers at packing angles approaching $\pi/2$ assists in more even distribution of stresses, consequently, in alleviating transverse strains and raising delamination threshold.

The derived dependencies perfectly correlate to experimental investigations of carbon-boron plastic rods of 20 mm diameter with different arrangements of the main and auxiliary layers (Table 7.11).

Relative content of filler ingredients is chosen 1:1 as it excludes scale factor effect when it's necessary to change location of the main components in the structure.

Experimental results on violation of compressed rod solidity have shown good compliance with theoretical calculations. Notice that experiments have shown higher delamination threshold and strength values of samples with intralayer hybridization as compared to the 3rd type interlayer hybrid rod (Table 7.7). This is because carbon filler fibers can fill spaces between boron fibers without any essential reduction of the filling degree. Moreover, the total filling degree of the rod material with the main reinforcement can be elevated.

Numerical computations of strains in compressed tubular rods with an external spiral layer laid perpendicularly to the axis have indicated that

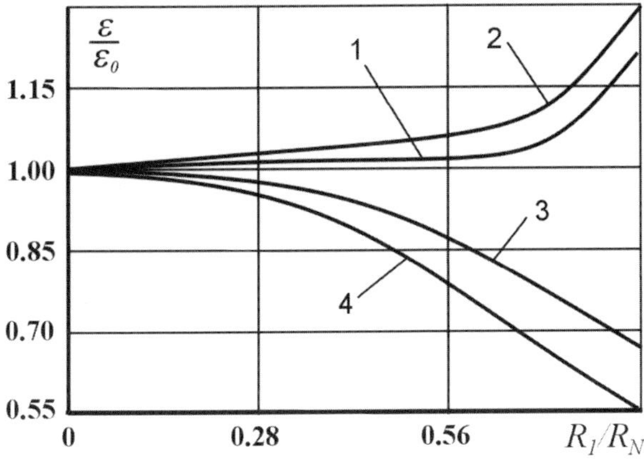

Fig. 7.11. Dependence of transverse strains in the tubular rod element with an external spiral layer upon hole radius ($R_N = 17.5\,\mathrm{mm}$). ε_0 – deformation of the solid rod; $1 - \varepsilon_r(R_1)$; $2 - \varepsilon_r(R_2)$; $3 - \varepsilon_\Theta(R_1)$; $4 - \varepsilon_\Theta(R_2)$

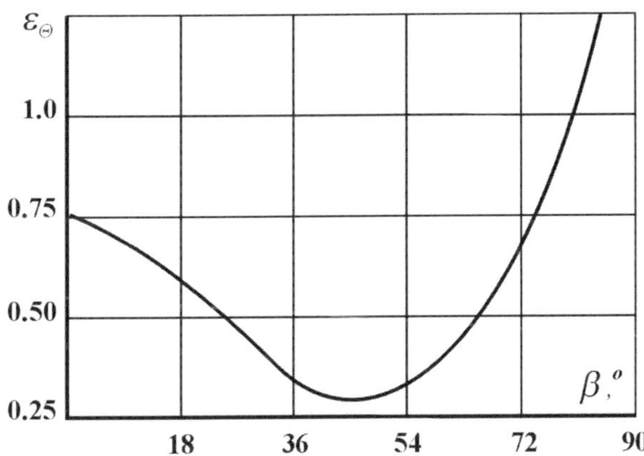

Fig. 7.12. Maximum strain ε_Θ dependence of a tubular carbon plastic rod with two spiral layers upon packing angle of the internal layer ($R_1 = 12.5\,\mathrm{mm}$, $R_2 = 17.5\,\mathrm{mm}$)

(Fig. 7.11) delamination of the main material can be due to attaining limiting strain values ε_θ.

To lower the strains layer L_1 strengthening the hole is introduced. The least transverse strain values are reached at packing angle $\beta = 45°$ (Fig. 7.12). Nevertheless, it should be born in mind that at the contact boundaries of the main and reinforcing layers arise stresses $\sigma_r(R_1 + 2h_1)$ able to exceed their interlayer bonding and tear off layer L_1. The stresses decrease as the packing angle reduces (Fig. 7.13). Strains ε_Θ can be also reduced by 20–25% using spiral layers.

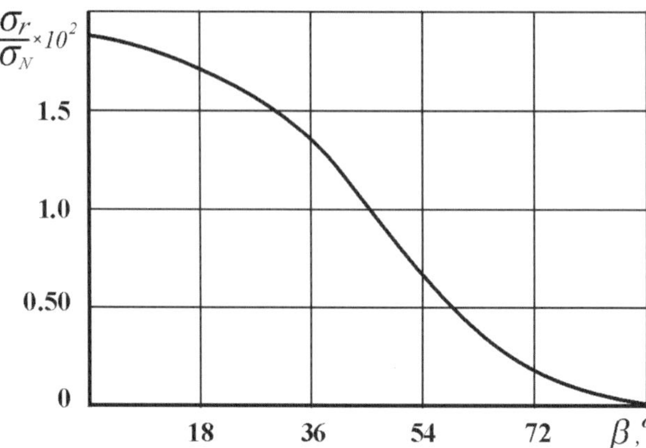

Fig. 7.13. Compressive stresses at contact boundaries between the internal spiral layers and the main direction

To analyze the influence of the arrangement of the main and spiral layers in tubular rod elements three types of the main material structure have been studied (Table 7.7). Carbon-boron plastic rods with spiral layers of organoplastic with different relative content of the main reinforcement (1:1, 3:1 and 1:3) have been investigated as an example.

The analysis suggests that in this case two types of structures close in efficiency to introduction of spiral layers exist. They are, first the structure having an external spiral layer with the packing angle close to $\pi/2$ and, second the structure with two spiral layers – the internal one with the packing angle about $\pi/4$ and the external layer with the packing angle $\pi/2$. The effect of the arrangement and relative content of ingredients of the main filler on deformability of tubular rods is a little less than the solid rods have. Note that the external disposition of more stiff filler turns preferable in this case as well.

An interesting proof to fairness of numerical results can be found in nature. Figure 7.14 illustrates a cross-section of the corn stern looking like a unidirectional shank reinforced by longitudinal fibrous stiff elements whose filling level increases from the center to periphery.

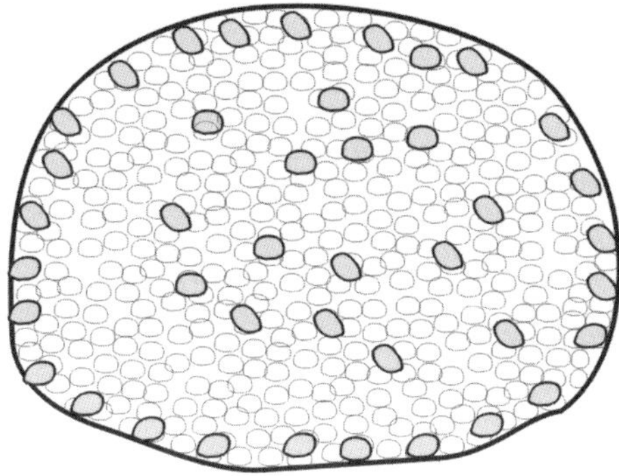

Fig. 7.14. A sketch of the cross-sectional structure of a corn stern

7.5 The Efficient Characteristics of the Material as Dependent on Parameters of Spirally Reinforced Rods

It has been proved during thermal treatment of rod elements with an outer spiral layer that with packing angles $0 < \beta_i < \pi/2$ fibers of the main material undergo twisting at cooling from polymerization down to room temperature (Fig. 7.15) which changes the rod length. Hence, as it follows from Fig. 7.16, by varying geometrical parameters of the spiral layers it's possible to regulate the efficient LTEC values of the rod.

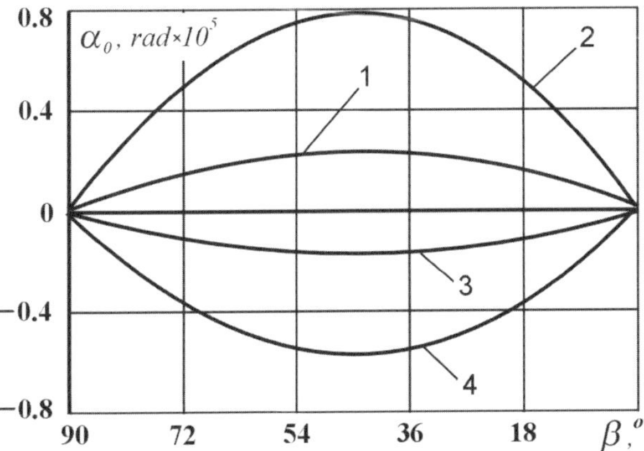

Fig. 7.15. Torsion angle of the rod versus packing angle of the winding layer at thermal shrinkage (layer thickness $2h = 0.03D$). *1, 3* – carbon plastic rod; *2, 4* – glass-reinforced plastic rod; *1, 2* – glass plastic as the auxiliary layer; *3, 4* – organoplastic as the auxiliary layer

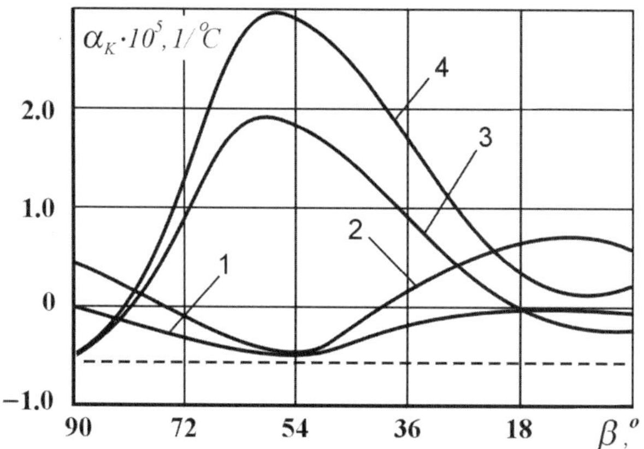

Fig. 7.16. Reduced LTEC value of the carbon plastic rod versus the auxiliary reinforcement packing angle. *1, 2* – glass-reinforced plastic auxiliary layer; *3, 4* – organoplastic auxiliary layer; *1, 3* – $2h/D = 0.03$; *2, 4* – $2h/D = 0.07$; *5* – rod of a unidirectional material

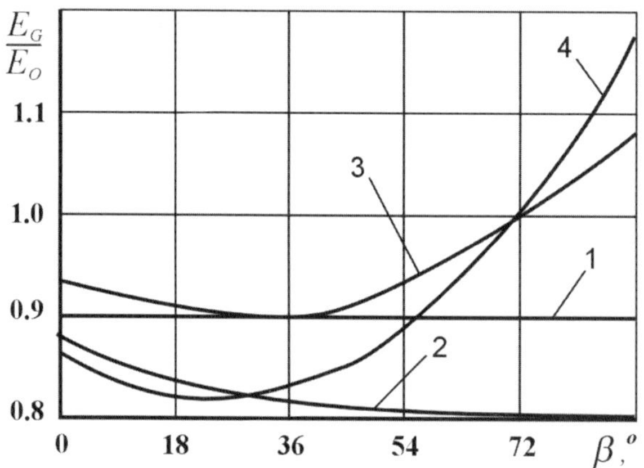

Fig. 7.17. Dependence of the efficient longitudinal elastic modulus of the spirally reinforced rod upon its spiral layer packing angle. E_0 – longitudinal elastic modulus of the main material

Above indicated property may be employed to achieve close to zero LTEC of carbon plastic rods through introduction of thin spiral layers.

As calculation results show, introduction of the spiral layer changes the efficient longitudinal elastic moduli (Fig. 7.17) and shear modulus (Fig. 7.18) of the rod material with an external winding and, consequently, its stability. From the comparison of Figs. 7.3 and 7.18 follows that the packing angle values ensuring an expanded delamination threshold of the rod material don't coincide with the value corresponding to its maximum shear modulus values.

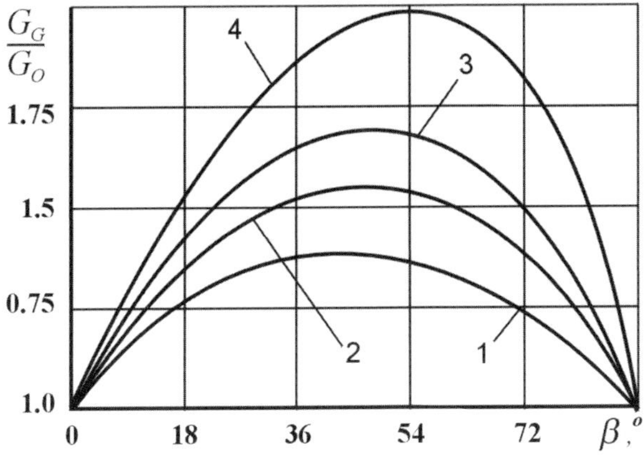

Fig. 7.18. The efficient shear modulus of the spirally reinforced rod relation to its spiral layer packing angle. G_0 – longitudinal shear modulus of the main material

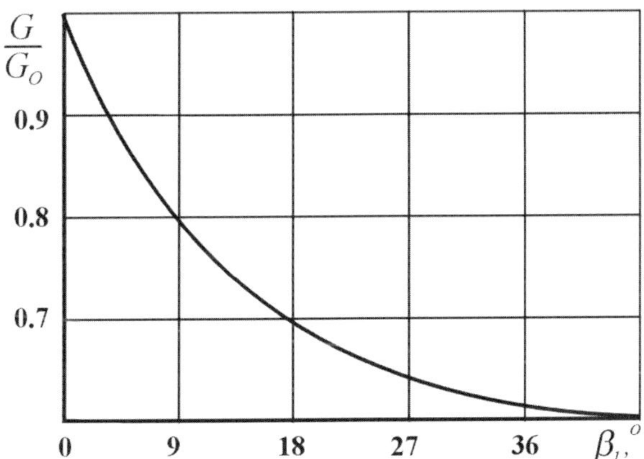

Fig. 7.19. Shear modulus of the tubular carbon plastic rod versus packing angle of the internal spiral layer ($R_1 = 12.5\,\mathrm{mm}$, $R_2 = 17.5\,\mathrm{mm}$). G_0 – Shear modulus of the unidirectional rod material

As the analysis has shown, the efficient characteristics of both solid and tubular rods are independent of the arrangement of the main components. At the same time, insertion of a strengthening hole in the tubular rod improves the efficient shear modulus and its maximum corresponds to packing angle $\beta_1 = \pi/4$ (Fig. 7.19). In this case, the efficient elastic modulus drops less than at introduction of the external layer which is critical when estimating stability of the rod element.

Proceeding from the above, a conclusion can be drawn that introduction of reinforcing layers changes the efficiency characteristics of the spirally reinforced rod elements. In this connection, the material thickness, packing angle and arrangement of the layers should be selected on the base of known structure of the rod and real conditions of its loading.

7.6 Conclusions

1. Introduction of spiral layers into the structure of hybrid composite rods varies their deformation behavior and efficient characteristics.
2. It's recommended to place layers possessing high elastic modulus from outside of the rod elements incorporating several types of the main reinforcement.
3. To lessen transverse strains the spiral layers should be arranged at the contact boundaries of the main layers of different materials as on the inner and outer surfaces of the rod element.
4. When designing the manufacturing process of spirally reinforced rods it's important to consider the effect of process factors on the product quality. Special attention should be paid to residual thermal stresses to avoid undesired violation of product solidity.

8 Optimal Design
of Hybrid Spirally Reinforced Rods

8.1 Statement of the Problem of Optimal Design

8.1.1 Choice of the Optimization Criterion

A peculiarity of hybrid spirally reinforced rod elements is the possibility to vary their structural and geometrical parameters, type of ingredients and reinforcing layers within a wide range. In this connection, the problem of optimizing their structure and formulation is very important. The optimal criteria for designing structural elements and the strategy of finding the solution are chosen proceeding from operational conditions and/or available design and technological bases of the structure. Many design situations are solved by the minimization problem of some quality function within a set space of optimized parameters and restrictions on a complex of limiting states.

The optimal criteria for estimating composite products can include such characteristics as its mass, LTEC, cost, improvement of, e.g. its stability, including stiffness and reduced maximum flexure, etc. The governing parameters in this case are the properties of the material components, including elastic and thermoelastic, their form and some others. Of importance for articles based on hybrid CM is the accepted reinforcement scheme and the original components. From the viewpoint of the technological process, the question of controlling residual thermal stresses is critical. In this case, the function dependent upon such stresses can be chosen as the optimal criterion of the material. Sometimes, minimization of the potential bending energy, e.g. for beams, or to limit the free oscillation frequency under certain boundary conditions of fixing, can be used as the criterion.

In a general case, when optimizing the structure of a composite material the optimal criterion is specified first of all by the loading type and operational conditions [248]. For spirally reinforced rods representing elements with spatial structures, thermal stability of the element length, stability or compressive strength can be taken as the quality criterion. The goal function can be one of the named criteria or a combination.

To ensure thermal stability of the rod elements, both intralayered hybridization of unidirectional rods using fibers with positive and negative LTEC, and spatial reinforcement can be employed. The quality criterion of these hybrid rods can be an integral LTEC of the product within the working temperature range (T_1, T_2)

$$\Psi(\overline{X}) = \frac{1}{T_2 - T_1} \int_{T_1}^{T_2} \alpha_Z(T) dT \qquad (8.1)$$

where $\alpha_Z(T)$ – the element LTEC at temperature T.

Composite materials are known to break down before the strength of their reinforcing fibers is exhausted. Thus far, the reinforcement structure that assists in reaching the strength limit of the main material fibers at preserved continuity of the product is considered to be optimal. For such structures the delamination threshold of the CM should be taken as the quality criterion. Violated continuity of the hybrid spirally reinforced rod may arise, as follows from Chap. 7, from exfoliation of one of the main material layers S_i $(i = 1, \ldots, N)$, breakage of the reinforcing layer L_i $(i = 1, \ldots, N)$, or the tearing of layers S_i, L_i or L_i, S_{i+1} from each other.

To avert the separation of layers (except for thin layer L_1 that reinforces the tubular rod hole) inequality (7.5) should be met. Since Poisson's ratios $\upsilon_{Z\Theta}$ of fibrous CM are intimately close, the inequality is satisfied when the main layers S_i $(i = 1, \ldots, N)$ are arranged such that the layer with a lower longitudinal elastic modulus is found inside that with a higher modulus. Spiral layers L_i $(i = 1, \ldots, N)$ can be introduced only between the main layers of different materials. In this connection, only rods with the indicated arrangement of the layers will be considered further, and the probability of their separation is neglected in optimization problems.

When stress $\sigma_r(R_1 + 2h_1)$ at the interfaces with layers L_i and S_2 surpasses their interlayer strength σ_1, then layer L_1 may detach. Therefore, the value

$$f_1 = \sigma_r(R_1 + 2h_1)/\sigma_1 , \qquad (8.2)$$

is a characteristic of the rod quality.

As is known, exfoliation of layer S_i from the main material occurs once the transverse strain reaches its limiting value. Consequently, for the layer solidity, the quality function is

$$f_i = \max_{S_i}(\varepsilon_\Theta)/\varepsilon_i , \qquad (8.3)$$

where ε_i the rating value of transverse strain in layer S_i.

By introducing coefficient δ_i:

$$\delta_i = \begin{cases} 0, \Rightarrow E_Z^{(i)} = 0; \\ 1, \Rightarrow E_Z^{(i)} \neq 0; \end{cases} \qquad (8.4)$$

we obtain from (8.2) and (8.3)

$$f_i = \delta_i \max_{S_i}(\varepsilon_\Theta)/\varepsilon_i + (1 - \delta_i)\sigma_r(R_1 + 2h_1)/\sigma_1 . \qquad (8.5)$$

Spiral layer L_i can fail as soon as stresses in the reinforcing direction $\sigma_L^{(i)}$ reach the tensile strength limit σ_L. In this case, the quality function is of the kind

$$\varphi_i = \sigma_L^{(i)}/\sigma_L \,. \tag{8.6}$$

It has been shown earlier that the least $\max(\varepsilon_\Theta)$ value in layer S_i corresponds to the highest stress value $\sigma_L^{(i)}$. Hence, for the continuity criterion of the rod element one should take the function

$$\Psi(\overline{X}) = \sum_{i=1}^{N} |f_i - \varphi_i| \,, \tag{8.7}$$

where f_i, φ_i are set by (8.5) and (8.6), respectively.

For reliable operation the elements of a structure should be stable. The calculation scheme of loading on metal rods is conditioned by their flexibility extent

$$\lambda = \mu L / i_{\min} \,; \tag{8.8}$$

where L – rod length; μ – the factor characterizing rod fixture; $i_{\min}$ – the minimum cross-sectional inertia radius. Thus, it has been proved that low-flexible isotropic rods of a brittle material fall out as the compressive stresses rises to the limiting strength. The critical load for high flexible rods ($\lambda > 100$) is calculated by Euler's formula

$$P_C = \frac{\pi^2 E_Z J}{(\mu L)^2} \,, \tag{8.9}$$

where J – cross-sectional inertia moment. For rods of medium flexibility one of the Yasinsky, Carman or Engesser–Shenly's formulas is used. In contrast to metal rods, the stability of composite ones, along with the longitudinal elastic modulus, experiences the effect of longitudinal shear modulus $G_{\Theta Z}$, which can be evaluated as

$$\eta = \frac{k'\pi^2 E_Z/G_{\Theta Z}}{\lambda^2} \cdot 100\% \,, \tag{8.10}$$

which corresponds to an error arising on finding P_C by Euler's formula (8.9). The relation of η to λ (for various ratios $E_Z/G_{\Theta Z}$) is presented in Fig. 8.1, from which it follows that Euler's formula is applicable for only high flexible composite rods with a small $E_Z/G_{\Theta Z}$ ratio. Furthermore, (6.1) was used to calculate the critical force as far as relation $E_Z/G_{\Theta Z} = 20 - 40$ is true for the rods and their flexibility is $\lambda = 20 - 100$. Consequently, the effect of $G_{\Theta Z}$ on the magnitude of P_C turns out to be equal to and in some circumstances exceeds the effect of E_Z. In this connection, it can be anticipated that notwithstanding lowering of the longitudinal elastic modulus of the rod material upon spiral layer introduction, simultaneous elevation of $G_{\Theta Z}$ will increment the critical load value P_C. As a result, it is necessary to determine an optimal structure of the spirally reinforced rod from the standpoint of

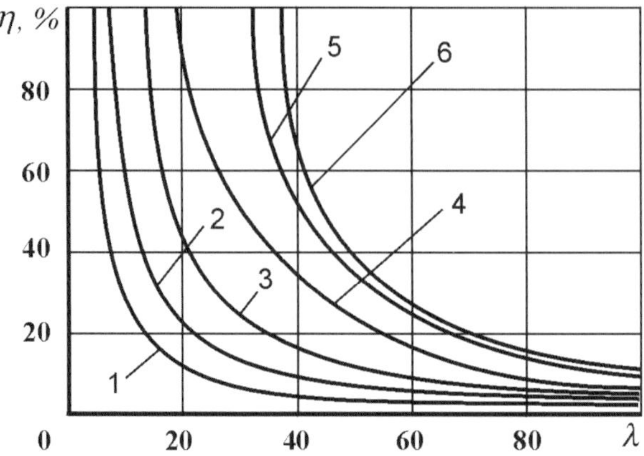

Fig. 8.1. Longitudinal shear modulus effect on the composite rod stability.
$1 - E_Z/G_{\Theta Z} = 3$; $2 - E_Z/G_{\Theta Z} = 8$; $3 - E_Z/G_{\Theta Z} = 10$; $4 - E_Z/G_{\Theta Z} = 20$; $5 - E_Z/G_{\Theta Z} = 40$; $6 - E_Z/G_{\Theta Z} = 60$

providing its high stability. To solve this problem it is expedient to denote the quality function as

$$\Psi(\overline{X}) = \frac{\lambda^2}{E_Z \pi^2} + \frac{1}{G_{\Theta Z}} , \tag{8.11}$$

whose minimum corresponds to the optimal solution.

It has been shown in 6 that an increase of $G_{\Theta Z}$ by 20–26% at the introduction of the spiral layer in the rod is accompanied by reduction of the longitudinal elastic modulus E_Z by 6–10%. Hence, minimization of function $\Psi(\overline{X})$ set in (8.11) may result in elevation of its characteristics by only 3–5%.

On the other hand, the critical force that arises as the low-flexible rods lose their stability and the compressive strength limit are very close in magnitude. This fact should be allowed for when designing structures incorporating spirally reinforced rod elements whose compressive strength limit is higher by 20–25% than in unidirectional designs. This is why some combination of quality functions (8.7) and (8.11) should be used as the optimal criterion in these problems.

If P_C and the strength limit of the rod are close enough, then the quality function can be given by the dependence

$$\Psi(\overline{X}) = \left| \frac{S}{\Psi_Y} - \max_{i=1,\ldots,N} (f_i, \varphi_i) \right| , \tag{8.12}$$

where Ψ_Y is calculated by (8.11) and f_i, φ_i are calculated at $P_Z = -1N$ by (8.5) and (8.6), and S is the cross-sectional area of the rod.

Since spiral winding is aimed at raising the delamination threshold and the longitudinal elastic modulus of the rod, as well as their P_C can be increased by only 3–5% for high-flexibility rods, the quality function should be taken as (8.7) with the limitation

$$\Psi_Y(\overline{X}) \le \Psi_Y^{(0)} , \tag{8.13}$$

where $\Psi_Y(\overline{X})$ is found from (8.11) and $\Psi_Y^{(0)}$ is the respective value of the unidirectional rod.

Thus, in accordance with loading conditions and operational requirements, the problem of optimizing spirally reinforced hybrid rods is reduced to computation of the minimum goal function values set by (8.1), (8.7), (8.12) or (8.7) under limitations (8.13).

8.1.2 The Control Parameter Vector Limitations

It has been stated in 7 that the elastic and thermoelastic properties of spirally reinforced rod layers and their dimensions exert a noticable effect on SSS and efficient characteristics of the element and, consequently, on functions (8.1), (8.2) and (8.7). Insofar as the spiral reinforcement is introduced in the unidirectional rod to perfect its characteristics, the material and relative amount of the main components are considered to be given for the optimization problem. As stated earlier, the optimal arrangement of the main reinforcement layers and the maximum number of spiral layers are set by inequality (7.5). Hence, the goal function is dependent upon the choice of both discrete parameters – reinforcing layer materials, and continuous ones – layer width and packing angle values. Because of the problem specifics, the discrete parameters are optimized by sorting out probable variants. Consequently, the optimization problem of the structural and geometrical parameters of the rod element is reduced to the minimal value of some quality function dependent upon control vector

$$\overline{X} = \{X_j\}_1^{2N} , \tag{8.14}$$

Here $X_{2i-1} = h_i$; $X_{2i} = \beta_i$; – number of main and reinforcing layers; h_i – layer L_i half width; β_i – layer packing angle.

On coordinates of vector $\overline{X}$ the corresponding limitations are imposed:

$$0 \le X_{2i} \le 0.1 D_H ;$$
$$2\pi/3 \le X_{2i} \le \pi/2 \tag{8.15}$$

As a result of analyzing the numerical calculations presented in Chap. 7, a limitation has been derived, which is necessary for solving each of the above-described optimization problems

$$\max\{\sigma_{RES}\} \ll \sigma_{RES}^{(m)} , \tag{8.16}$$

where σ_{RES} – residual stress crosswise in the rod, $\sigma_{RES}^{(m)}$ – its limiting value.

To solve the optimization problem (8.1) it is necessary to allow for the required width of the longitudinal elastic modulus, i.e. the inequality

$$E_Z \geq E_D \tag{8.17}$$

must be satisfied, where E_Z – efficient elastic modulus in the main reinforcement direction; E_D – its minimum admissible value.

Moreover, the introduced reinforcing layers must not reduce the delamination threshold of the rod, i.e. the inequality

$$\varepsilon_1(r, \Theta) \leq \varepsilon_1^{(0)} \tag{8.18}$$

must be true, where $\varepsilon_1(r, \Theta)$ – transverse strain in some point of the spirally reinforced rod; $\varepsilon_1^{(0)}$ – respective value in a unidirectional rod.

When determining the goal function by (8.7) for low flexible rods it is necessary along with limitation (8.16), to make provision for their stability, i.e. the inequality

$$P_C > P_B \tag{8.19}$$

should be satisfied, where P_C – the critical force corresponding to stability loss found by (8.1); P_B – breaking load on the rod element.

Since the goal function (8.12) takes account of (8.18) and (8.19), then only limitation (8.16) should be considered in our case.

Hence, the problem of optimizing the structural and geometrical parameters of the spirally reinforced rod brings us to one of the minimization problems:

1. Goal function of the type (8.1) depends upon control vector (8.14) under constraints (8.15) and (8.17).
2. Goal function of the form (8.7) depends on control vector (8.14) with limitations (8.14)–(8.16), (8.19).
3. Goal function of the type (8.7) depends on the vector of control parameters (4.14) with limitations (8.13), (8.15), (8.16).
4. Goal function of the form (8.12) depends on control vector (8.14) with limitations (8.15), (8.16).

8.2 Modification of the Random Search Method

The chief complexity of structural optimization is attributed to nonlinearity of the goal function and related constraints imposed on the control vector. Use of gradient procedures for the solution of these problems can essentially affect the optimization convergence process. These problems are often solved by various modifications of the haphazard procedure, e.g. random search with learning where, according to the Monte-Carlo principle [154], the studied dots are randomly generated. In [228, 248–250] a more effective variant of the method is proposed (the so-called evolution strategy. The convergence is attained in this case through a continuous optimization of the search during the goal function optimization process. The evolution strategy method includes a few stages.

During the first stage a reference point of the search is selected to satisfy problem limitations. In our case, this point is vector $\overline{X}_0$, which corresponds to the unidirectional rod, which means that the coordinates of vector $\overline{X}_0 = \{X_j^{(0)}\}$ are

$$X_{2i-1}^{(0)} = 0; \quad X_{2i}^{(0)} = \pi/2. \tag{8.20}$$

The second stage presupposes the choice of the initial value of the vector of standard deviations $\overline{S}$. It is also feasible to avoid here the cumbersome procedure of searching for an optimal value of vector components by assuming that

$$S_{2i-1} = 0.05 D_H; \tag{8.21}$$
$$S_{2i} = \pi/100. \tag{8.22}$$

During the third stage point $\overline{X}$ is selected to make allowance for

$$\overline{X} = \overline{X}^* + r\overline{S} \tag{8.23}$$

where $\overline{X}^*$ is the last fitting point; r – pseudorandom number from $[0;1]$.

The fourth stage includes estimation of a new point $\overline{X}$, keeping to the principle

$$\overline{X}^* = \begin{cases} \overline{X}^*, & \Rightarrow \Psi(\overline{X}) > \Psi(\overline{X}^*); \\ \overline{X}, & \Rightarrow \Psi(\overline{X}) \le \Psi(\overline{X}^*). \end{cases} \tag{8.24}$$

During the fifth stage it is evaluated whether vector $S = \{S_i\}_1^{2N}$ is the optimum.

To judge the effectiveness of the search let us introduce the value

$$W_L = \frac{N_P}{N_\Sigma}; \tag{8.25}$$

where N_P – number of successful attempts, N_Σ – total number of attempts.

It has been shown in [251] that W_L is practically independent of the object nature but is governed by the most favorable value $\overline{S}$. In this case, $W_L = 1/5$ indicates the fairness of selecting $\overline{S}$.

At the sixth stage the pitch length is adjusted

$$S_i = \begin{cases} S_i^*/1.2 & \Rightarrow W_L < 1/5; \\ S_i^* & \Rightarrow W_L = 1/5; \\ S_i^* \cdot 1.2 & \Rightarrow W_L > 1/5. \end{cases} \tag{8.26}$$

The optimization process (7th stage) is stopped in the case where $W_L = 0$ or reduction of S_i does not affect the goal function.

The described evolution strategy method is realized numerically in two stages. During the first stage N_1 new points are generated for the chosen initial values of vectors $\overline{X}_0$ and $\overline{S}_0$ from which the best one is selected, i.e. the point at which the limitations are satisfied and the goal function is the least. Furthermore, vector $S_2 = \{S_i^{(2)}\}$ is found:

$$S_i^{(2)} = 0.5 S_i^{(0)} . \tag{8.27}$$

At the second stage N_2 number of new vectors $\overline{X}$ found by (8.23) is generated and the search efficiency is estimated using (8.21). Values S_i are corrected following the rule (8.24). The second stage is repeated until the conditions of the process termination are satisfied. The numerical calculation presupposes process termination once the iteration number of the second stage reaches some N_3 magnitude. Thus, the optimization process can be stopped and started not from the initial readings of vectors $\overline{X}_0$ and $\overline{S}_0$ but from those intimately close to the optimum, which speeds up the optimization. SSS parameters and efficient characteristics are calculated by the method described in Chap. 6.

8.3 Estimation of the Optimal Structure of Thermally Stable Rods

Usually, to produce thermally stable rod elements, carbon fibers are employed as the main reinforcing material. This is accounted for, firstly, by their negative LTEC and secondly, the thermal stability of strength and elastic characteristics up to 2473 K and at low temperatures [246]. As the auxiliary reinforcement glass fibers are commonly used due to the sufficiently high thermal stability of their elastic modulus and less residual stresses, compared to organic carbon fibers.

Let us take the problem of defining structural and geometrical parameters of a 20 mm diameter solid carbon plastic rod having glass plastic reinforcing layers, to ensure minimal integral LTEC within 223–323 K. It is also required that the elastic modulus be 10% lower than the unidirectional plastic and residual temperature stresses will not surpass 0.1 of the main material strength limit at transverse tension or that of the spiral layer under compression in the reinforcement direction.

According to inequality (7.5), the spiral layer can be placed only on the external surface of the rod. Consequently, an optimal rod would contain one layer of the main material with an outer spiral layer, therefore the optimization problem can be formulated as follows:

Let us find the minimum of function

$$\Psi(\overline{X}) = \frac{1}{100} \int_{223}^{323} \alpha(T) dT , \tag{8.28}$$

which depends upon control vector

$$X = \{h_1, \beta_1\} , \tag{8.29}$$

by satisfying inequalities

$$0 \leq h_1 \leq 1.0 \, \text{mm} \, ;$$

$$60° \leq \beta_1 \leq 90° \, ;$$

$$E_Z \geq 0.9 E_Z^{(0)} \, ;$$

$$\sigma_r^{(S)} \leq 0.1 \sigma_+^{(S)} \, ;$$

$$\sigma_L \leq 0.1 \sigma_-^{(S)} \, ; \tag{8.30}$$

where $\sigma_+^{(S)}$ and $\sigma_-^{(S)}$ are breaking strength under tension and compression of layer S, respectively.

Numerical calculations have given the following results: width of the spiral layer $2h_1 = 0.23 \, \text{mm}$, packing angle of the layer $\beta_1 = 79°$ and the integral LTEC within 223–323 K is $\overline{\alpha} = 0.2 \times 10^{-8} \, K^{-1}$.

As a result, using the developed procedure of optimal design, structural and geometrical parameters of the carbon plastic rod with an external spiral layer of glass-reinforced plastic have been estimated whose LTEC turned out to be $\overline{\alpha} = 0.2 \times 10^{-8} \, K^{-1}$, within the 223–323 K temperature range, satisfying conditions (8.30).

8.4 Optimization of Compressed Spirally Reinforced Rod Elements

Let us consider the problem of an optimal arrangement of spiral layers in tubular carbon plastic rods having an inner diameter $D_1 = 25 \, \text{mm}$ and $D_0 = 35 \, \text{mm}$ outer diameter with a similar volume content of ingredients and of 500, 1000 and 1500 mm length. It is assumed further that temperature stresses are absent in the material.

A conclusion has been drawn from Chap. 7 that the efficient elastic and shear moduli of the rod material are independent of the mutual disposition of the main layers and its strength is higher than the rod with interlayer hybridization, where inequality (7.5) is met. For the spiral layer materials both organic and glass fibers can be used. It has been proved that when there are no thermal stresses in the material, the effectiveness of the reinforcing layer is a function of its elastic modulus. From inequality (7.5), it also follows that the layer can be placed on the inner, as well as the outer surface of the rod element, or even between the carbon and boron plastic fibers. This suggests that an optimal rod should possess the following structure: layer S_1 – the hole, $E_Z^{(1)} = 0$, $R_1 = 12.5 \, \text{mm}$, layer L_1 – organoplastic, layer width – $2h_1$, packing angle – β_1, layer S_2 – carbon plastic, radius R_2, layer L_2 – organoplastic with $2h_2$ width and β_2 packing angle, layer S_3 – boron plastic, radius R_3, layer L_3 – organoplastic with $2h_3$ width and β_3 packing angle to the rod axis. Radii R_2 and R_3 are found from the system of equations:

$$\begin{cases} R_2^2 - (R_1 + 2h_1)^2 = R_3^2 - (R_2 + 2h_2)^2; \\ R_3 + 2h_3 = 17.5. \end{cases} \tag{8.31}$$

Let us take a 500 mm long rod. In calculations the following characteristics of the materials will be accepted: ultimate compressive strength of a unidirectional carbon plastic $\sigma_C = 450$ MPa, and that of the boron plastic $\sigma_B = 840$ MPa. In the case of hinged, leaning rod ends, a rod of the indicated length shows flexibility equal to $\lambda = 23.25$. Since it is a low-flexibility rod, the quality function will be taken as (8.7), and it looks like

$$\Psi(\overline{X}) = \left| \frac{\sigma_r(R_1 + 2h_1)}{\sigma_1} - \frac{\sigma_L(R_1 + h_1)}{\sigma_L^{(1)}} \right| + \left| \frac{\varepsilon_\Theta(R_2)}{\varepsilon^{(C)}} - \frac{\sigma_L(R_2 + h_2)}{\sigma_L} \right|$$

$$+ \left| \frac{\varepsilon_\Theta(R_3)}{\varepsilon^{(B)}} - \frac{\sigma_L(R_3 + h_3)}{\sigma_L} \right| ; \tag{8.32}$$

where $\varepsilon^{(C)} = \varepsilon^{(B)} = 0.0012$ – limiting transverse strain of carbon and boron plastics; $\sigma_1 = 30.0$ MPa – interlayer strength of carbon and organoplastic based on epoxide binder; $\sigma_L = 1380$ MPa – ultimate tensile strength of organoplastic in reinforcing direction.

Function Ψ depends upon control vector

$$X = \{h_1, \beta_1, h_2, \beta_2, h_3, \beta_3\}, \tag{8.33}$$

on which with account of (8.14), (8.15), (8.19), limitations are imposed:

$$0 \le h_i \le 1.75 \, \text{mm} ;$$
$$60° \le \beta_i \le 90° ; \tag{8.34}$$

$$E_Z \ge 0.05(E_Z^{(C)} + E_Z^{(B)}) ; \tag{8.35}$$

where E_Z – efficient elastic modulus of the rod material; $E_Z^{(C)}, E_Z^{(B)}$ – elastic moduli of unidirectional carbon and boron plastics, respectively. The latter inequality means that the efficient elastic modulus of the spirally reinforced rod material may turn out to be less than the one rated on the basis of the additivity rule of the hybrid carbon-boron plastic by not more than by 10%.

The following numerical values of control vector components have been found $\beta_i = 90°$; $i = 2, 3$; $h_1 = 0$; $h_2 = 0.15$ mm; $h_3 = 0.17$ mm; $R_2 = 14.86$ mm; $R_3 = 17.16$ mm. These results give grounds for raising the delamination threshold of the rod material by 28%.

Let us consider a rod of 500 mm length where one end is rigidly fastened ($\mu = 2$). Its flexibility is $\lambda = 46.5$. The critical load corresponding to loss of stability of the unidirectional carbon-boron-reinforced plastic rod is calculated by the formula

$$\sigma_{CR} = \frac{E_Z \pi^2}{\lambda^2} \cdot \frac{1}{1 + \dfrac{1}{G_{\Theta Z}} \cdot \dfrac{E_Z \pi^2}{\lambda^2}} , \tag{8.36}$$

and is equal to $\sigma_{CR} = 742$ MPa. Its ultimate compressive strength is $\sigma_-^{(0)} = 650$ MPa. Since introduction of spiral layers increases compressive strength by

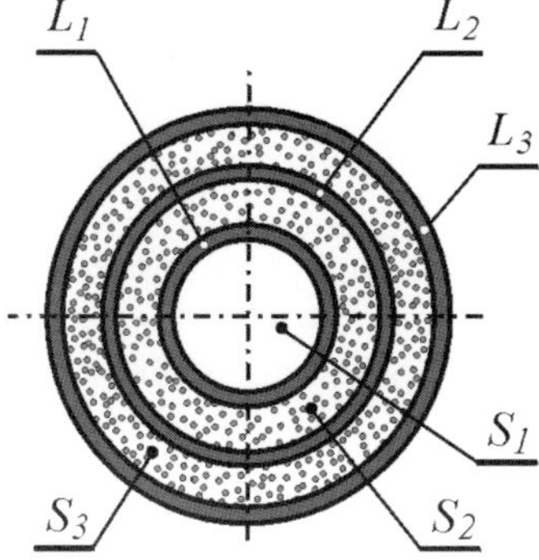

Fig. 8.2. Cross-sectional view of a hybrid tubular rod

15–20%, the quality function is taken as (8.12). As earlier, the control vector is of the form (8.33) with limitations (8.34). Numerical calculations yield the following results: $h_1 = 0$; $2h_2 = 0.12$; $\beta_2 = 77^\circ$; $2h_3 = 0.135$; $\beta_3 = 8^\circ$. The rod having such spiral layers displays a 14% higher delamination threshold compared to the unidirectional and $\sigma_{CR} \approx \sigma_{CR}^{(0)}$ (calculation error of σ_{CR} is less 1%).

A high-flexible rod of 1000 mm length (with a rigidly fastened end) shows $\lambda = 93$. Thus, for a unidirectional carbon-boron plastic rod we have

$$\frac{1}{G_{\Theta Z}} \cdot \frac{E_Z \pi^2}{\lambda^2} \approx 0.04 \,.$$

Consequently, it is improbable to increase σ_{CR} by more than 4%. Therefore, the function (8.7) with limitations (8.13) is chosen as the quality criterion of the rod. This allows one to obtain a rod of the same stiffness as the unidirectional one, but with a higher delamination threshold.

The control vector of the problem is of the kind of (8.33) with limitations (8.34) and (8.13). As a result of calculations an optimal structure of the rod is found, consisting of a layer of organoplastic L_1 of 0.1 mm thickness laid at a 67° to the axis, no spiral interlayer between the main carbon and boron plastic layers and an external 0.1 mm thick organoplastic layer packed at a 65° angle to the rod element axis.

The above-presented examples of the optimal design of rod elements have confirmed that the required structure of the spirally reinforced rod is governed by the operational conditions to be met by the selected optimization criterion.

8.5 Conclusions

1. The optimal design problem for spirally reinforced rod elements is reduced to minimization of the quality function given in a space of control parameters under some complex of limitations. The type of goal function

and limitations imposed on the control vector are governed in the first place by loading conditions.

2. To optimize the structure of the spirally reinforced rods being designed, it is expedient to use the method of evolution strategy with corresponding substantiation of the goal function choice.

3. The considered numerical examples show that the structures of optimal rod elements, the material and relative content of the main components can be similar or different and are specified by operational conditions.

9 Application of Composites Based on Spirally Reinforced Fillers and the Spiral Reinforcement Procedure in Design Units

9.1 Rod-type Products Based on Spirally Reinforced Fillers

Aluminum- and titanium-based materials displaying low density, production simplicity, high heat conductivity and acceptable cost are extensively used at present in modern engineering designs. However, their specific strength and rigidity are essentially inferior to composite materials. Use of composites in various structures lessens their weight, raises pay load in flying vehicles and cuts shipment costs [220, 252]. In aeronautic and space systems it is more expedient to use carbon-carbon composites with a spatial reinforcement [9, 253–255, 275, 276], as they withstand high thermomechanical loads.

Many studies have been devoted to investigations of carbon-carbon composite materials [144, 191, 256–260, 275–302]. For example, in [261] ablation characteristics of the chopped fiber based carbon-carbon composite produced by isothermal extrusion have been evaluated. Ablation rate and time of destruction onset were found to strongly depend upon fiber volume content. Increasing fiber content intensifies the ablation rate and results in surface roughness levelling. As a consequence, time of initial destruction extends by an order of magnitude compared to materials without reinforcement. It has been established in [262] that ablation rate is inversely proportional to material density. According to [276] porosity is an important feature of the carbon-carbon material. Materials with elevated porosity acquire an inclination to form very rough surfaces, which enlarges the winding surface area and elevates material removal. The head of the material experiences asymmetrical changes during ablation due to inhomogeneity of its properties across the depth. This fact may worsen the accuracy of the rated landing trajectories of flying vehicles [263]. Of great importance is the orientation of reinforcing carbon fibers about the flow. It was reported in [144] that ablation of materials along the flow is lower by 8% than that in the orientation perpendicular to the flow.

It is known that the carbon-carbon material used as heat-proofing in aeronautic, missile and space systems should be extremely dense (of about $2.1\,\mathrm{g/cm^3}$) [264], small-grained, with carbon fibers evenly distributed across its bulk, having minimum porosity and perfect adherence at the carbon matrix/fiber interface. Multidirectional structures obtained by netting (or weav-

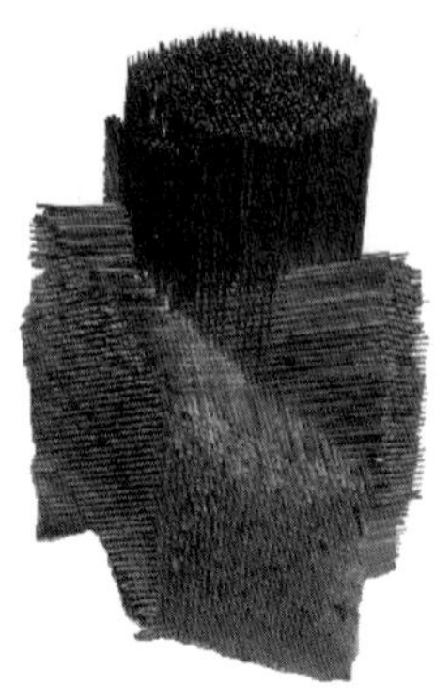

Fig. 9.1. The outward appearance of frameworks prepared for impregnation

ing), insertion or assembly of unidirectional rods are impregnated with a binder, meet the above requirements to the utmost [9, 265]. Most promising today are the materials based on a three-directional (3D) and four-directional (4D) texture (Fig. 9.1). They are also superior to sandwich structures in a number of parameters. Notice that [265, 266] the 4D texture outdoes the 3D one along with the 7D and recently elaborated 11D ones. It has been shown that as compared to the 3D texture, the 4D one is more isotropic, its filling degree is higher and found in the structure pores form an open interrelated spatial system. The latter circumstance assists in better saturation of the material.

The 3D texture is formed by an ordered netting of three families of straight-line fibers at 90° to each other. The most difficult procedure here is reinforcement laying in the third direction. Automated methods of netting such materials have been considered at length elsewhere [9]. The drawbacks of machine netting are a low filling stability of the framework and fiber deformation during drawing the material in the third direction. Currently, the three-directional framework is netted by hand as follows: metal spokes mounted instead of the third direction reinforcement are netted layer by layer with a reinforcing material in two directions (e.g. in X and Y coordinate directions), then upon fastening the longitudinal reinforcing material on the spokes by special hooks the reinforcement of the third direction is passed through the entire pile. Naturally, fibers of the layers in the XY plane became damaged during such netting, the third direction braids are in part broken and tension of the reinforcement in the XY direction is hampered, as this leads to spoke jamming. Consequently, it is hard to produce a high-quality material with stable performances in these conditions.

To significantly simplify the described technique of producing a 3D texture it is proposed to employ carbon plastic rods instead of metal spokes. To avoid fiber damaging the side surface of the rods are fitted with sleeving. Used with this aim, the materials are fine SHP threads, including nitron, kapron and others upon whose pyrolysis no admixtures are left in the material structure and gaseous matter can be easily removed prior to impregnation. Reinforcement laying around wrapped carbon plastic rods that are installed instead of spokes, eliminates reinforcement distortion within the XY plane and furnishes a chance of packing the reinforcement at a definite tension force. The outside winding of 0.04–0.05 mm thickness protects the third direction fibers against destruction in the course of netting. Moreover, it leaves a coke residue upon thermal treatment, which contributes to improving adhesion at the carbon fiber/carbon matrix interface. Application of the rods has made it possible to raise the productivity of manufacturing the 3D texture by 25–30%, to perfect their quality and the stability of the obtained material service characteristics.

As for the 4D texture, it is produced by netting four families of fibers laid at a 70.5° angle to one another. To produce a 4D texture the rods of small diameter are preliminary made of carbon fibers and then assembled into a framework. The assembly technique includes just a few stages at which any violations of the mutual orientation of fibers in each direction are excluded. It is natural that a series of requirements are imposed on the rods used in these structures, including as much filling as possible, dimensional stability, high strength and rigidity, along with the minimum winding layer thickness.

It appears that the chief element of the above-considered advanced multi-directional structures is the rod, based on high-modulus fibers of the VMN-4, VMN-5 or Colon grade. Binders for the rods should be selected from compositions that can, on the one hand be easily removed during pretreatment of the assembled framework and, on the other, ensure the strength and rigidity of the rod needed for the assembly. Furthermore, the fine-grained materials with a high filling degree were found to display the best thermomechanical characteristics. Consequently, to match up the demands it is necessary to diminish the rods cross-sectional size and carefully select its form. For a material reinforced in three directions rectangular configuration of the cross-section is preferable, for the 4D – a hexahedral one. However, due to the complexity of producing patterned rods, the most widespread configuration of their cross-section turned out to be round, for which the most influential parameter on the filling degree is the cross-sectional ellipticity.

Unfortunately, we have not so far worked out any methods for producing small-diameter (below 1.2 mm) carbon plastic rods able to guarantee the required process productivity and other technological parameters. Moreover, technological regimes of manufacturing carbon plastic rods have not yet been sufficiently studied. This is why the scientifically grounded commercial technology for producing carbon plastic rods is still not available and until now it is impossible to launch production of the carbon-carbon material at the needed scale.

9.1.1 Methods of Producing Rod-type Wares

Methods of pultrusion and netting prevail in the manufacture of rod-type products at present [267–274, 282, 298]. According to this method, fibrous filler is passed through a calibrated orifice, namely a spinneret, during which binder excess is removed, fibers are redistributed within the material structure and the rod cross-section is shaped. The produced rods are of low quality, since they are set free upon brief contact with the spinneret and start to warp, which results in their cross-sectional distortion. More perfect rods can be produced in long enough thermal chambers having cross-sections similar to the rod or in a series of supplementary devices able to improve impregnation quality, squeezing and reinforcement tension during molding and reduced fluffing of the reinforcing filler. Nevertheless, these devices can not avoid the chief drawback of the spinneret method, that is, reinforcement breakage during drawing through the spinneret, which naturally leads to variations in the filling degree and geometrical parameters of the obtained rods.

Along with pultrusion and netting, some other procedures are employed for the formation of intermediate adhesion-proof films. With this aim, the reinforcement is laid on a flat band, and wrapped in it, then molded in an auger extruder. This method is naturally more applicable for products with sufficiently large cross-sections. However, the presence of an intermediate film hampers the removal of the gaseous products of heat treatment, thus impairing the mechanical characteristics and essentially limiting the choice of binders.

More promising in this respect is the method of rod molding with application of spiral reinforcement. Notice that the compaction of the impregnated reinforcement reached in spinnerets is augmented during its winding by the auxiliary reinforcing fibers under tension. The method is frequently applied in manufacturing glass plastic bars used for reinforcing wood, glass-reinforced concrete and insulating structures instead of metal reinforcement.

Each of the above-considered methods has specific merits and disadvantages. Drawing through the spinneret does not allow the needed stability of the geometrical parameters of the rods, while use of a calibrated heatchamber induces certain inconveniences during operation, and so on. Most compliant to the requirements is the method of spiral reinforcement, during which the rods are molded due to the winding under tension of the main reinforcement followed by heat – treatment. In the course of the process the auxiliary reinforcement undergoes heat-induced shrinkage, which causes additional molding of the rod. The specifics of the method necessitate a stage of winding that supplements a typical flow chart of the manufacturing rods.

9.1.2 Elaboration of the Technological Process for the Spiral Reinforcement of Carbon Plastic Rods

The technique of producing spirally reinforced carbon-plastic rods includes the following stages: tension, impregnation, squeezing, winding, setting, control and cutting.

As investigations have shown, the main reinforcement tension value does not exert any decisive effect on the chief parameters of the rods. To eliminate their bending, some insignificant forces are applied, whose value is adjusted by slight braking of the spools with the main reinforcement.

Most critical within the schedule of the manufacturing process is the operation of impregnating the reinforcing material, as it influences the quality and mechanical properties of resultant products. Proceeding from the importance of this stage, a new, in principle hydrodynamic, procedure of impregnating fillers has been developed. It consists of combining the processes of tension and impregnation by using an oncoming binder flow driven under pressure within a closed cycle. The flow of the binder moving towards the reinforcing material impregnates it thanks to liquid friction forces, and stretches uniformly all the fibers of the reinforcing material. Tension forces and depth of binder penetration inside the fiber bunch are adjusted by varying the mutual velocities of the binder and reinforcement motion, binder viscosity and pressure in the working chamber. The tension force and impregnation quality are also affected by the length and shape of the working chamber. Microstructural investigations and examination of the binder diffusion front in the rods have proved that the proposed method makes it possible to decrease the porosity of the rods 1.4–1.6 times, increase binder penetration depth and diminish the amount of damage to individual fibers.

The impregnated and squeezed-out binder is compacted by the winding fibers laid on the surface at a definite tension. The resultant pressure is independent of the tension force value and can be calculated by the formulas presented in Chap. 5. Figure 9.2 illustrates variations in molding pressure q against the diameter of the treated rod under various tension forces of the winding fiber. It follows from Fig. 9.2 that increased diameter of the braid reduces q, providing the auxiliary reinforcement tension force is constant. Furthermore, molding pressure variation brings about corresponding variations in the filling degree of the rods. The dependence of filling degree on the winding fiber tension force for a carbon braid VMN-4 is presented in Fig. 9.3 at application of different impregnation compounds, namely epoxide and polyvinyl alcohol aqua solution. The region shown by a dotted line corresponds to the production of rods with deviations from linearity induced by material twisting at winding. The absolute twisting angle value is found within 1–5° and is conditioned by the tension force ratio of the winding and main reinforcements. To eliminate the twisting it is recommended to turn the braid in the opposite direction during impregnation.

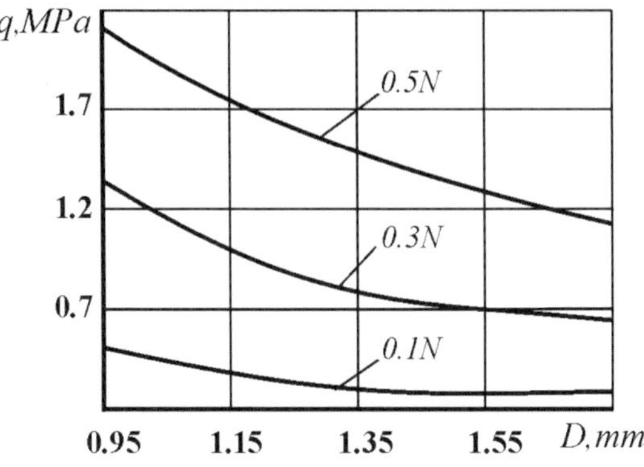

Fig. 9.2. Molding pressure variation of rods versus their diameter and winding fiber tension force

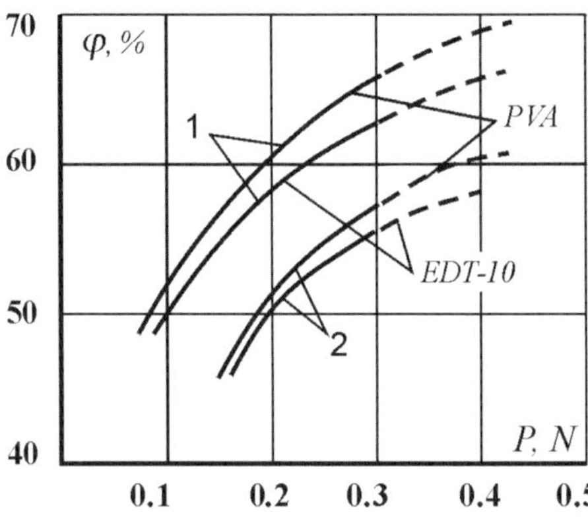

Fig. 9.3. Filling degree dependence of rods based on three folded in VMN-4 fibers upon winding fiber tension force. *1* – without winding layer consideration; *2* – with winding consideration

The existing heat-treatment regimes for materials based on the above-considered binders are inapplicable for carbon-plastic rods since their setting time should be minimal to reach the needed productivity. On the grounds of investigation results, setting regimes by thermal shock have been developed for rods using epoxy binder and polyvinyl alcohol. Setting time has been defined from the straight-line production conditions of the rods. For instance, for production of 1.2–3 running meters/min of rods in a 2–m long thermal chamber the time of molding is 0.75–1.75 min.

The performed investigations give grounds for refining, known as the art method, of drawing the reinforcing material through a calibrated spinneret and to develop a novel engineering process for manufacturing unidirectional carbon-plastic rods based on the spiral reinforcement procedure. In the course of investigations manufacturing equipment for a continuous molding of rods was developed. A circuit diagram of the processing line is illustrated in Fig. 9.4.

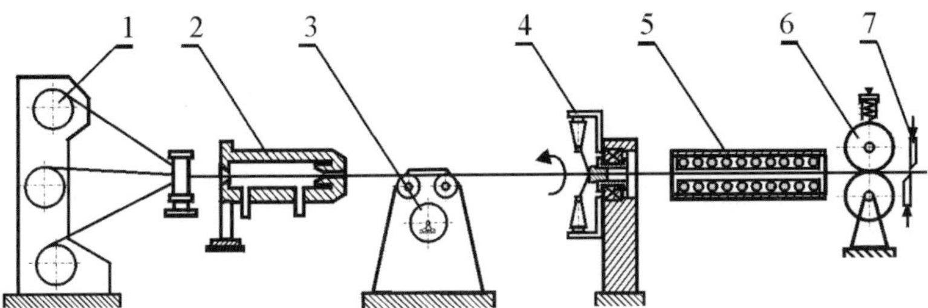

Fig. 9.4. Principal diagram of an automated line for the production of carbon-plastic rods. *1* – bobbin carrier; *2* – unit of hydrodynamic impregnation and tension; *3* – squeezing; *4* – winding; *5* – thermal treatment; *6* – heated air expulsion; *7* – drawing; *8* – cutting

9.1.3 Study of the Characteristics of Carbon-plastic Spirally Reinforced Rods

It has been noted earlier that the chief properties of the rods used for the assembly of multidirectional frameworks are dimensional stability, filling degree and relatively high mechanical parameters.

In fact, all methods of manufacturing rods suffer from either instability of dimensions, i.e., ellipticity of the cross-section, or deviation from their linearity. In contrast, the use of spiral reinforcement diminishes scatter to a definite extent in the named parameters. In Fig. 9.5 one can see empirical functions of diameter distribution of carbon-plastic rods obtained by drawing and winding. It is evident from the figure that, in the latter case, the scatter in diameter values and its absolute value are much lower. Experimental studies have shown that the dependence of the diameter values distribution of the rods obtained by the winding–on process parameters is in line with the normal law of distribution. On these grounds, a variation range for the pro-

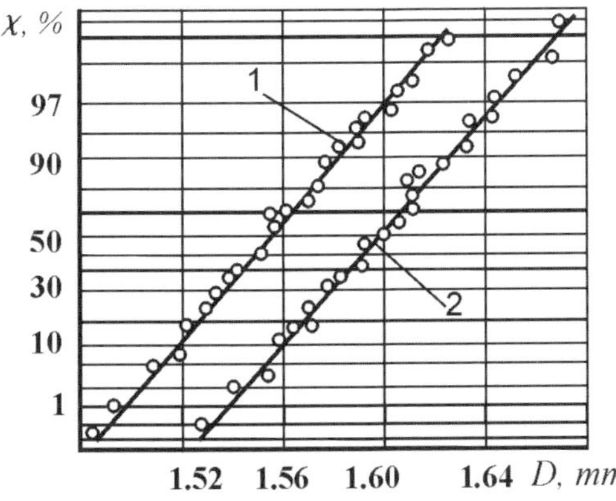

Fig. 9.5. Empirical functions of diameter distributions for carbon-plastic rods. *1* – winding; *2* – drawing

cess parameters and probable diameter deviations were chosen for different sizes of rods.

Undergage of the rod during production by spiral reinforcement results in increasing filling degree. During production of rods on a VMN-4 base by drawing, the degree of filling with the carbon fiber is $\varphi = 0.45$ and on spiral reinforcement it is $\varphi = 0.53$.

It has been found that the rod axis often becomes distorted during production as a result of differences in each braid tension and inhomogeneity in fiber distribution in the cross-section. For the drawing method, more typical is plane bending of the rod and on winding there occurs a combination of bending with twisting due to the action of the winding fiber, furnishing a chance to diminish the maximum bending value. By choosing the force parameters of the winding, including main reinforcement tension force ratio to the auxiliary one, and winding fiber kind, the deviation from linearity can be diminished. For example, when using Capron fibers as an auxiliary reinforcement at $0.2\,\mathrm{N}$ tension for winding a thrice-folded carbon braid VMN-4, the maximum bend reduces 1.23 times compared to drawn rods. Analysis of the service characteristics of rods on automated and manual assembly of multi-directional frameworks two modes of failure have proved to be dominant.

These occur firstly, due to the loss of stability on rod wedging inside the frame, and secondly, as a result of rod face crumpling on deflection from its movement trajectory. Both loading schemes are illustrated in Fig. 9.6. Test results of the rods according to the indicated loading schemes are cited in Table 9.1. It is evident that winding application raises critical compression force values by 36–42% and strength on crumpling by 34–36%.

The strength characteristics of spirally reinforced rods are conditioned by both structural parameters and production–process regimes. Under the

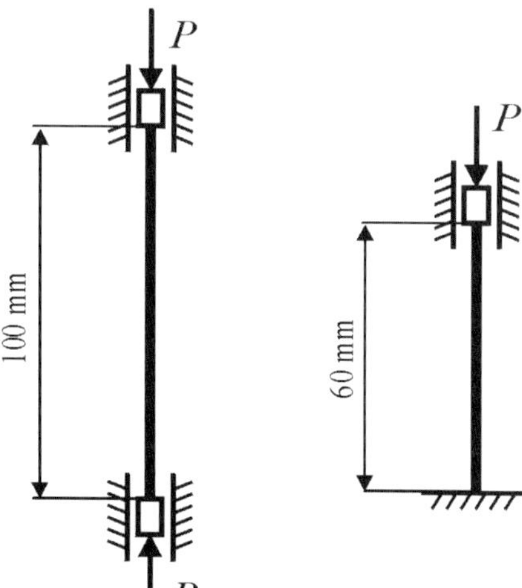

Fig. 9.6. Schematic diagrams of loading

Table 9.1. Mechanical Characteristics of Carbon-plastic Rods

Production process	D, mm	φ	Stability test			Test to crumpling		
			P_C, N	S, N	v, %	P_{bd}, N	S, N	v, %
Spinneret	0.75	0.46	7.4	1.0	13.5	8.0	0.9	11.3
	1.05	0.47	14.0	1.3	9.3	14.3	1.4	9.8
	1.19	0.55	31.0	2.5	8.0	42.0	3.4	8.1
Winding	0.71	0.52	10.5	0.7	6.7	10.9	0.7	6.4
	1.05	0.50	18.5	0.8	4.3	23.7	1.4	5.9
	1.18	0.56	42.1	2.1	5.0	56.4	1.7	3.0

chosen heat-treatment regime a large number of pores appear in the rod material, affecting its quality. In Fig. 9.7 variation of the material porosity is presented against temperature and exposure time. Apparently, there exists an optimum correlation of heat-treatment regime, generating the minimum number of pores. Analogous conclusions can be made from the analysis of the critical force dependence on the mentioned parameters (Fig. 9.8).

The performed investigations have given grounds for defining the region of optimum technological regimes of the straight line production of carbon-plastic rods of high filling degree (to 79%) and the respective geometrical parameters. Small-diameter rods of about 0.7 and 1.0 mm were produced

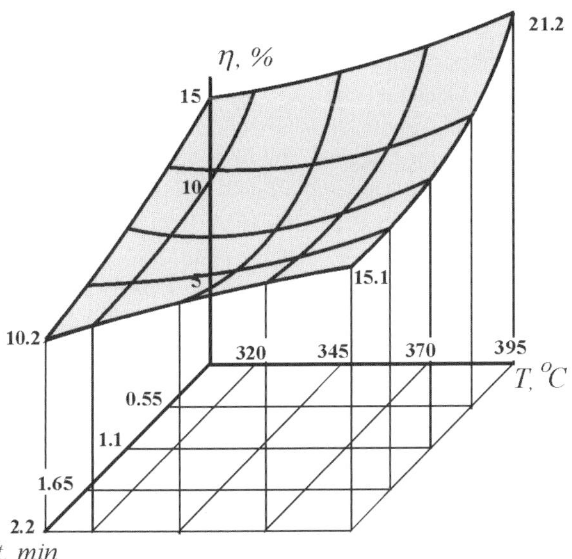

Fig. 9.7. Dependence of porosity of rods versus temperature and time of heat treatment

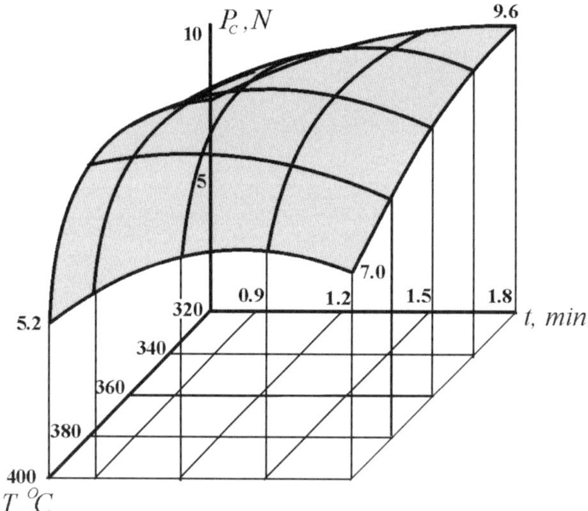

Fig. 9.8. Critical force variation in response to heat treatment parameters

by the method of spiral reinforcement based on a Colon carbon, which is practically impossible to achieve by drawing.

9.1.4 Metal-plastic Rods for Measuring Units

Experimental studies of characteristics and the elaborated production technique for carbon plastic rods have served as the basis for developing production processes for metal-plastic rods. These rods are employed as ablation sensors in measuring units aimed at determining the mass of removed heat-proof coatings on missiles and space vehicles in the dense atmospheric layers. A few modifications of ablation sensors are possible (Fig. 9.9) in which the bearing member of the rod is based on carbon, quartz or metal.

In the former case, the rod is formed from a carbon braid based on VMN-4 fibers or a braid of silica fibers, K11S6-180, wound with silica fiber. In the second case, a metal rod 0.6 mm in diameter, based on 29NK wire, is wound with a silica fiber, K11S6-180. The main requirements imposed on the thus obtained rods are: dimensional stability within 1.0 ± 0.05 mm, deviation from linearity – not more than 0.03 mm over 10 mm length, heat resistance up to 900°. To meet the latter condition the rods are produced using a specific binder, KE-6.

In the course of developing the production processes of the above-described rods, a series of negative features of design schemes c, d and e were singled out. It appeared difficult to eliminate heat-induced warping of the rods during operation and to keep to the needed dimensions of the measuring unit. Furthermore, only schemes a and b were analyzed. Then, realizing

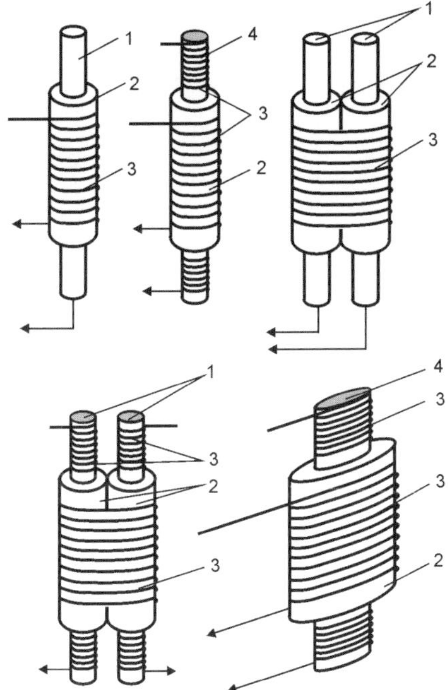

Fig. 9.9. Ablation sensors. *1* – carbon-plastic or metal-based rods; *2* – quartz-fiber winding; *3* – metal wire; *4* – quartz-fiber-based rods

that two measuring spirally should be mounted in case b, which enlarges the sensor and complicates its fabrication, scheme a was finally adopted. In this case, a layer of silica fibers is wound on a metal rod being simultaneously one of the conductors. The thickness of the layer is, on the one hand, conditioned by the design peculiarities and, on the other, by the needed electrical resistance of the insulation found between the inner metal rods and the outside surface (it is $20\,\mathrm{M\Omega}$ at $20°\mathrm{C}$). Then, a metal wire serving as the other conductor is laid on the rod. Bench and field tests have shown that the developed design of the measuring unit reduces measurement error 3.2 to 3.4 times in contrast to earlier-used measurements of material removal. Based on the designed measuring unit a new multichannel apparatus for automatic recording of the ablation rate of heat-proof coatings immediately during flight has been elaborated.

9.2 Composites with the Spirally Reinforced Filler in Strengthening Elements of Load-bearing Units

Local reinforcements are frequently used in modern aviation, missile and space structures. This essentially consists of mounting specific high-strength high-modular safety-firsts on the most loaded part of the structure. The reinforcing elements presuppose the application of expensive, composite materials and, consequently, the well-proven technology of adhesive joints. The structure of a glider is known to incorporate a stiffening casing of composite materials almost half of weight of aluminum. For stringers and frames open U and Z profiles are commonly chosen, or that of a closed box-like section. In the experiments, variants of stiffening elements, based on metal-composites, metals and composites were studied. Stringers and frames of unidirectional materials turned out to provide for a high specific strength of the structure, although with a danger of axial splitting. Therefore, orthogonally-reinforced materials, with a higher transversal stiffness, were used. The present-day technology of such types of stringers consists of creating a laminated sheet from different materials, including e.g., a undirected layer of carbon fibers and a layer of glass fabric, which is then cut into strips of a cross-section similar to the stringer. This process is, evidently, rather laborious and it is hard to produce high-quality products. Of special interest is the production of combined stringer and frame designs on a basis of boron fibers. With this aim high-modular boron fibers are placed in a definite place of a needed section (Fig. 9.10) to add strength and stiffness to the element. These reinforcing elements can be made by extrusion, bending, molding or assembly of structural profiles. This improves the weight characteristics of combined stringer designs 2.5–3 times, compared to aluminum structures. However, many problems need to be solved for production of these stiffening elements, which are connected with residual stresses emerging in the boron plastic and metal,

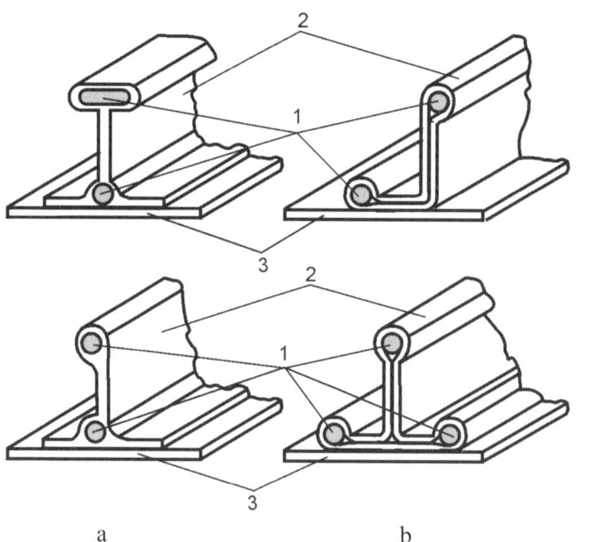

Fig. 9.10. Variants of stiffening elements of load-bearing structures made by a – extrusion and b – bending. *1* – boron-epoxide composite; *2* – metal stiffening element; *3* – casing

leading to distortion of their shape, the choice of a binder, which sets at the lowest possible temperature and perfection of the molding technology

In contrast to the above-described method, by the spiral reinforcement procedure it is possible to produce articles from any composite material and thus realize the spatial reinforcement scheme.

Products based on four types of materials have been produced as a result of investigations, namely unidirectional, spirally reinforced, coaxially-reinforced and coaxial-spirally reinforced. The production technique of profile elements on the base of these materials incorporates the following main stages: preparation of a semifinished product, article molding, heat treatment and finishing.

9.2.1 Semiproducts from Composites Based on Spirally Reinforced Fillers

It has been already shown that materials based on polyfibrous spirally reinforced fillers are advantageous over unidirectional layered composites [280, 281]. To these advantages belong elevated characteristics in the transversal direction, high shear resistance in the main reinforcement direction and good moldability. The latter feature gives grounds for manufacturing prepregs, based on specially designed materials for targeted goods, i.e. preparation of semiproducts. Their basis form is preliminary impregnated with a required binder polyfibrous braid wound with an auxiliary reinforcement. The braid commonly incorporates carbon, boron, organic or glass fibers, which raise filler properties in the main reinforcement direction.

Recognizing the high specific characteristics of organic fibers, a semiproduct has been produced with filler in the form of 5-, 10- and 25-times folded braids based on SHP fibers. The choice of geometrical and structural parameters is known to be governed by the molded semiproduct size and the loads exerted on it. Consequently, the share of the auxiliary reinforcement will diminish as the filler diameter and winding pitch increase. It should be emphasized that the optimal correlation of the main and auxiliary reinforcements in the filler bulk must be corrected during designing specific articles. Schemes of the spirally reinforced filler and its dimensions are presented in Fig. 9.11 and its production process layout is depicted in Fig. 9.12.

The most important technological parameters of manufacturing semiproducts are the winding reinforcement tension and winding pitch. These parameters exert an influence upon fiber packing density, article moldability, strength and stiffness.

It is expedient to use high-order material structures in materials based on spirally reinforced fillers. This means producing layer-by-layer coaxial reinforcement of the groups of unidirectional or spirally reinforced filler. This will raise the bearing capacity of the structure at transversal loading as well as during bending and compression. Evidently, the introduction of coaxial winding imposes a number of limitations on the technological process on the

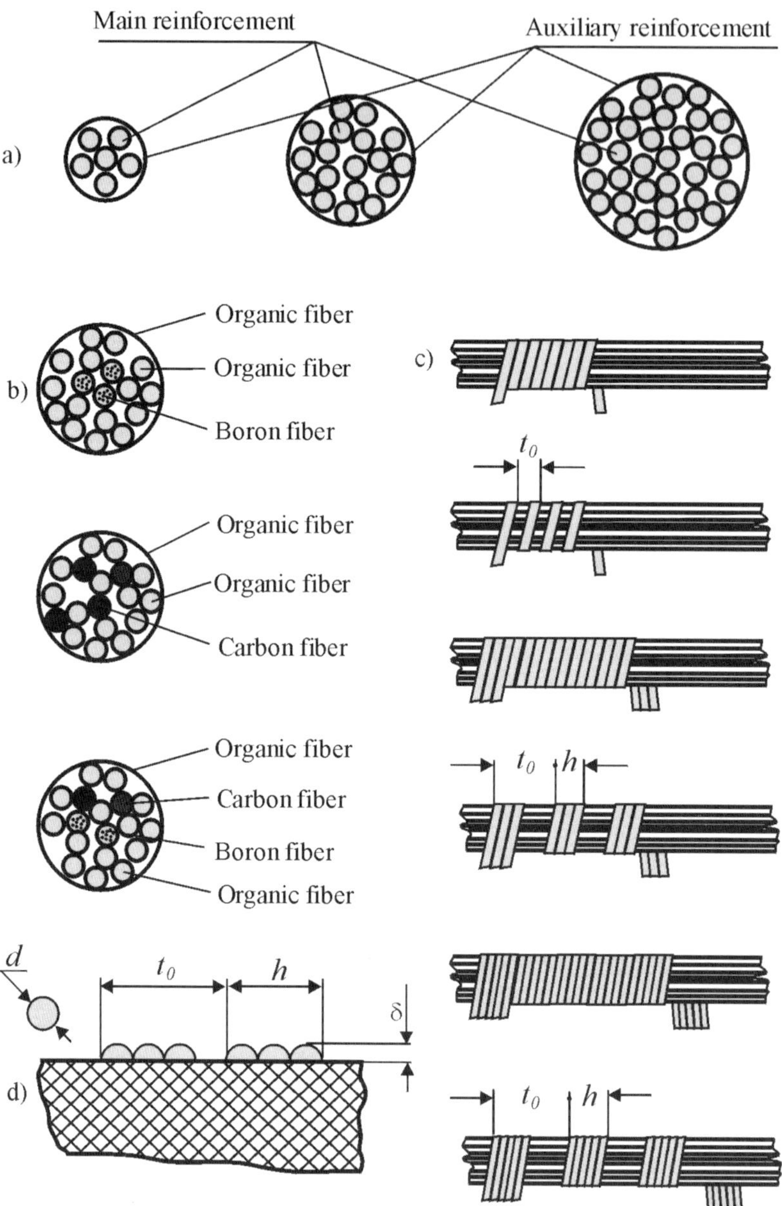

Fig. 9.11. Design schemes of used fillers (**a**) folding schemes of the original braid; (**b**) schemes of a polyfibrous spirally reinforced element; (**c**) variants of auxiliary reinforcement packing; (**d**) longitudinal section of the filler

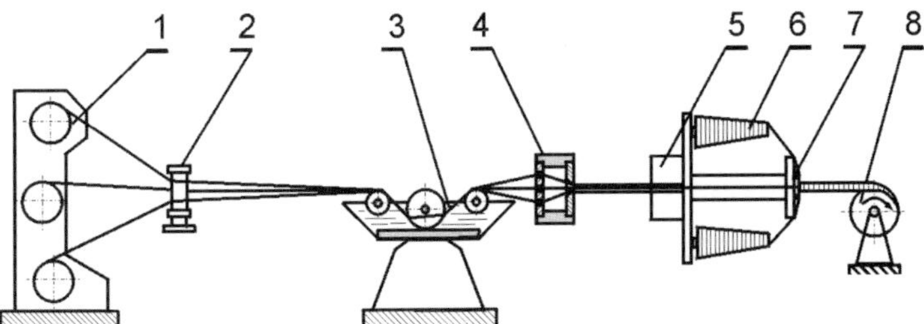

Fig. 9.12. Technological process layout for manufacturing spirally reinforced filler. *1* – bobbin carrier; *2* – tension device for the main reinforcement; *3* – unit for binder impregnation and squeezing; *4* – combined nozzle; *5* – axial winder; *6* – bobbin with the auxiliary reinforcement; *7* – tension device for the auxiliary reinforcement; *8* – drawing mechanism

one hand, and necessitates material designing for each chosen structure on the other.

Among the limitations one can name difficulties in realizing wet molding of the products. With this aim, it is necessary to exercise tension, impregnation, squeezing, winding and laying of a considerable amount of original spirally reinforced fillers simultaneously. For this reason the equipment turns out to be rather intricate and bulky, which complicates its maintenance and adjustment. So far, when manufacturing the products by coaxial reinforcement it is expedient to presaturate the fillers. Apart from this, there is a hazard of the semiproduct twisting during molding of the coaxial layers and this intensifies in cases where each of the layers consists of several elementary spirally reinforced braids. This suggests that most critical parameters of the semiproduct, including its stability, dimensions, moldability and, finally strength and stiffness are conditioned by the winding pitch, its direction, tension force of the winding fibers and amount of reinforcement. Proceeding from the above it is advisable to solve the optimization problem, the when designing structure and process parameters of the semiproduct.

Finished semiproducts usually gain a round or circular section and acquire the shape of a certain article during molding. Hence, a distinguishing feature of the production process of articles from semiproducts is noncoincidence between cross-sections of the former and the latter. However, a natural requirement of the semiproduct is to make its layers similar to the article in cross-sectional perimeter and any difference results in different filling degrees with the main reinforcement.

When the filling degree of an article is laid down in design, proceeding from attaining the required mechanical characteristics, then the semiproduct filling should be rated taking account of changes that occur during molding.

Let us consider the sequence of filling stages of a semiproduct for an article whose cross-section is given in Fig. 9.13. Dimensions of the i-th layer

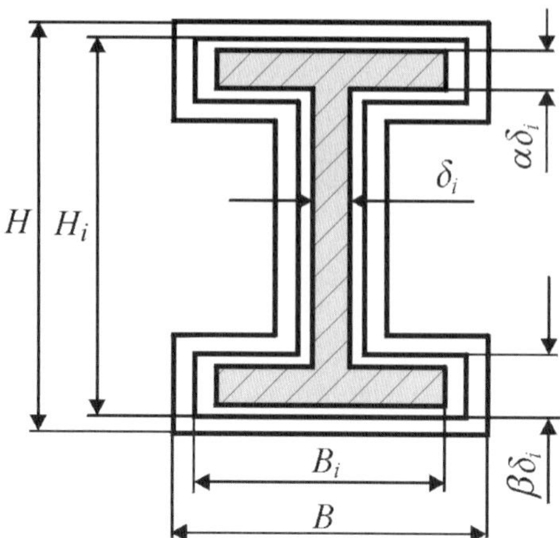

Fig. 9.13. Cross-sectional view of the reinforcing element

in article cross-section can be presented as follows

$$H_i = \frac{P_i}{2t_i}; \quad B_i = \frac{P_i m_i}{2t_i}; \quad \delta_i = \frac{P_i n_i}{2t_i}, \tag{9.1}$$

where $t_i = 2m_i - n_i + 1$; $n_i = \delta_i/H_i$; $m_i = B_i/H_i$; $P_i = 2(H_i + 2B_i - \delta_i)$ – perimeter of the ith layer.

The cross-sectional area of the ith layer is found from the corresponding parameters of the $(i-1)$th layer

$$\begin{aligned}
F_i &= \frac{P_i^2 n_i}{4t_i^2}\left[(\alpha_i + \beta_i)(m_i - n_i) + 1\right] \\
&\quad - \frac{P_{i-1}^2 n_{i-1}}{4t_{i-1}^2}\left[(\alpha_{i-1} + \beta_{i-1})(m_{i-1} - n_{i-1}) + 1\right].
\end{aligned} \tag{9.2}$$

Since the perimeters of the layers should remain constant at molding for both the article and semiproduct, then the cross-sectional area of the semiproduct corresponding to the ith layer is

$$F_i^n = \frac{P_i^2 - P_{i-1}^2}{4\pi}. \tag{9.3}$$

The filling degree of the ith layer of the semiproduct can be expressed through the filling degree of a finished article φ_i as follows

$$\varphi_i^n = \frac{F_i}{F_i^n}\varphi_i. \tag{9.4}$$

For articles whose cross-sectional area represents a double-T, channel or rectangle the main parameters of the layers should work for identity

$$\begin{aligned}
\alpha_i &= \alpha_{i-1} = \alpha; & \beta_i &= \beta_{i-1} = \beta; \\
m_i &= m_{i-1} = m; & n_i &= n_{i-1} = n; \\
t_i &= t_{i-1} = t.
\end{aligned} \tag{9.5}$$

Then, the relation for the ith layer cross-sectional area and filling degree of semipoducts will be

$$F_i = \frac{(P_i^2 - P_{i-1}^2)n}{4t^2}[(\alpha + \beta)(m - n) + 1];$$ (9.6)

$$\varphi_i^n = \frac{\pi n}{t^2}[(\alpha + \beta)(m - n) + 1].$$ (9.7)

Consequently, in obeying conditions (9.5), the filling degree of each semiproduct layer remains constant and depends linearly on parameters α and β. The dependence of the semiproduct filling degree on parameters m and n is illustrated in Fig. 9.14, where $\alpha = 0.6$ and $\beta = 1.4$. It follows from the diagram that function φ_i^n has an extreme line but the region of its practical values is bounded by the conditions of the actual existence of the section profile

$$\frac{n_i}{m_i} \leq 1;$$

$$n_i(\alpha + \beta) \leq 1.$$ (9.8)

Conditions (9.8) visualize that the article wall thickness should be less than its width and the total thickness of its flanges less than the height.

Thus, the real semiproduct can not acquire a filling degree surpassing

$$m = n = \frac{1}{\alpha + \beta}.$$

The highest filling degree values shown in Fig. 9.14 correspond to an article of rectangular cross-section and are calculated as

$$(\varphi_i^n)^{\max} = \frac{\pi\varphi_i m}{(1 + m)^2}.$$ (9.9)

Analysis of the presented data shows that with reducing factor n the semiproduct filling degree drops, while reducing factor m augments its filling. Note that the least filling degree of the semiproduct can be determined proceeding from process conditions, including provision for its integrity during production.

It is required to ensure the stiffness of the bearing surface when manufacturing multilayered semiproducts when laying each layer. This is attained when the filling degree of the semiproduct decreases with each layer from the center to the periphery, i.e.,

$$\varphi_1^n \geq \varphi_2^n \geq \ldots \geq \varphi_n^n.$$ (9.10)

Condition (9.10) is obeyed in cases where the values of α, β, m and n remain constant from layer to layer, or factors α, β and n are sequentially decreased while m is increased from the center to the periphery

$$\beta_1 \geq \beta_2 \geq \ldots \geq \beta_n;$$

$$\alpha_1 \geq \alpha_2 \geq \ldots \geq \alpha_n;$$

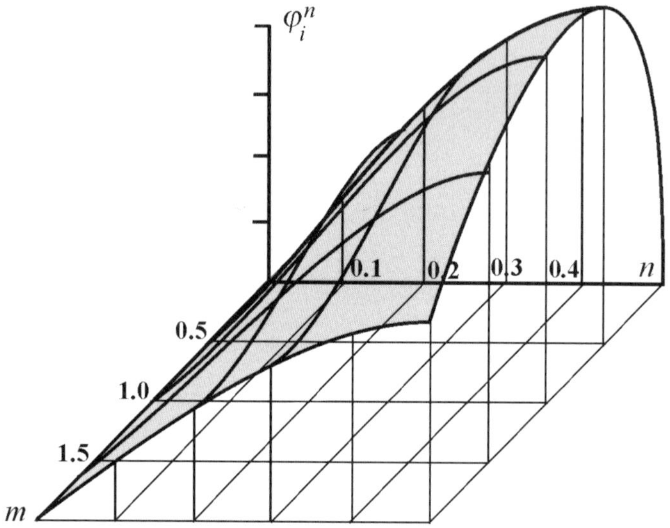

Fig. 9.14. Filling degree variation of a semiproduct for a complex-profile article dependent on parameters m and n ($\alpha = 0.4$, $\beta = 1.6$)

$$n_1 \geq n_2 \geq \ldots \geq n_n \, ;$$
$$m_1 \leq m_2 \leq \ldots \leq m_n \, . \tag{9.11}$$

The diameter of the mandrel upon which the semiproduct is formed is determined by the known parameters of the first layer of the article

$$d_0 = \frac{2(H_1 + 2B_1 - \delta_1)\sqrt{1 - \varphi_1}}{\pi} \, . \tag{9.12}$$

Thus, by setting the dimensions of the first layer of the profile and its filling degree, one can calculate the mandrel diameter d_0 using (9.12) and, taking account of (9.11) and limitations (9.8), the filling degree of each sequential layer of the semiproduct is found from (9.7).

9.2.2 Elaboration of the Manufacturing Process for Reinforcing Elements

Prepregs based on carbon braids VMN-5 and organic fibers SHP impregnated with, respectively, UP-2220 and T-71S, were used as the main reinforcement. They were wound with organic SHP fibers of 14.3 tex densities. Types and main dimensions of the obtained article cross-sections are depicted in Fig. 9.15.

For the manufacturing process of reinforcement, special equipment has been developed. It consists of a combined installation for spiral reinforcement of semiproducts, a drawing device and a set of molds. The process diagram for manufacturing semiproducts is illustrated in Fig. 9.16.

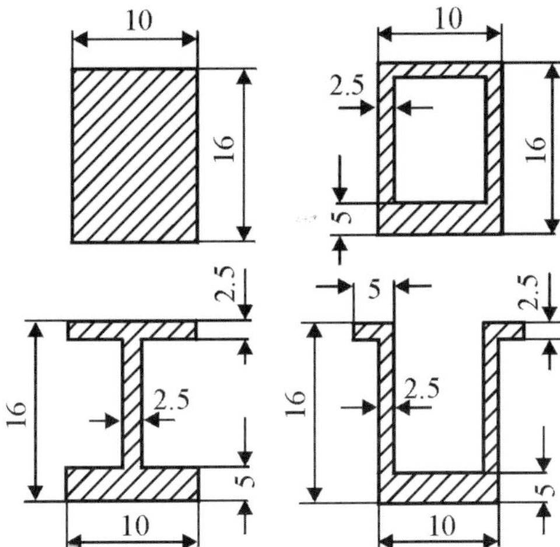

Fig. 9.15. Cross-sectional dimensions of obtained reinforcing elements

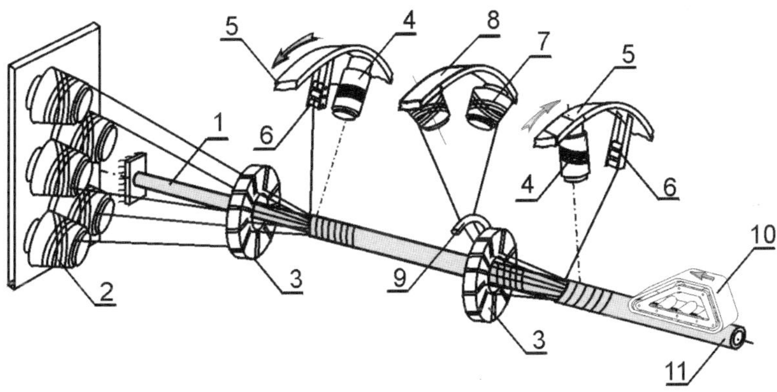

Fig. 9.16. Schematic diagram of the manufacturing process of semiproducts. *1* – mandrel; *2* – bobbin carrier; *3* – complex spinnerets; *4* and *7* – bobbins with winding material; *5* and *8* – ring pickups; *6* – guiding rollers; *9* – guide; *10* – pull-in device; *11* – finished semiproduct

According to the scheme the article is produced as follows. Bobbins with the prepared spirally reinforced prepreg are installed onto the bobbin-carrier of the installation and the number of braids in the filler is rated so as to ensure main reinforcement filling after molding to not less than 45–50%. For a high-quality impregnation of the auxiliary reinforcement the amount of binder in a prepreg is set to exceed by 6–7% the rated binder content needed for saturation of the main reinforcement only. Upon reeling off the slightly braking bobbins of the first bobbin carrier, the reinforcing material is fed through a heating distribution device into the first spinneret and thence to the first winder that forms the inside spiral layer of the auxiliary reinforcement. As soon as the reinforcement leaves the first winder, the filler is supplied from the bobbins of the second bobbin carrier to form the second layer of the longitudinal reinforcement, on which surface the outer layer of the auxiliary reinforcement is laid, using the second winder. The tension force of the auxiliary reinforcement is chosen so as to make the circuit of each semiproduct winding layer equal to its perimeter in the finished article cross-section. The produced semiproduct is cut into pieces similar to the article length and placed into a mold treated with a release agent PES-5.

The article was cured in line with the standard time and temperature regimes specific for each binder used in conditions of vacuum heat treatment. As gases cease to be isolated from the material structure it undergoes molding at less than 140 and 160°Cfor binders UP-2220 and T-71S,respectively. In Table 9.2 the chief technological parameters and structural characteristics of rectangular cross-section stringers on the base of coaxial spirally reinforced carbon plastic are cited.

The semiproduct for profiles of a box-type section are manufactured by the main reinforcement packing on a rectangular mandrel having dimensions similar to the inner profile cavity. The mandrel surface is made smooth enough and is covered by PES-5 adhesive. To facilitate mandrel removal upon article setting it is made wedge-like, with a 1/160 inclination. To ensure the difference in wall thickness of the obtained profile, the required number of additional bobbins with reinforcement is installed in the lower part of the back panel of the coaxial reinforcement. During operation a certain bias of the reinforcing elements from the angular portions of the mandrel to the plane center was observed, which hampered the semiproduct when laying into the mold. To eliminate this drawback, one of the side walls of the mold was made

Table 9.2. The Main Technological Parameters of Manufacturing Rectangular in Cross-section Stringers with Coaxial Disposition of Reinforcement

Process parameters	Values
Main reinforcement	VMN-5 braid in 2 folds
Auxiliary reinforcement	1 SHP fiber, 14.3 tex
Drawing velocity	120 mm/min
Number of coaxial layers	2
Filler winding pitch	1 mm
Winding pitch of the 1st coaxial layer	1 mm
Winding pitch of the 2nd coaxial layer	0,5 mm
Winding fiber tension force at filler production	0.5 N
Winding fiber tension force of the 1st coaxial layer	2.5 N
Winding fiber tension force of the 2nd coaxial layer	5.0 N
Tension force of the main reinforcement at winding	5–8 N/braid
Molding pressure	2 MPa

movable. Molding was exercised in two stages. First, over the side surfaces and then over the upper and lower surfaces upon fixation of the side wall and correction of the parallelism between the inner planes of the mandrel and the mold. It has been established that using this technique a box-type profile can be obtained from a carbon plastic with spirally reinforced filler, but with limited length. In spite of avoiding these drawbacks, the initial forces generated on squeezing of the inner mandrel from 320 mm long stringers reach 100–120 MPa, which is close to the ultimate strength of the mandrel material (about 10 KP).

Semi-products for double-T, channel and trough-shaped cross-sections are manufactured as hollow tubes, produced by laying the reinforcing material on a smooth cylindrical mandrel installed along the forming spinneret and winder axis, followed by pressing around spiral reinforcement layers. The mandrel must be rigidly fixed to avoid longitudinal displacement and the obtained tube is pulled off it using a drawing device. It has been found that a sufficiently high quality is attained when the prepreg is well dried and the volatile products content is less than 4–5%. This guarantees easy sliding of the reinforcement over the mandrel surface at 18–22°C. As the temperature rises to 30–40°C the binder in the prepreg becomes softened and makes sticking of the reinforcement to the mandrel quite probable. Since, when manufacturing thin-walled tubes the distributing devices and forming spinnerets are heated, the heat flows can be transferred to the mandrel. This is especially important when the equipment is switched off to repair fiber rupture or to change spools on the winders. To this end, the mandrels were made as hollow tubes inside which a cooling liquid was ran. To provide the difference in profile wall thickness at the preserved circular outer surface of the semiproduct, an additional reinforcement was introduced and a flat spot was made lengthwise of the mandrel. The removed segment area is equal to that of the additionally introduced reinforcement taking account of the rated filling degree.

To attain the needed profile shape the double-T and box-type products are produced in two stages. During the first stage the product is premolded, with which aim both side walls of the mold are made movable and mold shoes are fixed to the upper and lower lids of the mold. The side walls of the mold slide apart evenly to a distance equal to the obtained semiproduct diameter. The semiproduct is laid into the mold and is pressed to the upper lid, which has certain gaskets to avoid incomplete molding, after which the side walls draw together to a required size. The second stage includes heat treatment and removal of residual volatile products, after which the gaskets are taken away from under the upper lid and the product undergoes final molding. The sequence of the manufacturing stages for stringer molding is illustrated in Fig. 9.17.

sectional dimensions of all stringers can be inscribed into the 10×16 mm figure. Also, conditional specific characteristics of the obtained designs, taken as the strength and elasticity modulus relation to the weight of the samples of similar length, are given here.

Stringer samples of the following types were tested:

- Unidirectional carbon plastic based on epoxy-phenolic binder.
- Spirally reinforced organo-carbon plastic based on spirally reinforced filler produced by winding two carbon braids with an organic fiber of 14.3 tx density.
- Coaxially reinforced organo-carbon plastic obtained by double coaxial winding of carbon braids with the cross-sectional area of the main reinforcement ratio 1:1 in each of coaxial layers.
- The fourth type of the material designs represents a combination of the second and third ones which means that spirally reinforced braids of the main reinforcement are placed inside each of THE coaxial layers instead of unidirectional reinforcement.

It is evident that the highest absolute and specific characteristics are displayed by stringers with a double-T cross-section. In this case, the compressive strength ratio to strength of the stringer with an unbroken cross-section increases by 4% and the elasticity modulus by 14%. In contrast, their specific characteristics rise by 86 and 103%, respectively.

It has been established during experimental studies that the failure mode of the samples changes essentially in response to the reinforcement scheme. For example, breakage of the unidirectional carbon plastic is induced by exfoliation of the main reinforcement along its fibers. In contrast, destruction of the spirally reinforced carbon plastic is initiated within the edge contact stress area and propagates in the plane of the maximum tangential stresses. As a result of loading-induced cross-wise shear of the broken-down part of the sample, there occurs a step change in the tensile transversal stresses, leading to a lateral tearing off part of the spirally reinforced elements. The coaxial-reinforced material breaks through a multiple delamination that develops upon fiber rupture in coaxial windings. The process of fiber destruction in a combined material proceeds from fiber cut in the direction of the maximum tangential stresses, without any delamination.

As long as the experimental results are in complete agreement with derived earlier conclusions on the failure character of composites with spirally reinforced filler, it will be expedient to use them efficiently in designs operating under significant compressive loads.

9.3 Application of Composites Based on Spirally Reinforced Fillers in Design Casings

The structures produced by winding with fibrous materials were shown to have high specific characteristics and can therefore be justifiably used in airplane designs to economize on weight. A characteristic example is thin-walled shells for different purposes, which combine high carrying capacity and low weight. However, in contrast to the extensively studied molding of cylindrical, spherical and other types of closed-type casings, the production of conical products is still a laborious process. The most complex stage for all types of casings is considered to be the introduction of discrete hardening elements into their composition. This complicates the manufacturing process and sometimes even makes it improbable. In this connection, the production of structurally orthotropic conical casings with a large taper angle has lately become an acute problem. In the present work, a structure has been developed for a combined metal-composite conical casing with outer stiffening ribs and 30° vertex angle. The chief part of the casing, i.e. the hemisphere was made of AMG-6 metal, while the conical part and stiffening frame were made from carbon and glass fibers. The main requirement imposed on the design was weight restriction under improved bearing capacity in conditions of hostile environments, up to 150°Ctemperatures and 0.8 MPa outer pressures. Geometrical parameters of the casing are given in Fig. 9.18.

Structural features of the ribs exert a significant effect on the production process of casings as whole. Variants of mounting ribs of different profiles are depicted in Fig. 9.19.

Installation of the ribs in the casing body might lead to either adhesion-induced failure of the junction (a, b and c) or to formation of some additional layers (scheme d – one layer, schemes e and f – two layers). Most acceptable from the viewpoint of rib manufacturing and its installation in the casing body turns out to be scheme f, with the application of either continuous

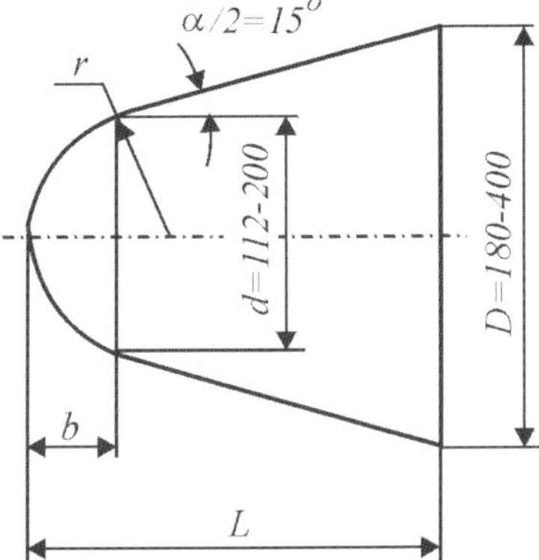

Fig. 9.18. Geometrical parameters of the articles and models

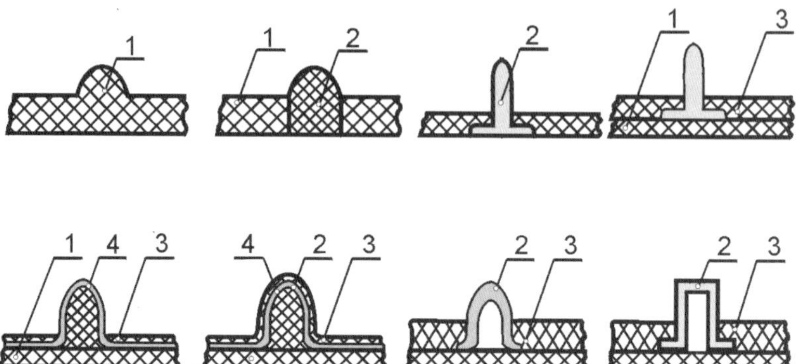

Fig. 9.19. Schematic diagram of installation of ribs. *1* – inner layer; *2* – stiffening element; *3* – outer layer; *4* – intermediate longitudinal layer

or hollow elements. Since practically all known methods of producing ribs are fairly laborious [199], a new procedure for manufacture of circular ribs of a spirally reinforced semiproduct has been developed, in a special mold whose working element is a small-pitch profiled spirally, between which coils a semiproduct is placed. Upon molding and thermal treatment, the thus-produced ribs are cut and junction points are glued together.

The next stage of article production is the development of a method of forming the conical part of the casing. In the course of analyzing the manufacturing processes, the longitudinal-transverse scheme of reinforcement packing was adopted as most perfectly complying with the loading regime and the relationship of the transversal to longitudinal layers, 2:1. One of the hardest tasks here is the laying of the transverse stratum on the conical element of the mandrel. A requisite of the process is maintaining the equilibrium state of the winding braid, since each wrap of the material is affected by the colliding constituent of the radial pressure, counterbalanced by the reinforcement friction force against the mandrel. This is why cones with large vertex angles are formed using specific technological means able to either balance the reinforcement on each portion of the mandrel through augmenting pressure or by compensating the colliding force, or to wind the reinforcement along geodesic lines.

It should be emphasized that all methods and devices indicated above are rather intricate and considerably complicate the production process. However, by using stiffer structural units as, e.g. spirally reinforced braids, it is possible to produce casings with large vertex angles of the cone. As a result of experimental analysis and technological studies, new and simpler methods of producing conical products with spirally reinforced fillers have been elaborated. In accordance with these methods, the first transversal reinforcement layer is laid on the mandrel beginning from the smaller diameter of the cone

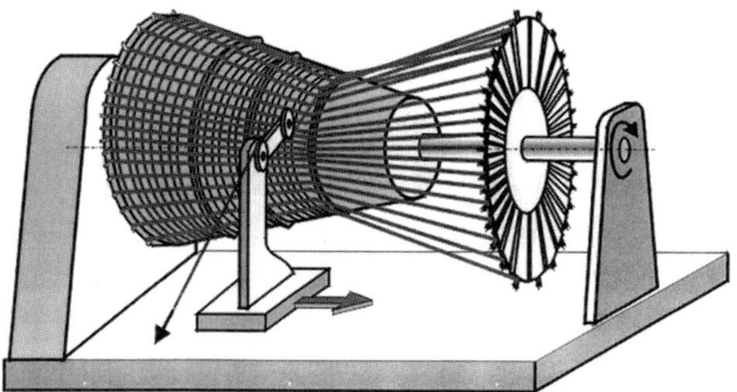

Fig. 9.20. Packing of the longitudinal reinforcement layer

and each sequential layer rests upon the preceding one and will not slip off along the generatrix thanks to the elevated rigidity of the braid. Preliminary produced ribs are installed on the mandrel over the first layer with the help of a special fastening system. The longitudinal reinforcement is then spread out on the combs, where the spring-loaded one is used as a movable comb, having resilient sectors to pull up the longitudinal reinforcement during forming (Fig. 9.20).

The outer layer is wound from the larger diameter of the cone and the spread out reinforcement rests upon strained longitudinal braids (Fig. 9.21). This sequence of stages ensures stability of the radial pressure and facilitates fixing of the ribs inside the casing body. Upon final forming, the article is placed in a vacuum thermal chamber for heat treatment. The outward appearance of the casing upon mandrel dismantling is presented in Fig. 9.22.

The casings are tested on a test bench whose functional diagram is given in Fig. 9.23. As the casing is loaded by the outer pressure, strains in the inner surface of the casing are recorded and the critical pressure is estimated. The results obtained during testing are cited in Table 9.6 for some dimension-types of casings on the basis of various materials. Proceeding from these

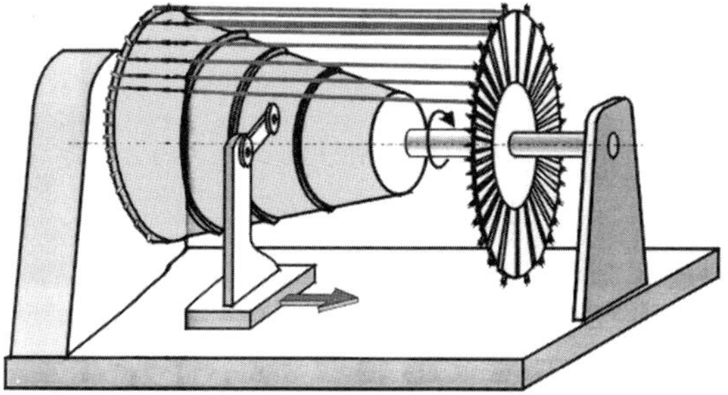

Fig. 9.21. Winding of the transversal layer

Fig. 9.22. External view of produced articles

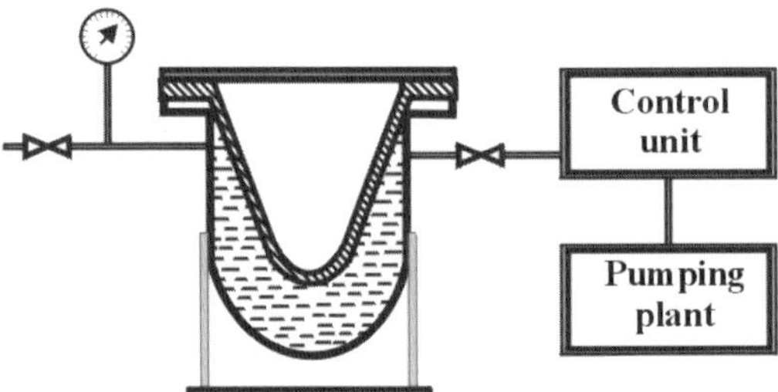

Fig. 9.23. Functional diagram of a testing bench

results, it can be stated that along with high technological effectiveness of the considered structures, spiral reinforcement of the fillers helps to raise their bearing capacity by 17–20%.

Table 9.6. Critical Pressure Values for Testing Conical Casings in Response to Outer Pressure

Casing dimensions, mm	Casing material	Number of ribs	Wall thickness, mm	q_{max}, MPa	S, MPa	v, %
$D = 180$	Glass plastic	–	2	1.9	0.15	7.8
$d = 112$	Carbon plastic	–	2	3.6	0.36	10
	CP SRF	–	2	4.2	0.34	8.2
	Glass plastic	2	2	3.9	0.24	6.2
	Carbon plastic	2	2	5.7	0.39	6.9
	CP SRF	2	2	6.8	0.44	6.4
$D = 180$	GP SRF	3	3	3.7	0.34	9.1
$d = 112$	CP SRF	3	3	4.6	0.4	8.6

GP SRF – glass plastic with the spirally reinforced filler;
CP SRF – carbon plastic with the spirally reinforced filler

9.4 Technical and Cost-effectiveness of Applying the Technology and Materials with Spirally Reinforced Fillers

Application of spirally reinforced fillers and the technology of spiral reinforcement when manufacturing structural elements have shown high effectiveness in both the engineering and economic aspects. For instance, molding of cross layers of a thin-walled conical casing with outer ribs with a spirally reinforced filler has made it possible to reject special devices for maintaining the equilibrium state of the reinforcement on the mandrel during winding and curing. Hence, expense on design, production, testing and run-in of the technological process and the devices thereof, their maintenance and control have been cut down drastically. Use of a simple and cheap installation, able to operate automatically for spiral reinforcement has reduced the time for preparation of the winding equipment, accelerated the winding process of the transverse layers and raised process stability through improving the strength of the spirally reinforced filler and reducing the probability of damage during winding. Moreover, substitution in part (6–8%) of costly carbon reinforcement by relatively cheap glass fibers without deteriorating the bearing capacity or incrementing the mass of the casing, has reduced the cost of the articles. The engineering effectiveness of applying materials based on spirally reinforced fillers consists of expanding the variety of engineering procedures aimed at realizing new design schemes for products with exterior location of stiffening ribs, in refining the manufacturing process, reducing labor-intensiveness and improving conditions of work.

As shown in Fig. 9.2, use of materials with the spirally reinforced filler in structures of the rib-type improves their compressive strength in the reinforcement direction at longitudinal shear by 30–60%. Application of the technique of spiral reinforcement during their manufacture broadens the range of the produced profiles by varying their cross-sectional form. So far, one can choose the most optimal cross-sectional form of the stiffening elements and locate them most efficiently, depending on the article type and its loading conditions. For instance, application of the double-T type carbon stiffening elements with the spirally reinforced filler in shell-type structures of airplanes have reduced their material capacity and mass by about 4%, under preserved bearing capacity.

The spiral reinforcement procedure has given grounds for realizing a continuous process of manufacturing semi-finished products for various kinds of wares. Upon molding and setting, a finished product is achieved with given structural and geometrical parameters. This eliminates the necessity of machining, with further recycling of part of the material as waste that, in turn, reduces the staff engaged in production, material capacity and power-consumption of the equipment and size of production areas and it improves working conditions.

For the production of carbon rods employed in heat-proofing materials with application of the spiral-winding method, a targeted automatic line has been elaborated and manufactured. The obligations of an operator on the automatic line presuppose just substitution of spools with the carbon material, control over the binder level and the packing of gaged and cut-to-size rods. As a result, three automatic lines can be served by one-person.

The presence of the spiral winding on cured rods gives grounds for polymerization of heat-set thermosetting binders under thermal shock conditions. This treatment increases the productivity of the manufacturing process by 2.5 to 3 times, depending on the rod diameter. Along with this, improved strength and stiffness characteristics allow for their application and the automated assembly of the casings. The efficiency of the process in manufacturing opens vistas for the further perfection of processing methods, ensuring evenly dense structure of the heat-proof materials, elevation of their protecting properties and bearing capacity.

The possibility of using spirally reinforced rods on carbon and siliceous basis as sensors suggest for their application in a measuring complex intended to register the ablation level in heat-shielding materials under the action of accelerated plasma flux. The complex enables the monitoring of failure processes of heat-proofing coatings in the course of their service-life, the study of ablation dynamics and estimating the optimal thickness of the protecting covering at different locations of the plasma flux action.

Results of the conducted investigations confirm that the developed materials with spirally reinforced filler and the techniques of their processing guarantee the technical and cost-effectiveness of their adoption in manufacturing.

9.5 Conclusions

1. Proceeding from the results of experimental studies a technological process of manufacturing rods on the basis of spirally reinforced fillers has been devised. The process makes it possible to augment the filling degree of the rods to 70%, reduce their diameter and improve strength by 30–40%, thus enabling automatic assembly of casings for multidirectional textures.
2. The developed design and technology for producing measuring elements with the use of spiral reinforcement have brought down the measurement error of ablation in the heat-proof coatings of airplanes 3.2–3.4 times. This fact has given grounds for developing automatic registration instruments to measure the ablation rate of protective coatings.
3. Experience in designing and manufacturing the stiffening elements of load-bearing units using spirally reinforced fillers has proved that the technique lessens the structure's weight, limits waste, reduces labor-

intensity and improves strength characteristics under compression by 30–60%.

4. The developed technological process for conical casings with external stiffening ribs, based on a spirally reinforced filler, made possible rejection of costly equipment, to raise technological effectiveness, cut down costs of expensive reinforcing materials by 6-8% and to improve their bearing capacity by 17–20%.

5. The proposed design and engineering solutions are worthwhile using for the creation of composite structures of different grades and purposes.

General Conclusions and Recommendations

1. A composite material structure based on hybrid spatial-reinforced filler has been developed. Bunches of the main reinforcement fibers are effectively separated inside the filler and they are regularly arranged within the material structure thanks to the main reinforcement winding with the fibers of the auxiliary reinforcement at a pitch that ensures mutual engagement of their coils. Avoidance of the material warping is reached by either double winding of the filler or by using a paired filler whose fiber groups of the main reinforcement are wound in different directions

2. The conducted investigations in microstructure of composites with the spiral-reinforced filler have visualized the interrelation between the material structures and technological parameters of their production. As a result, the calculation procedure has been devised for the main structural characteristics of the materials and their variation range, and to define the kind o the representative element.

3. It has been found out that introduction of interlayers into the composite structure alters its elastic properties and stresses in the bulk. To this end, recommendations have been proposed as for the choice of geometrical parameters, properties of the spiral-reinforced filler components, and for their manufacturing process aimed at elevating characteristics of the material under various types of loading.

4. The developed procedure of estimating elastic properties of the material and stresses in its structure takes into account orthotropic features and geometrical parameters of the auxiliary reinforcement layer. For the case of fine layers a simplified calculation method has been developed and its applicability limits have been evaluated.

5. By optimizing structure of the composites based on hybrid spiral-reinforced fillers under transversal loading it became possible to ascertain structural parameters of the material and its production technology regimes so as to obtain composites with most favorable correlation of elastic and strength characteristics.

6. Regularities observed at variations of structural and geometrical parameters of the spiral-reinforced fillers have been specified in the course of the main reinforcement winding with auxiliary fibers which have made

grounds for formulation of the manufacturing process and the equipment thereof.

7. Experimental investigations of physico-mechanical properties of the composites and fracture processes taking place in them corroborate well with theoretical conclusions that the suggested structure of the composites improves their characteristics at transversal loading, shear and longitudinal compression by 30 to 80% and their specific parameters outdo usual unidirectional composites.

8. It's proved that the optimum structural and geometrical parameters of the developed fillers can't be estimated unambiguously when applied in various structures, therefore it's worthwhile selecting them in each concrete case of an article proceeding from the parameter needed to improve in a given structure at certain loading conditions.

9. A mathematical model of a hybrid spiral-reinforced rod has been worked out which allows for structural and technological features of the material and loading conditions of the element within the whole structure. The conditions of an ideal mechanical and thermo-mechanical contact of the local orthotropic layer with solid bodies have been obtained proceeding from the supposition that the differential equations of the body equilibrium are met only on the median line of the layer. This served the base for a new design method of the stress-strain state (SSS), the effective elastic characteristics and linear thermal expansion coefficient (LTEC) of rod spiral-reinforced elements.

10. The generalized plane problem for a radial-inhomogeneous body has been solved at an axis-symmetrical and temperature loading using the perturbation theory.

11. An algorithm for optimum designing rod elements in compliance with required quality criteria has been derived. For a set of concrete structures optimum structural parameters of rod elements have been obtained. It has been experimentally established that such articles show by 25–30% higher longitudinal elasticity modulus in contrast to unidirectional rods and the delamination threshold is moved aside by 40%.

12. The interrelation between LTEC of the structure and its dimensional parameters has been defined and vistas in regulating the integral LTEC with account of materials used have been visualized.

13. The development of the advanced technological processes for designs incorporating composites on the base of hybrid spiral-reinforced fillers has proved to enlarge productivity and diminish material consumption, improve characteristics and expand listed products. The design and technological solutions put forward in this work are recommended for application at development of diversified composite structures of different purposes.

Addendum

Components of deformation and strength on contours (Section 3.5):

$$\frac{U_r^{(j)}}{R} = \sum_{k=0}^{N_j} \left[a_{j,k} r_j^{2k+1}(\kappa_j - 2k - 1)\cos 2k\Theta + B_{j,k} r^{-2k-1} \right.$$
$$\times (\kappa_j + 2k + 1)\cos(2k + 2)\Theta - C_{j,k} r^{2k+1}\cos(2k + 2)\Theta$$
$$\left. - d_{j,k} r^{-2k-1}\cos 2k\Theta \right];$$

$$\frac{U_\Theta^{(j)}}{R} = \sum_{k=0}^{N_j} \left[a_{j,k} r_j^{2k+1}(\kappa_j + 2k + 1)\sin 2k\Theta - B_{j,k} r^{-2k-1} \right.$$
$$\times (\kappa_j - 2k - 1)\sin(2k + 2)\Theta + C_{j,k} r_j^{2k+1}\sin(2k + 2)\Theta$$
$$\left. + d_{j,k} r_j^{-2k-1}\sin 2k\Theta \right];$$

$$\frac{\sigma_r^{(j)}}{2G_j} = \sum_{k=0}^{N_j} \left\{ - a_{j,k}(4k^2 - 1)r_j^{2k}\cos 2k\Theta + (2k+1)^2 r_j^{-2k-2}B_{j,k}\cos(2k+2)\Theta \right.$$
$$\left. - C_{j,k}(2k + 1)r_j^{2k}\cos(2k + 2)\Theta + (2k + 1)d_{j,k} r_j^{-2k-1}\cos 2k\Theta \right\};$$

$$\frac{\tau_{r\Theta}^{(j)}}{2G_j} = \sum_{k=0}^{N_j} \left\{ Pa_{j,k}2k(2k + 1)r_j^{2k}\sin 2k\Theta - (2k + 1)(2k + 2) \right.$$
$$\times B_{j,k} r_j^{-2k-2}\sin(2k + 2)\Theta + C_{j,k}(2k + 1)r_j^{2k}\sin(2k + 2)\Theta$$
$$\left. + (2k + 1)d_{j,k} r_j^{-2k-1}\sin 2k\Theta \right\}.$$

On contour L_2 in zone S_3:

$$\frac{\sigma_r^{(3)}}{2G_3} = 2\Gamma + 2\alpha_0 - \beta_0\cos 2\Theta - \Gamma'\cos 2\Theta$$
$$+ \sum_{k=0}^{N_3} \left\{ \alpha_{2k+2}R^{2k+2}\left[(2k + 4)\cos(2k + 2)\Theta \right. \right.$$
$$- \sum_{j=0}^{\infty}(2j - 2)r_{j,k}\cos 2j\Theta + \sum_{j=0}^{\infty}(2k + 2)S_{j,k}\cos(2j + 2)\Theta \Big]$$
$$\left. - \beta_{2k}R^{2k+2}\left[\cos 2k\Theta + \sum_{j=0}^{\infty} r_{j,k}\cos(2j + 2)\Theta \right] \right\};$$

$$\frac{\tau_{r\Theta}^{(3)}}{2G_3} = \Gamma' \sin 2\Theta + \beta_0 \sin 2\Theta + \sum_{k=0}^{N_3} \left\{ \alpha_{2k+2} R^{2k+2} \right.$$

$$\times \left[(2k+2)\sin(2k+2)\Theta + \sum_{j=0}^{\infty} 2j r_{j,k} \sin 2j\Theta \right.$$

$$\left. -(2k+2)\sum_{j=0}^{\infty} S_{j,k} \sin(2j+2)\Theta \right]$$

$$\left. +\beta_{2k+2}\left[-\sin 2k\Theta + \sum_{j=0}^{\infty} r_{j,k} \sin(2j+2)\Theta \right] \right\} ;$$

$$\frac{U_r^{(3)}}{R} = \Gamma(\kappa_3 - 1) + (\kappa_3 - 1)\alpha_0 - \Gamma' \cos 2\Theta - \beta_0 \cos 2\Theta$$

$$+ \sum_{k=0}^{N_3} \alpha_{2k+2} R^{2k+2} \left[(\kappa_3 + 2k + 1)\frac{\cos(2k+2)\theta}{2k+1} \right.$$

$$+ \sum_{j=0}^{\infty} r_{j,k}(\kappa_3 - 2j - 1)\frac{\cos 2j\theta}{2j+1} + (2k+2)\sum_{j=0}^{\infty} S_{j,k} \frac{\cos(2j+2)\theta}{2j+1} \Bigg]$$

$$+ \sum_{k=0}^{N_3} \beta_{2k+2} R^{2k+2} \left[\frac{\cos 2k\theta}{2k+1} - \sum_{j=0}^{\infty} r_{j,k} \frac{\cos(2j+2)\theta}{2j+1} \right] ;$$

$$\frac{U_\Theta^{(3)}}{R} = \sum_{k=0}^{N_3} \alpha_{2k+2} R^{2k+2} \left[(\kappa_3 - 2k - 1)\frac{\sin(2k+2)\theta}{2k+1} \right.$$

$$+ \sum_{j=0}^{\infty} r_{j,k} \frac{\kappa_3 + 2j + 1}{2j+1} \sin 2j\theta - (2k+2)\sum_{j=0}^{\infty} S_{j,k} \frac{\sin(2j+2)\theta}{2j+1} \Bigg]$$

$$+ \sum_{k=0}^{N_3} \beta_{2k+2} R^{2k+2} \left[\frac{\sin 2k\theta}{2k+1} - \sum_{j=0}^{\infty} r_{j,k} \frac{\sin(2j+2)\theta}{2j+1} \right] + (\Gamma' + \beta_0)\sin 2\Theta .$$

Set of equations for coefficients (3.48):

$$\sum_{n=j}^{j+1} (-1)^n \left[-a_{n,k}(4k^2 - 1)r_n^{2k} + (2k-1)^2 B_{n,k} r_n^{-2k} \right.$$

$$\left. -C_{n,k-1}(2k-1)r_n^{2k-2} + (2k+1)d_{n,k} r_n^{-2k-2} \right] = 0 ;$$

$$\sum_{n=j}^{j+1} (-1)^n \left[a_{n,k} 2k(2k+1)r_n^{2k} - (2k-1)2k B_{n,k-1} r_n^{-2k} \right.$$

$$\left. +(2k-1)C_{n,k-1} r_n^{2k-2} + (2k+1)d_{n,k} r_n^{-2k-2} \right] = 0 ;$$

$$\sum_{n=j}^{j+1}(-1)^n\Big[a_{n,k}(\kappa_n-2k-1)r_n^{2k+1}+B_{n,k-1}(\kappa_n+2k-1)r_n^{-2k+1}$$
$$-C_{n,k-1}r^{2k-1}-d_{n,k}r_n^{-2k-1}\Big]=0\,;$$

$$\sum_{n=j}^{j+1}(-1)^n\Big[a_{n,k}(\kappa_n+2k+1)r_n^{2k+1}+B_{n,k-1}(2k-\kappa_n-1)r_n^{-2k+1}$$
$$+C_{n,k-1}r^{2k-1}+d_{n,k}r_n^{-2k-1}\Big]=0\,;$$

$$B_{0,k}=d_{0,k}=0;\quad (j=0,1;\ k=0,\dots,N_j)\,.$$

$$\frac{G_2}{G_3}\left[\frac{P_1+P_2}{2}+\frac{\pi}{4\sqrt{3}}\beta_2-\beta_2+2\sum_{n=0}^{N_3}\alpha_{2n+2}r_{0k}\right]=a_{2,0}+d_{2,0}r_2^{-2}\,;$$

$$\frac{G_2}{G_3}\left[\left(4-\frac{\pi}{2\sqrt{3}}\right)\alpha_2-\frac{P_2-P_1}{2}\right.$$
$$\left.+\sum_{n=0}^{\infty}\alpha_{2n+2}(2n+2)S_{0k}-\beta_4-\sum_{n=0}\beta_{2n+2}r_{0k}\right]$$
$$=-3a_{2,1}r_2^2-B_{2,0}r_2^{-2}-C_{2,0}+3d_{2,1}r_2^{-4}\,;$$

$$\frac{G_2}{G_3}\left[\alpha_{2k}(2k+2)-\sum_{n=0}^{N_3}(2n-2)\alpha_{2n+2}r_{nk}+\sum_{n=1}^{N_3}(2n+2)S_{k-1}\alpha_{2n-2}\right.$$
$$\left.-\beta_{2k+2}-\sum_{n=0}^{N_3}\beta_{2n+2}r_{k-1,n}\right]=-(4k^2-1)a_{2,k}r_2^{2k}$$
$$+(2k+1)^2B_{2,k-1}r_2^{-2k-4}-C_{2,k-1}(2k-1)r_2^{2k-2}+(2k+1)d_{2,k}r_2^{-2k-2}$$
$$(k=2,\dots,N_3)\,.$$

$$\frac{G_2}{G_3}\left\{\frac{P_2-P_1}{2}+\frac{\pi}{2\sqrt{3}}\alpha_2+2\alpha_2+2\sum_{n=0}^{N_3}\alpha_{2n+2}[r_{1k}-(k+1)S_{0k}]\right.$$
$$\left.-\beta_2+\sum_{n=0}^{N_3}\beta_{2n+2}r_{0k}\right\}=6a_{1,2}r_2^2-2B_{2,0}r_2^{-2}+C_{2,0}+3d_{2,1}r_2^{-4}\,;$$

$$\frac{G_2}{G_3}\left\{2k\alpha_{2k}+2\sum_{n=0}^{N_3}\alpha_{2n+2}[kr_{kn}-(k+1)S_{k-1,n}]-\beta_{2k-2}+\sum_{n=0}^{N_3}\beta_{2n+2}r_{k-1,n}\right\}$$
$$=2(2k+1)a_{2k}r_2^{2k}-(2k-1)2kB_{2,k-1}r_2^{-2k-4}$$
$$+C_{2,k-1}(2k-1)r_2^{2k-2}+(2k+1)d_{2,k}r_2^{-2k-2}\,;$$
$$(k=2,\dots,N_3)\,.$$

$$(\kappa_3 - 1)\frac{P_1 + P_2}{4} + \frac{\kappa_3 - 1}{4\sqrt{3}}\pi\beta_2 + \sum_{n=0}^{N_3}\alpha_{2n+2}(\kappa_3 - 1)r_{0n} + \beta$$

$$= a_{2,0}(\kappa_2 - 1)(r_2 - d_{2,0}r_2^{-1}\,;$$

$$-\frac{P_2 - P_1}{2} - \frac{\pi}{2\sqrt{3}}\alpha_2 - \alpha_2(\kappa_3 + 1) + \sum_{n=0}^{N_3}\alpha_{2n+2}$$

$$\times\left[r_{1n}\frac{\kappa_3 - 3}{3} + (2n + 2)S_{0n}\right] + \beta_4 - \sum_{n=0}^{N_3}\beta_{2n+2}r_{0n}$$

$$= a_{2,1}(\kappa_2 - 3)r_2^3 + B_{2,0}r_2^{-1}(\kappa_3 + 1) - C_{2,0}r_2 - d_{2,1}r_2^{-3}\,;$$

$$-\alpha_{2k}\frac{\kappa_3 + 2k - 1}{2k - 1} + \sum_{n=0}^{N_3}\alpha_{2n+2}\left[r_{kn}\frac{\kappa_3 - 2k - 1}{2k + 1} + \frac{2n + 2}{2k - 1}S_{k-1,n}\right]$$

$$+\frac{\beta_{2n+2}}{2k + 1} - \sum_{n=0}^{N_3}\beta_{2n+2}\frac{r_{k-1,n}}{2k + 1}$$

$$= a_{2,k}r_2^{2k+1}(\kappa_2 - 2k - 1) + B_{2,k-1}r_2^{-2k+1}(\kappa_3 + 2k - 1)$$
$$-C_{2,k-1}r_2^{2k-1} - d_{2,k}r_2^{-2k-1}\,;$$
$$(k = 2, \ldots, N_3)\,.$$

$$\frac{P_2 - P_1}{2} + \frac{\pi}{2\sqrt{3}}\alpha_2 + \alpha_2(\kappa_3 - 1) + \sum_{n=0}^{N_3}\alpha_{2n+2}$$

$$\times\left[r_{1n}\frac{\kappa_3 + 2}{3} - (2n + 2)S_{0n}\right] + \beta_2 - \sum_{n=0}^{N_3}\beta_{2n+2}r_{0n}$$

$$= a_{2,1}r_2^3(\kappa_3 + 3) + B_{2,0}(1 - \kappa_2)r_2^{-1} + C_{2,0}r_2 + d_{2,1}r_2^{-3}\,;$$

$$\alpha_{2k}\frac{\kappa_3 - 2k + 1}{2k - 1} + \sum_{n=0}^{N_3}\alpha_{2n+2}\left[\frac{\kappa_3 + 2k + 1}{2k + 1} - \frac{2n + 2}{2k - 1}S_{k-1,n}\right]$$

$$+\beta_{2n+2} - \sum_{n=0}^{N_3}\beta_{2n+2}\frac{r_{k-1,n}}{2k - 1}$$

$$= a_{2,k}r_2^{2k+1}(\kappa_2 + 2k + 1) + B_{2,k-1}(2k - 1 - \kappa_2)$$
$$+C_{2,k-1}r_2^{2k-1} + d_{2,k}r_2^{-2k-1}\,;$$
$$(k = 2, \ldots, N_3)\,.$$

Coefficients for (4.7):

$$A^{(1)}_{2k+2} = -\delta(k)\frac{\pi}{2\sqrt{3}} + 2\operatorname{Re}\left\{ z^{-2k-2} + (k+1)\bar{z}z^{-2k-3} \right.$$
$$\left. + \sum_{j=0}^{\infty}[r_{j,k}(z^{2j} - j\bar{z}z^{2j-1}) + (k+1)S_{j,k}z^{2j}] \right\};$$

$$B^{(1)}_{2k+2} = \delta(k)\frac{\pi}{2\sqrt{3}} - \operatorname{Re}\left\{ z^{-2k-2}\sum_{j=0}^{\infty}r_{j,k}z^{2j} \right\};$$

$$A^{(2)}_{2k+2} = \delta(k)\frac{\pi}{2\sqrt{3}} + 2\operatorname{Re}\left\{ z^{-2k-2} + (k+1)\bar{z}z^{-2k-3} \right.$$
$$\left. + \sum_{j=0}^{\infty}[r_{j,k}(z^{2j} - j\bar{z}z^{2j-1}) - (k+1)S_{j,k}z^{2j}] \right\};$$

$$B^{(2)}_{2k+2} = \delta(k)\frac{\pi}{2\sqrt{3}} + \operatorname{Re}\left\{ z^{-2k-2}\sum_{j=0}^{\infty}r_{j,k}z^{2j} \right\};$$

$$A^{(3)}_{2k+2} = 2\operatorname{Im}\left\{ -(k+1)\bar{z}z^{-2k-3} + \sum_{j=0}^{\infty}[jr_{j,k}\bar{z}z^{2j-1} - (k+1)S_{j,k}z^{2j}] \right\};$$

$$B^{(3)}_{2k+2} = \operatorname{Im}\left\{ z^{-2k-2}\sum_{j=0}^{\infty}z^{2j}r_{jk} \right\};$$

$$\left\{\begin{matrix} A^{(4)}_{2k+2} \\ A^{(5)}_{2k+2} \end{matrix}\right\} = -\delta(k)\frac{\pi}{2\sqrt{3}}\frac{\operatorname{Re}}{\operatorname{Im}}\{\bar{z}\} + \frac{\operatorname{Re}}{\operatorname{Im}}\left\{ -\kappa_3\frac{z^{-2k-1}}{2k+1} - z\bar{z}^{2k-2} \right.$$
$$\left. + \sum_{j=0}^{\infty}\left[\left(\kappa_3\frac{z^{2j+1}}{2j+1} - z\bar{z}^{-2j}\right)r_{j,k} + (2k+2)S_{j,k}\frac{\bar{z}^{2j+1}}{2j+1}\right] \right\};$$

$$\left\{\begin{matrix} B^{(4)}_{2k+2} \\ B^{(5)}_{2k+2} \end{matrix}\right\} = \delta(k)(\kappa_3 - 1)\frac{\pi}{4\sqrt{3}}\frac{\operatorname{Re}}{\operatorname{Im}}\{z\} + \frac{\operatorname{Re}}{\operatorname{Im}}\left\{ \frac{\bar{z}^{2k-1}}{2k-1} - \sum_{j=0}^{\infty}r_{j,k}\frac{z^{2j+1}}{2j+1} \right\};$$

$$\left\{\begin{matrix} C^{(4)} \\ C^{(5)} \end{matrix}\right\} = \frac{\operatorname{Re}}{\operatorname{Im}}\{-v_3 z\}; \quad \delta(k) = \left\{\begin{matrix} 0, & k \neq 0 \\ 1, & k = 0 \end{matrix}\right\}.$$

References

1. United States Patent 3, 565, 127. Inextensible filamentary structures and fabrics woven there from, Doyle C. Nicely, Samutl J. Davis, Patented Feb. 23, 1971.
2. Jeronimidis, G. (1978) A biological way of preventing crashes, New scientist, 79, No. 1111, 112.
3. Katsube, N. (1995) Estimation of effective elastic module for composites, J. Solids and Struct., 32, No. 1, 79–88.
4. Kendall, K. (1976) Interfacial cracking of a composite. Part 2. Bending, J. of Mater. Sci., No. 11, 1263–1266.
5. Chamis, C.C., Sinclair, G.H. (1979) Mech. of entropy hybrid composites – properties, analysis and design. Techn. Proc. 34th Ann. Conf. Reinf. Plast. Compos. Inst. Reinf. Future, New Orleans. La, New York.
6. Guild, F.J. and Summerscales, J. (1998) Quantitative microstructural analysis for continuous fiber composites, Chapter 6 in J Summerscales (ed.), Microstructural Characterization of Fiber-Reinforced Compos., Woodhead Publishing, Cambridge, 179–203.
7. Poss, A.L. (1975) Designing with three directional composites, Mech. Eng., No. 4, 32–37.
8. Wagner, H.D. (1996) Solution of the thermal residual stress problem in composites with anisotropic interphases, Physical Review B, 53 (9), 5055–5058.
9. Zhou, X.-F., Wagner, H.D. (1999) Stress concentrations caused by fiber failure in two-dimensional composites, Compos. Sci. and Technol., 59 (7), 1063–1071.
10. Mechanics of Composite Materials, Ed. J. Sendetski (1978), Moscow, Mir, 564
11. Shun-Fa Hwang and Ching-Ping Mao (2001) Failure of delaminated interply hybrid composites plates under compression, Compos. Sci. and Technol., 61, 11, 1513-1527.
12. Tomblin, J.S., Barbero, E.J. (1997) Statistical Microbuckling Propagation Model for Compressive Strength Prediction of Fiber Reinforced Composites, ASTM STP, 1242.
13. Wagner, H.D. (1989) Statistical concepts in the study of fracture properties of fibers and composites, in application of fracture mechanics to composite mater. (K. Friedrich, Ed., R.B. Pipes, Compos. Mater. Series Ed.), Compos. Materials. Series 6, Elsevier Sci. Publishers B.V., Amsterdam, 1989, pp. 39–77.
14. Wagner, H.D., Roman, J., Marom, G. (1982) Hybrid effects in the bending stiffness of graphite/glass reinforced composites, J. Mater. Sci., 17, No. 5, 1359–1363.
15. Xie, Y.J., Yan, H.G., Liu, Z.M. (1996) Buckling optimization of hybrid-fiber multilayer-sandwich cylindrical shells under external lateral pressure, Compos. Sci. and Technol., 56, 12, 1349–1353.

16. Yokozeki, Tomohiro, Aoki, Takahira and Ishikawa, Takashi (2002) Fatigue growth of matrix cracks in the transverse direction of CFRP laminates, Compos. Sci. and Technol., 62, 9, 1223–1229.

17. Fuwa, M., Bunsell, A.R., Harris, B. (1975) Tensile failure mechanisms in carbon fiber reinforced plastics, J. Mater. Sci., 10, No. 12, 2062–2070.

18. Huesgen, R. (1997) Flexural behavior of ductile hybrid FRP rebars in singly reinforced concrete beams, MSc Thesis, Department of Civil and Architectural Eng., Drexel University, Philadelphia, PA, 1997.

19. Kendall, K. (1976) Interfacial cracking of a composite. Part 3. Compression, J. of Mater. Sci., No. 11, 1267–1269.

20. Slower, E.A., Lin, T.S. (1967) On mechanical behavior of fiber reinforced crystalline materials, J. Mech. and Phys. of Solids, 9, No. 4, 242-246.

21. Thorat, H.T., Lakkad, S.C. (1983) Fracture toughness of unidirectional glass/carbon hybrid composites, J. Compos. Mater., 17, No. 1, 2–4.

22. Wagner, H.D., Marom, G. (1982) On composition parameters for hybrid composites materials, Compos., 13, 18–20.

23. Anifantis, N.K. (2000) Micromechanical stress analysis of closely packed fibrouscomposites, Compos. Sci. and Technol., 60, 8, 1241–1248.

24. Gudmundson, P., Alpman, J. (2000) Initiation and growth criteria for transverse matrix cracks in composite laminates, Compos. Sci. and Technol., 60, 2, 185–195.

25. Miska, K.M. (1978) Hybridizing expands properties, cuts cost of advanced composites, Mater. Eng., 88, No. 2, 35–37.

26. Barbero, E.J. (2000) Compressive strength of composites for aircraft structures, World Market Series Business Briefing-Global Aerospace Technol., London, 96–100.

27. Tandon, G.P. (1995) Use of composite cylinder model as representative volume element for unidirectional fiber composite, Compos. Mater., 29, No. 3, 388–403.

28. Tsai, C.-L., Daniel, I.M. (1994) Method for Thermo-Mechanical Characterization of Single Fibers, Compos. Sci. and Technol., 50, 7–12.

29. Wu By, E.M. (1978) Phenomenological deterioration criteria for anisotropic media, Mech. of Compos. Mater., V.2, Editor J. Senedtski, Moscow, Mir.

30. Zweben, C. (1977) Tensile strength of hybrid composites, J. of Mater. Sci., 12, No. 7, 1325–1337.

31. Halm, H.T. (1976) Residual stress in polymer matrix composite laminates, J. Compos. Mater, 10, No. 4, 266–278.

32. Wisnom, M.R. (1993) Analysis of shear instability in compression due to fiber waviness, J. Reinf. Plast. and Compos., 12, No. 11, 1171–1189.

33. Barbero, E.J., Raftoyiannis, I. (1993) Local buckling of FRP beams and columns, ASCE J. of Mater. in Civil Eng., 5(3), 339–355.

34. Barbero, E.J., Tomblin, J. (1996) A damage mechanical model for compression strength of composites, I.J. Solid Struct., 33(29), 4379–4393.

35. Schipperen, J.H. A. (2001) An anisotropic damage model for the description of transverse matrix cracking in a graphite–epoxy laminate, Composite Struct., 53, 3, 295-299.

36. Wood, A. (1979) Compoestos hybridos, Rev. Plast. Mod., 38, No. 277, 98–101.

37. Dorey, G., S., G.R., Mutchings, J. (1978) Impact properties of carbon fiber / Kevlar – 49 fiber hybrid composites, Compos., 9, No. 1, 25–32.

38. Morozov, E.V. (2000) Thermoelasticity of spatially reinforced composite plates, Composite Struct., 48, 1-3, 129-133.

39. Zhou, X. -F. and Wagner, H.D. (2000) Fragmentation of two-fiber hybrid micro composites: stress concentration factors and interfacial adhesion, Compos. Sci. and Technol., 60, 3, 367–377.

40. Barbero, E.J., GangaRao, H.V. S. (1992) Structural applications of composites in infrastructure, Part II, SAMPE J., 28(1), 9–16.

41. Bleay, S.M., Humberstone, L. (1999) Mechanical and electrical assessment of hybrid composites containing hollow glass reinforcement, Compos. Sci. and Technol., 59, 9, 1321–1329.

42. Delneste, L. (1983) Designing for trees in plastics, Eng. Mater. and Des, 27, No. 9, 52–60.

43. Gaudenzi, P. (1997) On delamination buckling of composite laminates under compressive loading, Compos. Struct., 39, 1–2, 10, 21–30.

44. Mc Crath, J.A., Rheaume, W.A., Campman, A.R. (1967) The weaving of three – dimensional fabrics for the aerospace industry, 12th Nat. SAMPE Symp., 1967, p. X/P1.

45. Scudra A.M., Bulavs F. Ya. (1982) Strength of Reinforced Plastics. Moscow, Khimiya, p. 213.

46. Berglund, L.A., Asp, L.E., Talreg, R. (1996) Prediction of matrix-initiated transverse failure in polymer composites, Compos. Sci. and Technol., 56, 9, 1089–1097.

47. Delneste, L., Perez, B. (1983) An interlastic finite element model of 4D carbon – carbon composites, Carbon Compos., AIAA J., 21, No. 8, 1143–1149.

48. Luciano, R., Barbero, E.J. (1995) Formulas for the stiffness of composites with periodic microstructure, Int. J. of Solids Struct., 31–21), 2933–2944.

49. Lomakin V.A. (1976) Theory of Elasticity of Nonuniform Bodies, Moscow, MGU, p. 367.

50. Ogihara, Shinji, Reifsnider, K.L. (2002) Characterization of nonlinear behavior in woven composite laminates, Appl. Compos. Mater., 9 (4), 249–263.

51. Wei, C.Y., Li, R, Ding, L (1988) Tensile properties of GF/CF intra-hybrid reinforced plastics, FRP, Compos., No. 1, 6–10.

52. Rabotnov Yu. N. (1979) Strength of laminated composites, News of AS USSR, Mech. of Solid Body, 1,113–119.

53. Berglund, L.A., Asp, L.E., Talreg, R. (1996) Effects of fiber and interphase on matrix-initiated transverse failure in polymer composites, Compos. Sci. and Technol., 56, 6, 657–665.

54. Fukuda, H.and Kawata, K. (1983) A Monte Carlo simulation of the strength of laminated hybrid composites, Trans. Japan Soc. Aero. Space Sci., 25, No. 70, p. 213.

55. Wagner, H.D. (1986) Elastic Response of FibrousCompos. Mater. with Weak Bonding, Competes Rendus de l'Academie des Sci.s (Paris) Serie II, 14, 1283–1288.

56. Seong Sik Cheon, Tae Seong Lim, Dai Gil Lee (1999) Impact energy absorption characteristics of glass fiber hybrid composites, Composite Struct., 46, 3, 267-278.

57. Vinson, J.R., Chou, T.W. (1975) Compos. Materials and their Use in Structures, London, 1975, XII, 438.

58. Adali, S., Verijenko, V.E. (1997) Minimum cost design of hybrid composites. Cylinders with temperature dependent properties, Compos. Struct., 38, 1–4, 8, 623–630.

59. Papanicolaou, G.C., Michalopoulou, M.V., Anifantis, N.K. (2002) Thermal stresses in fibrous composites incorporating hybrid interphase regions, Compos. Sci. and Technol., 62, 14, 1881–1894.

60. Lekhnitsky S.G. (1981) Torsion of Anisotropic Rods. Moscow, Nauka, p. 375.

61. Mazumdar, S.K. (2002) Composites Manufacturing: Materials, Product, and Process, CRC Press.

62. Phillips, M.G. (1981) Composition parameters for hybrid composites materials, Compos., 12, No. 2, 113–116.

63. Miyairi Hirao, Nagai Masahiro, Muramatsu Atsuyoshi (1979) Fatigue strength of FRP hybrid construction GFRP/GFRP hybrid construction, Proc. of 19th Jap. Congress of Mater. Research, 1979, Keota, p. 189–194.

64. Rhodes, J. (1996) A semi-analytical approach to buckling analysis for composite structure, Compos. Struct., 35, 1, 93–99.

65. Kantorovich L.V. (1934) About a method of solving differential equations in particular derivatives. Reports of AS SSSR, 2, No. 9, 532–535.

66. Influence de la distribution spatiale' des fibres au sein d'un minicomposite sur., Le comportement mecanique en traction: Journees Sci. et techn. "Micromec. et mec. Endommagement compos. ", 1995, No. 1, 108–109.

67. Piggott, M.R., Harris, B. (1981) Compression strength of hybrid fiber – reinforced plastics, J. Mater. Sci., 16, No. 3, 687–693.

68. Marissen, R. (1983) ARALL (Aramidfaserstarker Aluminium – Schicht – verbundverkstoff) – Ein neuer Hybrid – Verbundwerk – stoff mit besonderen Schwingfestigkeitseigenschaften, Werkstofftechnik, 14, No. 8, 278–283.

69. Duvaut, G., Terrel, G., Léné F., Verijenko, V.E. (2000) Optimization of fiber reinforced composites, Compos. Struct., 48, 1–3, 83–89.

70. Hill, R.G. (1964) Theory of mechanical properties of fiber – strengthened materials, J. Mech. Phys. Sol., 12, No. 4, 199–212.

71. Khatri, S.C. and Koczak, M.J. (1996) Thick-section AS4-graphite/E-glass/PPS hybrid composites: Part I. Tensile behavior, Compos. Sci. and Technol., 56, 2, 181–192.

72. O'Brien, T.K., Salpekar, S.A. (1993) Scale effects on the transverse tensile strength of graphite/epoxy composites, Compos. Mater.: Testing and Design (Eleventh Volume), ASTM STP 1206, Camponeschi, E.T., Jr, Ed., American Society for Testing and Mater., Philadelphia, 1993, pp. 23–52.

73. Wagner, H.D., Nairn, J.A., Detassis, M. (1995) Toughness of Interfaces from Initial Fiber-Matrix Debonding in a Single(FiberCompos. Fragmentation Test, Appl. Compos. Mater., 2 (2), 107–117.

74. Xing, J., Hsiao, G.C., Chou, Tsu-Wei (1981) A dynamic explanation of the hybrid effect, J. Compos. Mater., 15, No. 9, 443–461.

75. Freger, G.E., Bakst, E.E., Freger, D.G. (1996) Technologia wykonywania zbrojonych ukladow silowych zlizonej konstrukcji. Seminarium pt: Wybrane technologie obrobki plastycznej, Poznan, Maja 20–21, 1996, p. 75 – 78.

76. Gunnink, J.W., Vlot, A.T., de Vries, J., van der Hoeven, W. (2002) Glare technological development 1997–2000, Appl. Compos. Mater., 9 (4), 201–219.

77. Hatiashwili G.M. (1983) Almanzi-Mitchell Tasks for Uniform and Composite Bodies, Tbilisi, Mentsiereba, p. 236.

78. Le Riche, R. and Gaudin, J. (1998) Design of dimensionally stable composites by evolutionary optimization, Compos. Struct., 41, 2, 97–111.

79. Chamis, C.C., Lark, R.F., Sinelair, G.H. (1981) Mechanical property characterization of entropy hybrid composites, Test Meth. and Des. Allowables Fibrous Compos. Symp., No. 8, 261–280.

80. Engineering Plastics, Ed. Trostyanskaya E.B. (1974), Moscow, Khimiya, p. 304.

81. Piggott, M.R. (1981a) A theoretical framework for the compressive properties of aligned fiber composites, J. Mater. Sci., 16, No. 10, 2837–2845.

82. Chou, T.W., Fukuda, H. (1981) Stiffness and strength of hybrid composites, Compos. Mater: Mech., Mech. Prop. and Fabr. Jap.–US Conf., Tokyo, Jan.12–14, 1981, Barking: 1981, 78–88.

83. Galveston, G., Sillwood, G.M. (1976) Synergistic fiber strength in hybrid composites, J. of Mater. Sci., 11, No. 10, 1877–1885.

84. Averston, G., Kelly, A. (1978) Tensile first cracking strain and strength of hybrid composites and laminates, New fibers and their composites, 18–19 May, Philos. Trans. Roy. Soc. London: 1979, A 294, No. 1411, 519–534.

85. Yavin, B., Wagner, H.D. (1993) Micromechanical measurements of interfacial adhesion in E-glass/epoxy composites, Adv. Compos. Letters, 2 (2), 47–50.

86. Wang, J., Karihaloo, B.L. (1997) Matrix crack-induced delamination in composite laminates under transverse loading, Compos. Struct., 38, 1–4, 8, 661–666.

87. Aiello, M.A. and Ombres, L. (1996) Maximum buckling loads for asymmetric thin hybrid laminates under in-plane and shear forces, Composite Struct., 36, 1–2, 10, 1–11.

88. Pelekh B.N., Fleishman F.N. (1984) Effective Modulus of disperse filled structures with thin interphase layers, Reports of AS USSR, A, 6, 53–56.

89. Wagner, H.D., Marom, G., Harris, B. (1993) Microphenomena in Advanced-Compos., Elsevier Appl. Sci., London and New York, 1993.

90. Marom, G., Fischer, S., Tuler, F.R., Wagner, H.D. (1978) Hybrid effects in composites: conditions for positive or negative effect versus rule – of mixtures behavior, J. of Mater. Sci., 13, No. 7, 1419–1426.

91. Favre, J.-P. (1977) Improving the facture energy of carbon fiber – reinforced plastics by delaminating prom, J. Mater. Sci., 12, No. 1, 43–50.

92. Timoshenko S.P. (1967) Vibration in Engineering, Moscow, Nauka, p. 444.

93. Daniel, M., Ishai, O. (1994) Eng. Mech. of Compos. Mater., Oxford University Press, New York.

94. Maistre, M.A. (1976) Development of a 4-D reinforced carbon – carbon composites, A1AA Pap., No. 607, 1–5.

95. Wei, C.Y. and Zeng, H. (1987) Hybrid effects of hybrid reinforced composites, Appl. of Eng. Plastics, No. 3, 25–32

96. Morley, I.J. (1979) Composite Materials: designing for structural integrity, Contemporary Physics, 20, No. 3, 257–292.

97. Adali, S., Dufiy, K.J. (1990) Optimal design of asymmetric hybrid laminates against thermal buckling, Therm. Stresses, 13, No. 1, 57–71.

98. Lu, X., Liu, D. (1992) Interlayer shear slip theory for cross-ply laminates with nonrigid interfaces, AIAA J.., 30, No. 4, 1063–1073.

99. Manderrs, P.W., Bader, M.G. (1981) The strength of hybrid glass / carbon fibercomposites. Part 1. Failure strain enhancement and failure mode, J. Mater. Sci., 16, No. 8, 2233–2245.

100. Manderrs, P.W., Bader, M.G. (1981) The strength of hybrid glass / carbon fibercomposites. Part 2. A statistical model, J. Mater. Sci., 16, No. 8, 2246–2256.

101. Pegoretti, A., DellaVolpe, C., Detassis, M., Wagner H.D., Migliaresi, C. (1996) Thermomechanical behavior of the interfacial region in carbon-fiber/epoxy composites, Compos., Part A, 27A, 1067–1074.

102. Huang, Y., Hwang, K.C., Hu, K.X., Chandra, A. (1995) A unified energy approach to a class of micromechanics models for composite materials, Acta Mechanica Sinica, 11, 1, 59–75.

103. Park, Rohchoon, Jang, Jyongsik (1998) The effects of Hybridization on the mechanical performance of aramid/polyethylene intraply fabric composites, Compos. Sci. and Technol., 58, 10, 1621–1628.

104. White, M.J., Heightional, C.N., Dirlihg, R.B. (1979) Thermostructural design of a carbon – carbon heatshield for a Jovian entry, A1AA/NASA Conf. Adv. Texhnol. Future Space Syst., 1979, Hampton, Collect. Techn. Pap. 147–149.

105. Arrington, M., Harris, B. (1978) Some properties of mixed fiber CFRP, Compos., 9, No. 3, 149–152.

106. Scola, D.A., Roylance, M.E. (1978) The effect of process variables on the dry and wet shear strength of fiber reinforced polysulfone composites, 23rd Nat. SAMPE Symp. and Exhibit, Anaheim, Calif., 1978, Vol. 23, Azusa, Calif., 1978, 950-979

107. Stowell, E.A., Lin, T.S. (1967) On mechanical Behavior of Fiber-Reinforced Crystalline Mater., J. Mech. and Phys. of Solidas, 9, No. 4, 242-246.

108. Takahashi, J., Kemmochi, K., Watanabe, J., Fukuda, H., Hayashi, R. (1995) Development of ultra-high temperature testing equipment and some mechanical and thermal properties of advanced carbon/carbon composites, Adv.Compos. Mater., 5, No. 1, p. 73.

109. Wagner, H.D., Eitan, A. (1993) Stress concentration factors in two-dimensional composites: effects of material and geometrical parameters, Compos. Sci. and Technol., 46 (4), 353–362.

110. Barbero, E.J. (1998) Prediction of compression strength of unidirectional polymer matrix composites, J. Compos. Mater., 32, No. 5, 483–502.

111. Grigolyuk E.I., Filshtinsky L.A. (1970) Perforated Plates and Shells, Moscow, Nauka, 67.

112. Huang, Y., Hu, K.X. (1995) A generalized self-consistent mechanical method for solids containing elliptical inclusions, ASME J. of Applied Mech., 62, 566–572.

113. Olivier Siron, O., Pailhes, J., Lamon, J. (1999) Modeling of the stress/strain behavior of a carbon/carbon composites with a 2.5 dimensional fiber architecture under tensile and shear loads at room temperature, Compos. Sci. and Technol., 59, 1, 1–12.

114. Pardoen, G.C. (1975) Improved structural analysis technics for orthogonal weave carbon – carbon materials, A1AA Pap., 13, No. 6, 756–761.

115. Schmitt-Thomas, Kh. G., Yang, Zhen-Guo, Malke, R. (1997) Failure behavior and performance analysis of hybrid-fiber reinforced PAEKCompos. at high temperature, Compos. Sci. and Technol., 58, 9, 1509-1518.

116. Wagner, H.D. (1995) Residual stresses in micro composites and macro composites, J. of Adhesion, 52, 131–148.

117. Kliger, H.S. (1978) Designing with carbon fiber hybrid composites, 23rd Nat. SAMPE Symp. and Exhibit. – Anaheim., Calif., 1978, 23, Azusa, Calif., 1978, p. 10–20.

118. Theocaris, P.S., Paipetis, S.A., Stassinakis, C.A. (1978) Effect of geometry and imperfect in composite systems with limiting shear properties, Fiber Sci. and Technol., 11, No. 5, 335-352.

119. Fukuda H., Kawata, K. (1984) Comparison of interply and intraply hybrid composites from the viewpoint of stress concentration, Theoretical and Applied Mech. (NCTAM), 32, p. 459.

120. Deng, Shiqiang, Ye, Lin, Mai, Yiu-Wing (1999) Influence of fiber cross-sectional aspect ratio on mechanical properties of glass fiber/epoxy composites, I. tensile and flexure behavior, Compos. Sci. and Technol., 59, 9, 1331–1339.

121. Fu, Shao-Yun, Xu, Guanshui, Mai, Yiu-Wing (2002) On the elastic modulus of hybrid particle/short-fiber/polymer composites, Compos., Part B: Eng., 33, 4, 291–299.

122. Fukuda, H. (1984) An advanced theory of the strength of hybrid composites, J. Mater. Sci., 19, No. 3, p. 974.

123. Shum, D.K. M. and Huang, Y. (1990) Fundamental Solutions for Microcracking Induced by Residual Stress,Eng. Fracture Mech., 37, 1, 107-117.

124. Wagner, H.D., Gallis, H.E., Wiesel, E. (1993) Study of the interface in kevlar 49/epoxy composites by means of them microbond and the fragmentation tests: effects of materials and testing variables, J. of Mater. Sci., 28 (8), 2238–2244.

125. Wagner, H.D. (1995) Interfaces in micro composites and macro composites: The issue of thermal residual stresses, Compos. Interfaces, 2(5), 321–336.

126. Abu-Farsakh, G.A., Abdel-Jawad, Y.A., Abu-Laila, Kh. M. (2000) Micromechanical characterization of tensile strength of fiber composite materials, Mech. of Compos. Mater. and Struct., 7, 1, 105–122.

127. Chiu, C.H., Tsai, K. -H., Huang, W.J. (1999) Crush-failure modes of 2D triaxially braided hybrid composite tubes, Compos. Sci. and Technol., 59, 11, 1713–1723.

128. Huang, Y., Hu, K.X., Wei, X., Chandra, A. (1994) A generalized self-consistent mechanical method for composite materials with multiphase inclusions, J. of the Mech. and Physics of Solids, 42, 491–504.

129. Incardona, S., Migliaresi, C., Wagner, H.D., Gilbert, A.H., Marom, G. (1993) The mechanical role of the fiber-matrix transcrystalline interphase in carbon fiber reinforced, J. Polymer Micro Compos., Compos. Sci. and Technol., 47 (1), 43–50.

130. Malmeister A.K., Tamuz V.P., Teters G.A. (1980) Resistance of Polymeric and Composite Materials. Riga, Zinatne, p. 571.

131. Rossettos, J.N., Sakkas, K. (1991) Effect of fiber modulus variations on stress concentration in hybrid composites, AIAA J., 29, No. 3, 482–484.

132. Vanaja, A., Rao, R.M. (2002) Fiber Fraction Effects on Thermal Degradation Behavior of GFRP, CFRP and hybrid composites, J. of Reinforced Plastics and Compos., 21, No. 15, 1389–1397.

133. Chou Tsu-wei, Kelly, A. (1980) Mechanical properties of composites, Ann. Rev. Mater. Sci., Vol. 10 Palo Alto, Calif., 229–259.

134. Chou Tsu-wei, Nomura, S. (1980) On the thermomechanical behavior of short fiber and hybrid composites, Adv. Compos. Mater. Proc. 3rd Int. Conf., Paris, Aug.26–29, 1980, Vol. 1. – Oxford, 1980, 69–80.

135. Cosmodamiansky A.S. (1975) Plane Task of Theory of Elasticity for Plates with Holes, Openings and Bulges. Kiev, Vischa Shkola.

136. Hayashi, T. (1972) On the improvement of mechanical properties of composites by hybrid composition, Proc. 8th Int. Reinf. Plast. Conf., 1972, 149–152.

137. Hill R.G. (1965) Theory of mechanical properties of fiber – strengthened materials, J. Mech. Phys. Sol., 13, No. 4, 189–198.

138. Hill, R.G. (1952) The elastic behavior of a crystalline aggregate, Proc. Phys. Soc., 65, No. 389, 349–354.

139. Abu-Farsakh, G.A., Barakat, S.A., Abed, F.H. (1999) A macromechanical damage model of fibrous laminated composites, Appl. Compos. Mater, 6 (2), 99–119.

140. Schwartz, P., Rosensaft, M., Wagner, H.D. (1985) The Effects of Filament Diameter Variability on the Failure of Kevlar 49/Epoxy Strands, J. of Mater. Sci. Letters, 4, 1409-1411.

141. Che, F.C., Young, K. (1977) Inclusions of arbitrary shape in an elastic medium, J. ofMath. Physics, 18, 1412–1416.

142. Shun-Fa Hwang and Ching-Ping Mao (1999) The delamination buckling of single-fiber system and interply hybrid composites, Composite Struct., 46, 3, 279-287.

143. Kim Youngchan, Julio, D.F., Barbero, E.J. (1996) Progressive failure analysis of laminated composite beams, J. Compos. Mater, 30, No. 5, 536–558.

144. Developments in Reinforced Plastics. (1982) Vol. 2: Properties of Laminates, Ed. G. Pritchard. – London; New York: Appl. Sci. Publ., IX, 183 p.

145. Fukuda, H., Itohiya, G., Watanabe, J., Takahashi, J., Kemmochi, K. (1998) Effect of thermal shock on interlaminar shearing strength of woven C/C composites, Proc. 8th Japan-U.S. Conf. on Compos. Mater. 1998, Baltimore, pp. 427–436.

146. Kirk, J.N., Munro, M., Beaunont, R.W. (1978) The fracture energy of hybrid carbon and glass fiberCompos., J. Mater. Sci., 13, No. 10, 2197–2204.

147. Park, R., Jang, J. (1998) Hybridization effects on the mechanical performance of aramid/polyethylene intraply fabric composites, Compos. Sci. and Technol., 58, 1621–1628.

148. Rubinstein, A.A. (1998) The fracture analysis of Compos. by a micromechanical approach, Compos. Sci. and Technol., 58, 11, 1785-1792.

149. Anstice, P.D., Beanmont, Peter, W.R. (1983) Direct observations of the failure and toughness of hybrid composites, J. of Mater. Sci. Letters, 2, No. 10, 617–622.

150. Freger, G.E. (1983) Calculation of elasticity characteristics of spirally reinforced materials with account of anisotropy of interlayers, Mech. of Reinforced Plastics, Riga, Polytechnic Institute, 51–60.

151. Marom, G., Harel, H., Neumann, S., Friedrich, K., Schulte, K., Wagner H.D. (1989) Fatigue behavior and rate-dependent properties of aramid fiber/carbon fiber hybrid composites, Compos., 20, 537–544.

152. Zhao, Y.H., Weng, Q.J. (1990) Effective elastic module of ribbon-reinforced composites, J. Appl. Mech., 57, No. 1, 158–167.

153. Harris, B., Bunsell, A.R. (1979) Impact properties of glass fiber / carbon fiber hybrid composites, Compos., No. 9, 197–201.

154. Manderrs, P.W., Bader, M.G., Chou Tsu-Wei (1982) Monte Carlo Simulation of the strength of composite fiber bundles, Fiber Sci. and Technol., No. 17, 183–204.

155. Fukuda, H., Chou, T.W. (1983) Stress concentrations in a hybrid composites sheet, J. Appl. Mech., 50, No. 4, p. 845.

156. Bolotin, V.V. (1976) Stochastic models of cumulative damage in composite materials, Eng. Fracture Mech., 8, 103–113.

157. Davalos, J.F., Qiao, P., Barbero, E.J. (1996) Multiobjective material architecture optimization of pultruded FRP I-beams, Compos. Struct., 35, 3, 271–281.

158. Donguy, P., E., R.A. (1977) Ultra-simple carbon – carbon nozzle demonstration, AJAA Pap., No. 1254, 1–6.

159. Di Sciuva, M., Gherlone, M., Lomario, D. (2003) Multiconstrained optimization of laminated and sandwich plates using evolutionary algorithms and higher-order plate theories, Composite Struct., 59, 1, 149–154.

160. Perry, J.L., Adams, D.F. (1976) Mechanical tests of a three – dimensionally – reinforced carbon – carbon composite material, Carbon, 14, No. 1, 61–70.

161. Moh, J.S., Hwu, C. (1997) Optimization for buckling of composite sandwich plates, AIAA J., 5, 35, 863–868.

162. Chaudhari, R.A., Garala, H.J. (1995) Analytical /experimental evaluation of hybrid commingled carbon/glass/epoxy thick-section composites under compression, Compos. Mater, 29, No. 13, 1695–1718.

163. Horgan, C.O. (1982) Saint-Venant s principle in anisotropic elasticity theory, Coilog. Int. CNRS, No. 295, 19–22.

164. Sprecher, N. (1982) New practical payoffs for high-tech RPs., Mod. Plast. Int., 12, No. 5, 90–92.

165. Hofer, K.E., Porte, R. (1978) Influence of moisture on the impact behavior of hybrid glass / graphite / epoxy composites, Elastom. and Plast, 10, No. 3, 271–281.

166. Hitchen, S.A. and Kemp, R.M. J. (1996) Development of novel cost effective hybrid ply carbon-fiber composites, Compos. Sci. and Technol., 56, 9, 1047–1054.

167. Chamis, C.C., Lark, R.F. (1977) Hybrid composites – state of the art review, analysis, design, application and fabrication, Collect. Techn. Pap. Struct. and Mater., Vol. A, 311–331.

168. Bunsell, A.R., Harris, B. (1974) Hybrid carbon and glass fiber composites, Compos., 5, No. 4, 157–164.

169. Sekine, H., Yagishita, Y. (1998) Computational simulation of interlaminar crack extension in angle-ply laminates due to transverse loading, J. Compos. Mater., 32, No. 8, 744–755.

170. Goree, J.Q., Kaw, A.K. (1990) Effects of interleaves on fracture of laminated composites. Part. I. Analysis, J. Appl. Mech., 57. No. 1, 168–174.

171. Gunyajev G.M. (1981) Structure and Properties of Polymer Fiber Composites. Moscow, Khimiya, p. 232.

172. Harris, H.G., Somboonsong, W., Ko, F.K. (1997) A new ductile hybrid fiber reinforced polymer (FRP) reinforcement for concrete structures, Proceedings of the 1997 International Conference on Eng. Mater., June 8-11, Ottawa, Canada, Vol. I, pp 593–604.

173. Ko, F.K., Krauland, K., Scardino, F. (1983) Weft Insertion Warp Knit for Industrial Applications: hybrid composites, J. of Industrial Fabrics, 1, No. 3, 26.

174. Wood, J.R., Wagner, H.D., Marom, G. (1996) The compressive fragmentation phenomenon: using micro composites to evaluate thermal stresses, single fiber compressive strengths, Weibull parameters and the interfacial shear strengths, Proceedings of the Royal Society, London, A452 (1996), 235–252.

175. Short, D., Summerscales, J. (1979) Hybrid – a review. Part 1. Techniques, design and construction,Compos., 10, No. 4, 215–221.

176. Copeland, C.W., Zadoo, D.P. (1990) Hydrothermal influence on the flexural properties of kevlar-graphite/epoxy hybrid composites, J. of Reinf. Plastics and Compos., No 6, 602-603.

177. Rubicki, E., Kanninen, M. (1977) Fracture and fracture mechanic of non-metallic hybrid composites, hybrid and select metal-matrix composites state – Art Rev., New York, 1977, p. 53–65.

178. Sutherland, L.S., Shenoi, R.A., Lewis, S.M. (1999) Size and scale effects in composites: Literature review, Compos. Sci. and Technol., 59, 2, 209–220.

179. Adsit, N.R., Carnahan, K.R., Green, G.E. (1972) Mechanical behavior of three-dimensional composite ablative materials, Compos. Mater.: Testing and Design. ASTM STP 497, Philadelphia, 107–120.

180. Tanimoto, T. (1994) Suppression of interlaminar damage in carbon/epoxy laminates by use of interleaf layers, 2nd Int. Workshop Interfaces, Santiago, Sept. 27–29, 31, No. 8, 1071–1078.

181. McAllister, L.E., Taverna, A.R. (1972) Development and evaluation of Mod 3 carbon/carbon composites, 17th Nat. SAMPE Symp., 1972, p. 111–3(7.

182. Alva Vishnu, S., Alien, H.G. (1996) Through-the-thickness tension strength of 3-D braided composites, Compos. Mater., 30, No 1, 51–68.

183. Halpin, J.C., Jerine, K., Whitney, J.M. (1971) The laminate analogy for 2 and 3 dimensional composite materials, J. Compos. Mater., No. 1, 36–49.

184. Qiao, P., Davalos, J.F., Barbero, E.J. (1998) Design optimization of fiber reinforced plastic composites shapes, J. Compos. Mater., 32(2), 177–196.

185. Vollmar, S., Pompe, W. (1978) Zur Einfluss von Zwischenschichten auf die mechanischen Eingenschaften von Teilchenverbunden, Plaste und Kautschuk, No. 2, 78–83.

186. Detassis, M., Frydman, E., Vrieling, D., Zhou, X.-F., Nairn, J.A., Wagner, H.D. (1996) Interface toughness in fiber composites by the fragmentation test, Compos., Part A, 27A, 769–773.

187. Freger, G.E. (1984) Research of composite materials based on spirally reinforced fillers, Mech. of Compos. Mater., 3, 412–416.

188. Busche, M.G. (1969) Boost composites shear strength with 3-D reinforcement, Mater. Eng., 67, No. 4, 62–64.

189. Chou, T.W., Fukuda, H. (1982) Monte Carlo simulation of the strength of hybrid composites, J. Compos. Mater, 16, No. 5, 371-375.

190. Barbero, E.J., Luciano, R. (1995) Micromechanical formulas for the relaxation tensor of linear viscoelastic composites with transversely isotropic fibers, Int. J. of Solids Struct., 32(13), 1859–1872.

191. Hua, C.T., Chu, J.N., Ko, F.K. (1990) Damage Tolerance of 3-D Braided Commingled PEEK/Carbon Composites, ASTM, 23.

192. Ko, F.K., Fang, P., Chu, H. (1988) 3(D Braided Commingled Carbon Fiber/PeekCompos., 33rd International SAMPE Symposium and Exposition, Vol. 33, March 7–10, 1988, pp. 899–911.

193. Handbook of Fiberglass and Advanced Plastic Composites, Ed. G. Lubin, New York, Van Nostrand Reinhold, 1969, p. 894.

194. Ko, F.K., Pastore, C. (1989) Analysis of Three Dimensional Complex Shaped Composites, Transactions of the ASME, presented at Gas Turbine and Aeroengine Congress and Exposition, June 4–8, 1989, Toronto, Ontario, Canada.

195. Luo, H.A., Wang, Q. (1997) Stress concentrations in an intermingled hybrid composites, J. Appl. Mech., 64, No. 4, 738–742.

196. Ozden, O.O. (1996) Micromechanical Analysis of hybrid composites, Reinforced Plastics and Compos., 16, 828(836

197. Moser, T.L., Scheider, W.C. (1981) Strength integrity of the space shuttle orbiter tiles, A1AA Pap., No. 2469, 1-9.

198. Jigun I.G., Polyakov V.A. (1978) Properties Spatially-Reinforced Plastics, Riga, Zinatne, p. 79.

199. United States Patent 4, 357, 193. Method of fabricating a composite structure, McCann W., Olsen G., Patented Sept. 2, 1982.

200. Hollman, M., Pitkethly, M.- J., Doble, J. B (1999) Characterizing the fiber/matrix interface of carbon fiber-reinforced composites using a single fiber pull-out test, Compos., 21, No. 5, 389–395.

201. Wood, J.R., Wagner, H.D., Marom, G.A. (1994) Model for compressive fragmentation, Adv. Compos. Letters, 3 (4), 133–138.

202. Lee, D.G., Kim, Y.G., Oh, J.H. (1997) Optimum bolted joints for hybrid composites materials, Composite Struct., 38, 1–4, 8, 329–341.

203. Muto, N., Arai, Y., Shin, S.G., Matsubara, H., Yanagida, H., Sugita, M., Nakatsuji, T. (2001) hybrid composites with self-diagnosing function for preventing fatal fracture, Compos. Sci. and Technol. 61, 6, 875–883.

204. Lourie, O., Wagner, H.D., Levin, N. (1997) Effective width of interface in a stressed model polymer composites measured by micro-FTIR, Polymer, 38 (22), 5699–5702.

205. Muskhelishwili N.I. (1954) Some Major Tasks of Mathematical Theory of Elasticity, Moscow, AS USSR, p. 648.

206. Joshi, S.P., Lyengar, G.R. (1990) Analysis of hybrid laminated composite plates, J. Aerosp. Eng., 3, No. 1, 78–90.

207. Adnan, H.H. (1977) Thermomechanically induced interfacial stresses in fibrous composites, Fiber Sci. and Technol., 10, No. 3, 195–209.

208. Alfredo Balacó de Morais (2000) Transverse module of continuous-fiber-reinforced polymers, Compos. Sci. and Technol., 60, 7, 997–1002.

209. Fukuda H., Chou, T.W. (1982) Monte Carlo simulation of the strength of hybrid composites, J. Compos. Mater., 16, No. 5, p. 371.

210. Greszcuk, L.B. (1971) Interfibre stresses in filamentary composites, A1AA J., No. 7, 1274–1280.

211. Hua, C.T., Ko, F.K. (1989) Properties of 3-D Braided Commingled PEEK/CarbonCompos., Proceedings, 21st International SAMPE Technical Conference, September 25-28, 1989, Atlantic City, NJ.

212. Madi, U.R., Smith, D.L. (1984) A finite element model for determining the constitutive relation of a compression member, 3-rd Int. Conf. Space Struct., 1984, London, New-York, p. 625–629.

213. Gellhorn, E.V., Menges, G. (1982) Wickeln-verfahrens-optimierung durch Vakuumtrankung, Gummi – Asbest – Kunstst., 35, No. 11, 630–635.

214. Chawla, K.K., Aragao, E.E., Monteiro, R.R. (1980) Mechanical behavior of the hybrid composite system: polyester matrix/glass and jute fibers, Adv. Compos. Mater. Proc. 3rd Int. Conf., Paris, Aug. 26–29, Vol. 1. – Oxford, 1980, 414–424.

215. Freger, G.E., Kestelman, V.N. (1990) Auswahl technologischer Parameter zur Herstellung von spiralförming anisotropen Hybridfüllstoffen, Plaste und Kautschuk, 37, No. 10, 349–351.

216. Chu, J.N., Ko, F.K. (1992) Dynamic mechanical properties of three-dimensional braided graphite/PEEK composites, SAMPE Quarterly, July, 1992.

217. Kuo, Wen-Shyong, Ko, Tse-Hao, Lo, Tzu-Sen (2002) Failure behavior of three-axis woven carbon/carbon composites under compressive and transverse shear loads, Compos. Sci. and Technol., 62, 7–8, 989–99.

218. Xiaoyu, J., Xiangan, K. (1999) Micro-mechanical characteristics of fiber/matrix interfaces in composite materials, Compos. Sci. and Technol., 59, 5, 635–642.

219. Freger, G.E. (1982) Investigation of tense-deformation state of reinforced materials under shear, Problems of Strength, 11, 116–119.

220. Rapid growth VS for advanced composites, Des. End. (Can.), 1981, 27, No. 9, 18–19.

221. Mehan, M.L., Schadler, L.S. (2000) Micromechanical behavior of short-fiber polymer composites, Compos. Sci. and Technol., 60, 7, 1013–1026.

222. Freger G.E., Kats M.L., Ignatiev B.B. (1980), Investigation of tense state of reinforced materials with low filling, Applied Mech., 7, 132–137.

223. Fitzer, E., Huttner, W. (1981) Structure and strength of carbon – carbon composites, J. Phys. D. Appl. Phys., 14, No. 3, 347–371.

224. Fleishman N.P., Starovoitenko N.V. (1970) Plane Task of Elasticity Theory for plates with curve board support, Mater. Resistance and Theory of constructions, Kiev, 97–103.

225. António, C.A. C., Marques, A.T. and Soeiro, A.V. (1995) Optimization of laminated composite structures using a bi-level strategy, Compos. Struct., 33, 4, 193–200.

226. Greszcuk, L.B., Hawley, A.V., Couch, W. (1979) Mechanical properties of glass fiber composites, Hydroinaut, 13, No. 4, 105–112.

227. Bolotin, V.V., Novichkov Yu. N. (1980) Mech. of Multilayer Constructions, Moscow, Mashinostroeniye, 376.

228. Hartmann, D. (1978) A generally applicable structural optimization method and its application to the design of cylindrical shells, Eng. Optimization, 3, No. 4, 201–213.

229. Freger, G.E. (1983) Stress-deformation state of spirally reinforced composites under transversal loading, Mech. of Compos. Mater., 6, 989–995.

230. McGrath, J.E., Grubbs, H., Woodward, M.H., Rogers, M.E., Gungor A., Joseph, W.A. Mercier, R., Brennan, A. (1994) Development of new polymeric matrices for high performance composites, Compos. Struct., 27, 7–16.

231. Tirosh, J., Katz, E., Lifschitz, G. (1979) The role of fibrous reinforcements well bonded or partially of composite materials, Eng. Fract. Mech., 12, No. 2, 267–277.

232. GangaRao, N. Raghupathi (1997) Evaluation of the hybridization of wood crossties with glass fiber-reinforced composites (GFRC), Proceedings of the Compos. Institute's International Compos. EXPO-97, Jan. 27–29, 1997, p. 1–6.

233. Van Fo Fi, G.A. (1971) Theory of Reinforced Materials with Coatings. Kiev, Naukova Dymka.

234. Ishikawa, T., Chou Tsu-Wei. (1982) Elastic behavior of woven hybrid composites, J. Compos. Mater., 16, No. 1, 2–19.

235. Kagava, Y., Okuhara, H., Watanabe, Y. (1981) Some properties of composite metals reinforced with helical fiber, Compos. Mater.: Mech., Mech. Prop. and Fabr. Jap. (US Conf., Tokyo, Jan. 12-14, 1981, Barking, 1981, 213–222.

236. Summerscales, J. (1998) Microstructural Characterization of Fiber-ReinforcedCompos., Woodhead Publishing, Cambridge.

237. Fukuda, H., Kawata, K. (1983) A Monte Carlo simulation of the strength of laminated hybrid composites, Trans. Jap. Soc. Aeronaut. and Space Sci., 25, No. 70, 203–215.

238. Kendall, K. (1976) Interfacial cracking of a composite. Part 1. Interlaminar shear and tension, J. of Mater. Sci., No. 11, 638–644.

239. Kessler, A., Bledzki, A. (2000) Influence of the fiber/matrix-interphase on the post-impact properties of glass/epoxy-laminates, Adv. Compos. Mater., 9, No. 2, 109 – 118.

240. Ambarzumjan S.A. (1974) General Theory of Anisotropic Shells. Moscow, Nauka, 446

241. Mai, Y.W., Andonian, R., Cotterell, B. (1980) On polypropylenecellulose fiber – cement hybrid composites, Adv. Compos. Mater., Proc. 3rd Int. Conf., Paris, Aug. 26–29, 1980, Vol. 2, Oxford, 1980, 1687–1699.

242. Benerji P., Batterfield R. (1984) Methods of Boarder Elements in Applied Sci., Moscow, Mir, 494.

243. Holstein, D., Jüptner, W., Höfling, R., Aswendt, P. and Schmidt, C. -D. (1997) Deformation analysis of thermally loaded composite tubes, Composite Struct., 40, 3–4, 257-265.

244. Levita, G., Di Landro, L., Marchetti, (1997) A. Interface strength in composites having epoxy matrix toughened with reactive rubber, Plast. Rubber and Compos. Process. and Appl., 26, No. 6, 260–255.

245. Perry, J.L., Adams, D.F. (1975) Sharp impact experiments on graphite/epoxy hybrid composites, Compos., No. 7, 166–172.

246. Gunyaev, G.M. (1979) Use of the rule of mixtures in designing polyfibrous composites, Compos. Mater. Reports, Soviet–Jap. Symp, Moscow, 309–325.

247. Sonparote, P.W., Lakkad, S.C. (1982) Mechanical properties of carbon / glass fiber reinforced hybrids, Fiber Sci. and Technol., 16, No. 4, 309–312.

248. Mohan, R.K., Shridhar, M.K., Roa, R.M. (1983) Composites strength of jute-glass hybrid fiber composites, J. Mater. Sci. Lett., 2, No. 3, 99–102.

249. Lekhnitsky S.G. (1977) Theory of Elasticity of Anisotropic Body. Moscow, Nauka, p. 416.

250. Detassis, M., Pegoretti, A., Migliaresi, C., Wagner, H.D. (1996) Experimental evaluation of residual stresses in single fiber composites by means of the fragmentation test, J. of Mater. Sci., 31, 2385–2392.

251. Reda, D.C., Raper, R.M. (1979) Measurements of transition front asymmetries on large-scale, ablating graphite noise tips in hypersonic flights, A1AA Pap., No. 268, 12–18.

252. Hofer, K.E., Stander, M., Bennet, L.C. (1987) Duration and enhancement of the fatigue behavior of glass / graphite / epoxy hybrid composites after accelerated adding, Polymer Eng. and Sci., 18, No. 2, 120–127.

253. Horsch, F., Sprenger, K.H. (1983) Erhentnisse aus der Verarbeitung und Prufung von Glas, Aramid – Sowie Kohlenstoffaserergeweben und daraus hergestellter Laminate, Plastverarbeiter, 34, No. 1, 41–44.

254. Dirling, R.B. (1977) On the relation between material variability and surface roughness, AIAA/ASME 18th, Dun. And Mater. Conf., AIAA Aircraft Compos. Emerging Methodology, Struct. Assur., San Diego, Calif., 1977, Vol. A, No. 1, 246–250.

255. Fish, J. and Shek, K. (2000) Multistage analysis of Compos. Mater. and Struct., Compos. Sci. and Technol., 60, 12–13, 2547–2556.

256. Fukuda, H. (1991) Micromechanical strength theory of hybrid composites, Adv. Compos. Mater., 1, No. 1, p. 39.

257. Takeda, Nobuo, Neil, McCartney L., Ogihara, Shinji (2000) The application of a ply-refinement technique to the analysis of microscopic deformation in interlaminar-toughened laminates with transverse cracks, Compos. Sci. and Technol., 60, 2, 231–240.

258. Park R., Jang, J. (1997) Stacking Sequence Effect of Aramid/UHMPE hybrid composites by Flexural Test Method, Polymer Testing, 16(6), 549.

259. Phillips, L.N. (1976) On the usefulness of glass fiber – carbon hybrids, Reinf. Plastics Congr., Brington, 1976, Proc., p. 207–211.

260. Wisnom, M.R. (1995) The effect of transverse compressive stresses on tensile failure of glass fiber-epoxy, Compos. Struct., 32, 1–4, 21–626.

261. Auerbach, G. (1979) Effect of fiber fraction on ablation properties in short fiber graphite composites, AGAA Pap., No. 374, 108–117.

262. Williams, J.M., Imprescia, R.J. (1975) Improvement in oxidation resistance of the leading edge thermal protection for a space shuttle, J. Spacecraft and Rockets, 12, No. 3, 151–154.

263. Reifsnider, K., Case, S., Duthoit, J. (2000) The mechanics of composites strength evolution, Compos. Sci. and Technol., 60, 12-13, 2539–2546.

264. Murray, T.A. (1981) Lightweight composites take to the air, Des. Eng. (USA), 52, No. 7, 31–35.

265. Maksimov R.D., Plume E.Z. (1980) Elasticity of hybrid composites material based on organic and boron fibers, Mech. of Compos. Mater., 3, 399–403.

266. Davies, I.J., Hamada, H. (2001) Flexural properties of a hybrid polymer matrix composites containing carbon and silicon carbide fibers, Adv. Compos. Mater., 10, No. 1, 77 – 96.

267. Vinson, J.R. (1981) On the state of technology and trends in composite materials in the United States., Compos. Mater.: Mech., Mech. Prop. and Fabr. Jap., US Conf., Tokyo, 12–14 Jan., 1981, Barking, 1981, 353–361.

268. Cross, S.L. (1980) Prospects improve for automotive composites, SAE-Australas, 40, No. 3, 135–139.

269. Bank, L.C., Gentry, T.R., Nuss, K.H., Hurd, S.S., Lamanna, A., Duich, S. and Oh, B. (2000), Construction of a pultruded composite structures: a case study, ASCE J. of Compos. for Constr., 4., No. 3, 112–119.

270. Bank, L.C., Yin, J., Moore, L.E., Evans, D., Allison, R (1996), Experimental and numerical evaluation of beam-to-column connections for pultruded structures, J. of Reinf. Plastics and Compos., 15, No. 10, 1052–1067.

271. Barbero, E.J. (1991) Pultruded structural shapes - from the constituents to the structural behavior, SAMPE J., 27(1), 25–30.

272. Ko, F.K., Morán-López, J.L., Sanchez, J.M. (1993) Advanced textile structural composites, advanced topics in materials science and engineering, 1993, Plenum Press, NY, pp. 117–137.

273. Ko, F.K. (1990) Textile Structural Composites for Building Constructions, Textile Composites in Building Construction, Part 3, 43–60.

274. Ko, F.K., Kutz, J. (1988) Multiaxial Warp Knit for AdvancedCompos., 4th Annual ASM/ESD ACCE, September 13–15, 1988, Dearborn, MI.

275. Akiva, U., Itzhak, E., Wagner, H.D. (1997) Elastic constants of 3–dimensional orthotropic composites with platelet/ribbon reinforcement, Compos. Sci. and Technol., 57, 173–184.

276. Auerbach, G., Liberman, M.L., Pierson, H.O. (1977) Effect of porosity in graphite materials on ablation in arc-heated jets, J. Spacecraft and Rockets, 14, No. 1, 19–24.

277. Brennan, A.B. (1991) Thermal analysis of hybrid composites, Seiko Thermal Analysis Conference, Charlotte, NC, February, 1991.

278. Dixon, R.R. (1977) Properties of unidirectional two-fiber hybrid composites, Soc. Plast. Eng., No. 1, 344-346.

279. Fisher, S., Marom, G., Tuler, F.R. (1979) Hybrid effects in composites: a comparison between interlaminar and translaminar configurations, J. Mater. Sci., 14, No. 4, 863–868.

280. Froger, G.F. (1983) Investigation of tense state of spirally reinforced composites with high degree of filling, Mech. of Reinforced Plastics, Riga, Polytechnic Institute, 61–68.

281. Freger, G.E., Ignatev, B.B., Chesnocov, V.V., Karvasarskaja, N.A., Panfilov, A.M. (1987) Thermal stresses in rod constructions from spirally reinforced composites, Mech. of Comp. Mater., No. 2, 305–309.

282. Hardaker, K.M., Richardson, M.O. (1980) Trends in hybrid composites technology, Polym. – Plast. Technol. and Eng., 15, No. 2, 169–182.

283. Harlow, D.G. (1983) Statistical properties of hybrid composites. 1 Recursion analysis, Proc. Roy. Soc., London, A 389, No. 1796, 67–100.

284. Harlow, D.G., Phoenix, S.L. (1978) The chain-of-bundles probability model for the strength of fibrous materials. 1 Analysis and conjectures, J. Compos. Mater, 12, No. 4, 195–214.

285. Huang, Y., Gong, X.Y., Suo, Z. and Jiang, Z.Q. (1997) A model of evolving damage bands in materials, International J. of Solids and Struct., 34, 3941–3951.

286. Huang, Y., Hu, K.X. and Chandra, A. (1994) Several variations of the generalized self-consistent method for hybrid composites, Compos. Sci. and Technol., 52, 19–27.

287. Lancin, M. (1991) Relationship between the microstructure of the interface and the mechanical behavior of Compos. Mater., Rev. Phys. Appl., 1, No. 6, 1141–1166.

288. Marom, G., Arridge, R.G.C. (1976) Stress concentrations and transverse modes of failure in composites with a soft fiber-matrix interlayer, Mater. Sci. and Eng., No. 23, 23–32.

289. Naik, N.K., Ramasimha, R., Arya, H., Prabhu, S.V. and ShamaRao, N. (2001) Impact response and damage tolerance characteristics of glass–carbon/epoxy hybrid composites plates, Compos., Part B: Eng., 32, 7, 565–574.

290. Ogi, Keiji, Smith, P.A. (2002) Modeling creep and recovery behavior of a quasi-isotropic laminate with transverse cracking, Adv. Compos. Mater., 11, No. 1, 81–93.

291. Parratt, N.J., Potter, K.D. (1980) Mechanical behavior of intimately – mixed hybrid composites, Adv. Compos. Mater. Proc. 3rd Int. Conf., Paris, Aug., 26–29 1980, Vol. 1, Oxford, 1980, p. 313–326.

292. Ramsden, J.M. (1980) Towards the plastic airplane, Flight, 118, No. 3715, 182–185.

293. Rechenberg, I. (1970) Optimerung Technischer Systeme nach Prinzipien der Biologischen Evolution, TU, Berlin, 257 p.

294. Renton, W.J. (1977) Introduction to hybrid and selected metal-matrix composites, hybrid and select metal-matrix composites state, Art Rev., New York, 1977, p. 1–11.

295. Renton, W.J. (1981) An overview of hybrid composites applications to advanced structures, Compos. Mater.: Mech., Mech. Prop. and Fabr. Jap.-US Conf., Tokyo, Jan. 12–14, 1981, Barking, 1981, p. 362–382.

296. Stewart, R.W., Verijenko, V.E., Adali, S. (1997) Analysis of the in-plane properties of hybrid glass/carbon woven fabricCompos., Composite Struct., 39, 3–4, 319–328.

297. Summerscales, J., Short, D. (1978) Carbon fiber and glass fiber hybrid reinforced plastics, Compos., 9, No. 3, 157–166.

298. Vaughan, J.G., Roux, J.A., Raju Mantena (1992) Characterization of Mechanical and Thermal Properties of AdvancedCompos. Pultrusions, Proceedings of the 1992 NSF Design and Manufacturing Systems Conference, January 8–10, 1992, p. 1141–1145.

299. Wagner, H.D. (1989) Dependence of fractures stress upon diameter in strong polymeric fibers, J. of Macromolecular Sci.-Physics, B28 (3 & 4), 339–347.

300. Wagner, H.D., Roman, I., Marom, G. (1982) Hybrid composites: designing for a positive hybrid effect, polymers and plastic materials (Polim. Vehomarim Plast. In Hebrew), 12, 8–15.

301. Wagner, H.D., Steenbakkers, L.W. (1989) Microdamage analysis of fibrous composites monolayers under tensile stress, J. of Mater. Sci., 24, 3956–3975.

302. Wu, Z.J., Ye, J.Q., Cabrera, J.G. (2000) Interfacial effects on stress transfer in fiber-reinforced composites, Mech. of Compos. Mater. and Struct., 7, No. 4, 315–330.

Springer Series in
MATERIALS SCIENCE

Editors: R. Hull   R. M. Osgood, Jr.   J. Parisi   H. Warlimont